Aquatic Plants of Pennsylvania

AQUATIC PLANTS OF PENNSYLVANIA

A Complete Reference Guide

TIMOTHY A. BLOCK AND ANN FOWLER RHOADS

Illustrations by Anna Aniśko

Morris Arboretum of the University of Pennsylvania

PENN

UNIVERSITY OF PENNSYLVANIA PRESS

Philadelphia

Published by
University of Pennsylvania Press
Philadelphia, Pennsylvania 19104-4112
www.upenn.edu/pennpress

Printed in Canada on acid-free paper
10 9 8 7 6 5 4 3 2 1

A Cataloging-in-Publication record is available from the Library of Congress

ISBN 978-0-8122-4306-2

Frontispiece: *Orontium aquaticum*, photo by Roger E. Latham

Contents

List of Illustrations vii
Preface ix
Acknowledgments xi
Chapter 1. Evolution and Ecology 1
Evolution and Form of Aquatic Plants 1
Role of Plants in Aquatic Ecosystems 7
Types of Aquatic Ecosystems 11
Management of Aquatic Ecosystems 15
Chapter 2. Identification Keys for the Aquatic Plants of Pennsylvania 21
Key to Growth Habit Groups 22
Key to Emergent Plants 22
Key to Floating-leaf Plants 24
Key to Thalloid or Tiny Floating Plants 25
Key to Submergent Leafy-stemmed Plants 26
Key to Submergent Rosette-forming Plants 27
Key to Submergent Plants with Finely Divided Leaves 28
Chapter 3. Emergent Plants 29
Arrow-arum, tuckahoe 29
Arrowhead 31
Beggar-ticks/bur-marigold 42
Bulrush 47
Bur-reed 61
Cat-tail 71
Chinese lobelia 74
Flowering-rush 76
Goldenclub 77
Grasses 79
Water horsetail 88
Iris 90
Lizard's-tail 94
Mud-plantain 96
Pickerel-weed 100
Bayonet rush 102
Smartweed 104
Spike-rush 111
Three-way sedge 121
Twig-rush 123
Water-hemp ragweed 125
Water-horehound 127
Water-parsnip 128
Water-plantain 129

Water-primrose 132
Water-purslane 136
Water-willow 137
Watercress 141
Chapter 4. Floating-leaf Plants 143
Floating-heart 143
American lotus 147
Water-chestnut 149
European water-clover 151
Water-lily 153
Watershield 161
Chapter 5. Thalloid and/or Tiny Floating Plants 163
Aquatic liverworts 163
Charophyte algae 164
Duckweed, watermeal, water-flaxseed, and bogmat 166
Eastern mosquito fern 178
Riverweed 180
Chapter 6. Submergent Leafy-stemmed Plants 183
Aquatic mosses 183
Hydrilla 185
Pondweeds 186
Water star-grass 221
Waternymph, naiad 223
Water-starwort 229
Waterweed 233
Waterwort 237
Chapter 7. Submergent Rosette-forming Plants 241
Mudmat 241
Pipewort 243
Quillwort 245
Water lobelia 251
Water-celery, eel-grass 253
Chapter 8. Submergent Plants with Finely Divided Leaves 255
Bladderwort 255
Coontail, hornwort 269
Fanwort 272
Mermaid-weed 273
Water-crowfoot 275
Water-marigold 278
Water-milfoil 280
Glossary 291
References and Sources 295
Appendix: Taxonomic List of Plants Included 299
Index 305

Tables and Figures

Tables

1.1. Representation of Land Plant Lineages in the Aquatic Flora of Pennsylvania 1
1.2. Endangered, Threatened, and Rare Aquatic Plants of Pennsylvania 16
1.3. Non-native, Invasive Aquatic Plants in Pennsylvania 18
3.1. Comparison of Key Characters of Bur-reed Species 67
8.1. Comparison of Vegetative Characteristics of *M. sibiricum*, *M. spicatum*, and *M. verticillatum* 287
Appendix Taxonomic List 299

Figures

1.1. Evolutionary relationships of major groups of aquatic plants 2
1.2. Pollination in water-celery 5
1.3. Lake surface covered with white male flowers of water-celery 6
1.4. Aquatic food chain 8
1.5. Great Blue Heron 8
1.6. Diagram of the littoral zone 10
1.7. Zonation in littoral zone vegetation 10
1.8. Shed skin of a dragonfly 10
1.9. Distribution of lakes in Pennsylvania 11
1.10. Succession in a glacial kettle lake 12
1.11. A glacial kettle lake 12
1.12. Freshwater tidal marsh 13
1.13. Tidal riverbank 14
1.14. Northeastern bulrush 15
1.15. An infestation of European water-chestnut 18
1.16. Water surface nearly covered by native fragrant water-lily 19
2.1. Stems jointed, tiny scale-like leaves in whorls at the node 22
2.2. Leaf expanded at the base, containing spore sac 22
2.3. Leaf venation with and without prominent mid-vein 24
2.4. Leaf arrangement in waterweeds 26
2.5. Vein reticulation 27
2.6. Quillwort leaf with spore sac in the base 27
2.7. Bladeless leaf sheaths in a spike-rush 27
2.8. Leaf form and arrangement 28
5.1. Mixed population of water flaxseed, duckweed, and watermeal 166
6.1. Pondweed stipules 187
6.2. Leaf sections of pondweed showing lacunar band 188
6.3. Pondweed stem with glands 189
6.4. Pondweed achenes 189
8.1. Bladderwort traps 255

Preface

Purpose of the Book

This book is intended to foster an appreciation for the diversity of native aquatic plants and their role in aquatic ecosystems. We have included information to help readers distinguish major groups of aquatic plants by growth form and habitat relationships. The identification keys, descriptions, and illustrations will assist users in identifying plants to species, a necessary step in assessing the status of a body of water.

The book also includes information about endangered, threatened, and rare aquatic plants as well as invasive exotic species that have colonized lakes, ponds, and streams. Species descriptions include information on ethnobotanical uses (Moerman 2000) and wildlife value (Martin et al. 1961). Management issues and concerns are also discussed.

How to Use the Book

The book includes identification keys, which consist of a series of paired statements. By selecting the statement that best describes the sample, you can proceed through a sequence of steps to identification.

Species descriptions were drawn from our own extensive experience with the aquatic plants of Pennsylvania and from references and sources listed in the back of the book. References that pertain to a single species are cited in the description of that plant. All species descriptions are accompanied by line art and photographs for further help in confirming the identity of a specimen. Most of the photographs are by the authors; the few exceptions are noted. The botanical illustrations by Anna Aniśko were drawn from live material or herbarium specimens under the direction of the authors.

Distribution maps for each species were prepared using herbarium specimen records from the Pennsylvania Flora Database (2010) plus records from a 2001–2002 study of glacial lakes of northwestern Pennsylvania (Grund undated) and additional observations we made during an eight year study of lakes of eastern Pennsylvania 2000–2007.

Species with similar patterns of growth are grouped together for ease of comparison. A taxonomic list of the plants treated in the book is included in the Appendix.

Acknowledgments

Funding to support costs of publication was received from:

William L., III and Elizabeth P. McLean
Sandra L. McLean
John A. H. and Elizabeth J. Shober
Mt. Cuba Center, Inc., Rick Lewandowski, Director
University of Pennsylvania, Business Services Division

We are also grateful to the following:

The Wild Resource Conservation Program of the Pennsylvania Department of Conservation and Natural Resources, which funded our field work on aquatic plants through annual grants for the years 2000–2007;

Pennsylvania Natural Heritage Program botanists Steve Grund and John Kunsman for access to their collection data;

Bonnie Isaac, Herbarium of the Carnegie Museum of Natural History in Pittsburgh, James Bissell, Cleveland Museum of Natural History, Alina Freire-Fierro, Herbarium of the Academy of Natural Sciences of Philadelphia, and Daniel F. Brunton for access to specimen information from their collections;

Barbara Lathrop, Pennsylvania Department of Environmental Protection Clean Lakes Program, who collaborated with us on vegetation and water sampling at many lakes;

Chris Firestone, Bureau of Forestry, Pennsylvania Department of Natural Resources, who reviewed an early version of the manuscript and provided many helpful suggestions;

the owners and managers of numerous lakes who provided access to us for the purpose of conducting surveys of aquatic plants;

and, the John and Janet Haas Family, and the McLean Contributionship for their ongoing support of botanical research at the Morris Arboretum of the University of Pennsylvania.

Chapter 1. Evolution and Ecology

EVOLUTION AND FORM OF AQUATIC PLANTS

It's not always easy to decide what to include in a list of aquatic plants. We treat 194 macrophytes (large plants) here, including 178 flowering plants, 2 horsetails, 2 ferns, 5 quillworts, 3 mosses, 2 liverworts, and 2 Charophyte algae. Phytoplankton and filamentous algae are not included. Macrophytes range from plants that grow with their roots under water but most of their stems, leaves, flowers, and fruits above the water surface—the emergent flora—to species that are completely submergent.

The difficulty in choosing what to include arises because of the flexibility of some species, which can grow as emergents for part of the time, but survive long periods without standing water. In selecting the species to include it was our intention to focus on plants that typically grow completely beneath the water surface, floating on water, or with at least their roots in standing water. This does not exclude occasional periods of low water when plants that are normally aquatic may be stranded on a muddy shore.

Most of the aquatic plants treated in this book are flowering plants. They reproduce sexually to produce seeds which are contained in a matured ovary (fruit). These plants also have specialized vascular tissues which make possible the distribution of water and dissolved nutrients throughout the plant body, along with an external cuticle to retain moisture. These are features that allowed plants to expand from their aquatic origins to colonize land early in their evolutionary history. Algae have continued to diversify in aquatic environments including both fresh and salt water, and, in fact, dominate some aquatic habitats.

Continuing evolution of the land flora resulted in secondary adaptation to aquatic environments. This "return to water" has occurred in all major groups of land plants: bryophytes, quillworts, horsetails, ferns, and flowering plants (see Figure 1.1 on page 2). Thirty-six families of flowering plants are represented in this book, including members of the basal angiosperms, monocots, and dicots (Table 1.1).

Table 1.1. Representation of Land Plant Lineages in the Aquatic Flora of Pennsylvania (see Appendix for a full list of species)

TAXONOMIC GROUP	FAMILIES	SPECIES
Bryophytes		
liverworts	1	2
mosses	3	3
Quillworts	1	5
Horsetails	1	2
Ferns	2	2
Flowering plants		
basal angiosperms	2	6
monocots	14	110
dicots	20	62

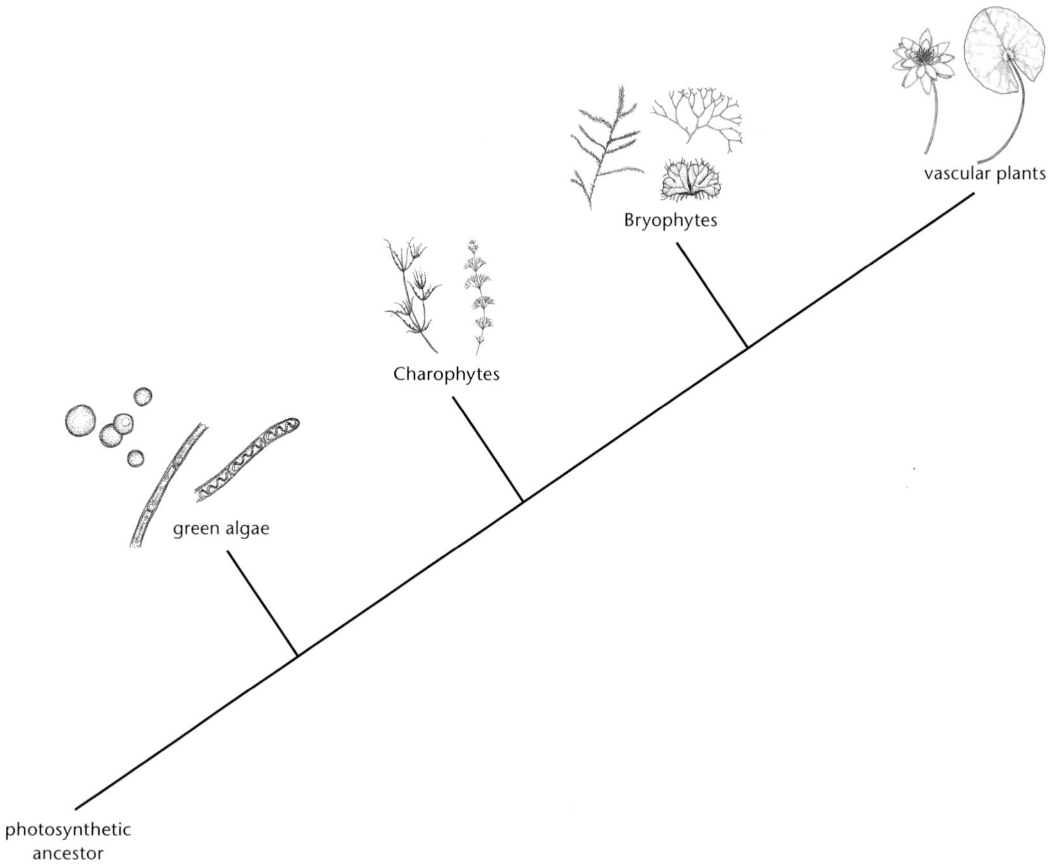

Figure 1.1. Evolutionary relationships of major groups of aquatic plants.

Plants have evolved a variety of growth forms to take advantage of the full range of aquatic habitats. Emergent plants occupy lake and stream margins where their roots can be under water but the stems and leaves are largely above the surface. Rooted plants with floating leaves such as water-lilies (*Nymphaea* and *Nuphar* spp.) and watershield (*Brasenia*) are limited to water depths to about 1.5–2 m. The leaf blades are attached by long petioles to rhizomes imbedded in lake or streambed sediments.

Free-floating plants such as the duckweeds (*Lemna* spp.), watermeals (*Wolffia* spp.), and water flaxseed (*Spirodela punctata*) are independent of water depth, but winds and waves usually push them toward the lake or stream margins except on small ponds, where they may cover the entire surface.

Rooted submergent species such as waterweed (*Elodea* spp.) are limited to depths where light penetration is sufficient to support photosynthesis. This can vary from less than 1 m in very turbid water to 3–4 m or more in exceptionally clear lakes. Other submergent species such as Eurasian water-milfoil (*Myriophyllum spicatum*), hydrilla (*Hydrilla verticillata*), and many pondweeds (*Potamogeton* spp.) typically produce a long stem that only branches when it approaches the water surface. It has been shown that low light promotes shoot elongation and inhibits branching. High light availability has the opposite effect, inhibiting further increase in stem length and stimulating branching. The result is to place the bulk of the leaves just below the water surface where photosynthesis can be most efficient.

Modifications for Life Under Water

An aqueous environment poses challenges different from those encountered on land. Water is denser than air and provides support through greater buoyancy. Consequently, submergent and floating aquatic plants have less need for tissues that provide the stiffening that allows terrestrial species to stand erect. At the same time roots, stems (including rhizomes), and petioles of aquatic plants usually have porous, gas-filled columns of tissue (aerenchyma) that contribute to buoyancy and allow diffusion of carbon dioxide and oxygen throughout the plant.

The vascular tissues of submergent aquatic plants are also greatly reduced. Consequently, these plants tend to be limp and to collapse in a soggy heap when removed from the water. In order to prepare herbarium specimens, it is necessary to float the plants in a tray of water, slide a piece of mounting paper under them, and carefully lift them out of the water so the stems and leaves are spread in a lifelike manner as they are when floating in a column of water.

Terrestrial plants have evolved various coverings (cuticle, surface hairs, etc.) to prevent excess water loss through evapotranspiration, a problem not encountered under water. Submergent aquatics typically have thin leaves with little or no cuticle; as a result they shrivel quickly when removed from water. Floating leaves, on the other hand, typically have a cuticle on surfaces exposed to the atmosphere.

In order to resist the force of moving water, aquatic plants have evolved strong anchoring features such as rhizomes that are buried in lake and stream bottom sediments, or various forms of holdfasts by which species like water moss (*Fontanalis* spp.) and riverweed (*Podostemum ceratophyllum*) cling to rocks and other surfaces. Flexibility is another important characteristic in water plants; many species have long flexible stems or leaf petioles that can move with the water currents or waves. The petioles and peduncles of water-lilies are longer than the distance from the water surface to the rhizome to which they are attached, to allow for wave action; in addition, the petioles are firmly attached near the middle of the leaf blade. The rounded shape and smooth margins of the leaf blades also reduce resistance. The underwater leaves of many species such as the water-crowfoots (*Ranunculus* spp.), coontail (*Ceratophyllum* sp.), bladderworts (*Utricularia* spp.), and water-milfoils (*Myriophyllum* spp.) are finely divided, thus offering less resistance to water currents and also increasing the surface area for absorption of carbon dioxide.

Stomata are pores that allow for the movement of gasses such as carbon dioxide and oxygen in and out of leaves. In terrestrial plants stomata are typically on the lower leaf surfaces; floating leaves of aquatic plants have their stomata on the upper surface, providing access to atmospheric carbon dioxide. Submersed leaves have few or no functional stomata because carbon dioxide and oxygen dissolved in water enter the leaf, or in some cases, the roots, by diffusion.

Emergent aquatics are more like terrestrial plants in many ways; however, their roots must be able to withstand constant inundation and the stress of wave action. Many are rhizomatous, providing a network of anchoring structures. Porous tissues (aerenchyma) that allow oxygen to diffuse through stems, rhizomes, and roots are often present. Even the leaves of emergent species show adaptations such as blades with an arrowhead shape that offer less resistance to the forces of wind and water.

Some emergent species have evolved leaf surfaces that shed water. This characteristic is most notable in American lotus (*Nelumbo lutea*), goldenclub (*Orontium aquaticum*), and northern mannagrass (*Glyceria borealis*). The hydrophobic quality is the result of tiny rounded, wax-covered projections called papillae that cover the leaf surface (Neinhuis and Barthlott 1997). Engineers in Germany have used the leaf surface of lotus as a model to design a paint that sheds water and dirt.

Variability in Form

Aquatic plants are notoriously variable in form, which can make identification challenging. Variation in water depth is a major cause. For example, common bur-reed, which is normally an emergent plant with stiffly erect leaves, will grow in deeper water, but does not flower, and the leaves are less rigid and bend over and float at the tip. Several arrowheads (*Sagittaria graminea* and *S. rigida*) grow vegetatively as short sterile rosettes of narrow, pointed leaves in deep water. *Sagittaria graminea* can even flower under water, but in the absence of flowers it is impossible to tell the rosettes of these species apart visually.

Plants like the water-crowfoots (*Ranunculus* spp.), false-mermaid (*Proserpinaca* spp.), and most of the water-milfoils (*Myriophyllum* spp.) and pondweeds (*Potamogeton* spp.) have both underwater and floating or emersed leaves. The floating or emersed leaves are simpler and sturdier compared to the submersed leaves of the same plant. In addition, many of the pondweeds vary as to whether floating leaves are produced. Species such as *Potamogeton bicupulatus* and *Potamogeton diversifolius* can grow as submergents in deeper water, but often produce floating leaves in shallow water.

Reproduction

Sexual reproduction—Aquatic angiosperms (flowering plants) may produce their flowers under water, at the water surface, or elevated above the water surface. Pollination strategies vary accordingly. For species that hold their flowers above the water surface, insect or wind pollination can proceed as for terrestrial plants. Underwater flowers, such as those of the waternymphs (*Najas* spp.) and some pondweeds must depend on water currents to transport pollen, much like wind-pollinated plants on land. For annuals like the waternymphs (*Najas* spp.) and waterworts (*Elatine* spp.) seed production is essential to maintain populations from year to year. The pollination success rate for perennial species is not quite as critical, since the plants usually live from year to year. In addition, most aquatic perennials also have effective asexual or clonal growth mechanisms (see below).

Some aquatic plants have evolved unique pollination strategies utilizing the water surface as a stage. Species in the Frog's-bit Family (Hydrocharitaceae) are especially interesting in this regard.

Water-celery (Vallisneria americana) has one of the most unusual pollination methods. Water-celery is dioecious, meaning that male and female flowers are produced on separate plants. Male flowers are released from an inflorescence at the base of the plant to float freely to the water surface where they are dispersed by wind. Meanwhile female flowers on long peduncles just reach the water surface where their tips produce a dimple that attracts the floating male flowers to where the pollen can be transferred directly from anther to stigma (Figure 1.2). We have seen lakes white with male flowers at the leeward end from a large blooming population of water-celery (Figure 1.3). Once pollination has occurred, the peduncle coils tightly, pulling the developing fruit farther under the water surface where it can mature.

The waterweeds (*Elodea canadensis and E. nuttallii*) and hydrilla (*Hydrilla verticillata*), also in the Frog's-bit Family, have similar mechanisms. Pollen is released onto the water surface from male flowers, explosively in the case of

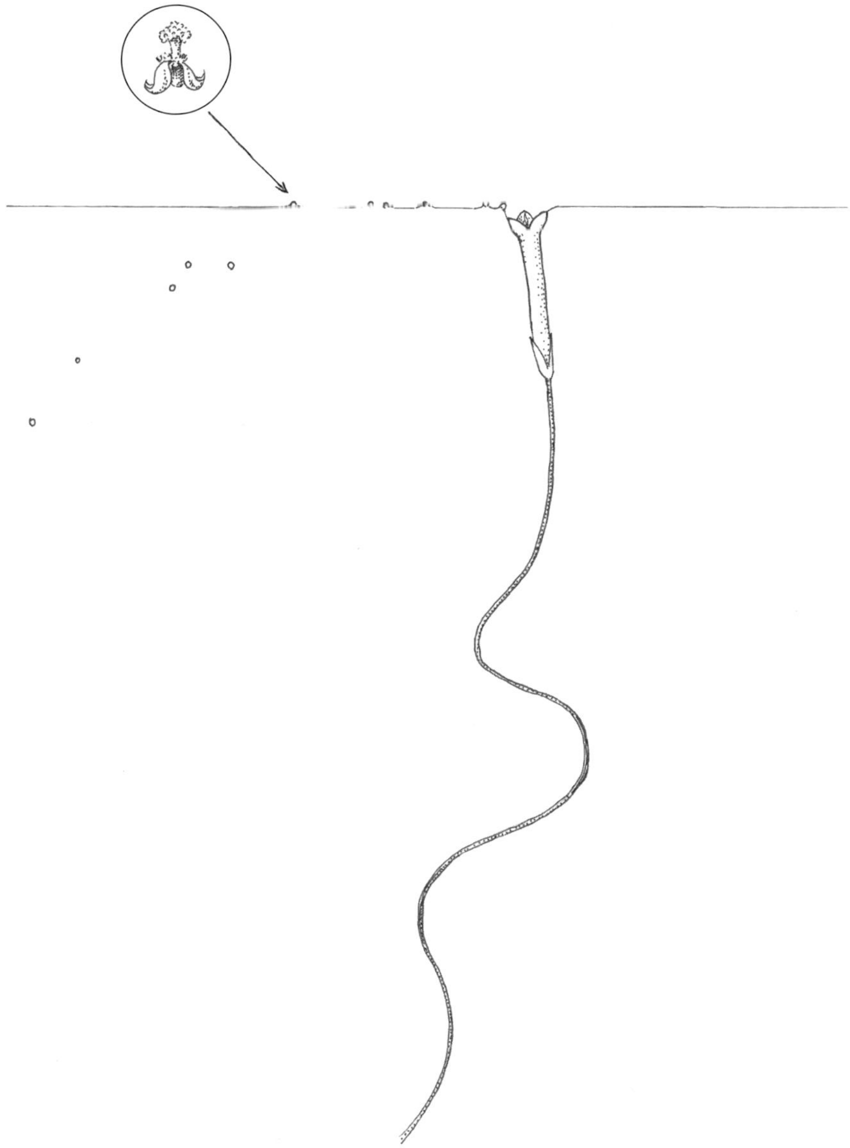

Figure 1.2. Pollination in water-celery (*Vallisneria americana*) occurs when free-floating male flowers drift into contact with female flowers positioned at the water surface.

Figure 1.3. Lake surface covered with white male flowers of water-celery (*Vallisneria americana*) in Sullivan County, PA.

hydrilla, where it is carried by wind or water currents to the stigmas of the female flowers which are positioned at the surface.

Asexual reproduction—Asexual, or clonal, reproduction is very common in aquatic plants. Many perennial species produce rhizomes, modified stems that grow horizontally in the substrate, sending up shoots, leaves, or flowers at intervals. Water-lilies, many pondweeds, and most emergent species are in this category. Clonal colonies can expand to cover large areas, forming dense patches of genetically identical plants. In addition, pieces of rhizome that become detached can start a new colony at another location.

Fragmentation is not limited to rhizome segments. Pieces of the stems of many aquatic plants break off and float; these may continue to grow and even flower and fruit as floating fragments. In addition, through the formation of adventitious roots, these fragments can start new colonies. Water-milfoils (*Myriophyllum* spp.), waterweed (*Elodea* spp.), coontail (*Ceratophyllum* spp.), and bladderworts (*Utricularia* spp.) are examples of plants that reproduce this way.

The formation of dormant buds, or turions, is another way aquatic plants can reproduce. Turions are usually formed at the end of the growing season, resulting in a vegetative structure that can survive the winter or other periods of unfavorable growing conditions. Turions are generally dense and sink to the bottom. When the water warms in the spring they resume active growth. Species that form turions include duckweeds (*Lemna* spp.), waterweeds (*Elodea* spp.), bladderworts (*Utricularia* spp.), and some pondweeds (*Potamogeton* spp.).

Dispersal

Lakes and ponds are islands of habitat; although streams connect some to watersheds, others are completely isolated. Even streams contain a variety of growing conditions based on variables such as the degree of shading and speed of water movement. Each body of water, connected or not, seems to have a slightly different set of plants. The differences are partly due to the water chemistry (more on that later), but the random movement of seeds and plant fragments (propagules) is also involved.

How do aquatic plants move? Flowing water connects some lakes and ponds to watersheds and carries seeds or plant fragments downstream. Some seeds, such as those of the cat-tails (*Typha* sp.) are dispersed by wind. But waterfowl and other animals are important vectors of many aquatic species.

Charles Darwin was one of the first to study the potential for birds to disperse seeds (Browne 1995). More recent studies have documented that many waterfowl feed on the fruits and seeds of aquatic plants and serve as effective dispersal agents (Charalambidou and Santamaria 2002). Seeds can also be carried externally on the feet or feathers of birds. Small plants such as the duckweeds and watermeals can also adhere to birds or other animals such as beaver, otter, muskrats, or turtles and be carried from one site to another.

Humans too, spread plants from site to site. Some of this is inadvertent, through seeds or plant fragments that cling to the exterior of boots, boats, motors, oars, or fishing gear. Some of it is more deliberate, such as dumping the contents of an aquarium into a local stream, canal, or lake. Cultivated water gardens can also be a source of plant introductions resulting from overflow during heavy rains, the careless handling of garden waste, or bird-mediated dispersal. The deliberate introduction of species to lakes or ponds, usually for ornamental value, is yet another source.

ROLE OF PLANTS IN AQUATIC ECOSYSTEMS

Photosynthesis

Green plants play a basic role in aquatic ecosystems because of their ability to carry out photosynthesis. Plants, including macrophytes, phytoplankton, and filamentous algae, plus cyanobacteria, are the primary producers that drive aquatic food chains (Figure 1.4). Powered by the sun, they fix carbon, producing energy-rich sugars and starches. Other members of the aquatic community that eat plants are termed herbivores. They include fish, turtles, snails, insects, and some birds. Carnivores in turn depend on the herbivores for food. Predatory birds such as osprey, eagles, herons, and egrets are at the top of the food chain in these systems (Figure 1.5). Directly or indirectly all are dependent on plants.

Sugars and starches are not the only product of photosynthesis; the process also releases oxygen. In aquatic systems the oxygen produced by green plants is important in meeting the respiratory needs of plants and animals.

Limiting factors—The need for light to drive photosynthesis limits the depth to which plants can grow in a lake; this is especially true for rooted submergent species. Light does not travel as far through water as through air. In addition, plankton and suspended sediments create turbid conditions that limit light penetration. The clearer the water, the greater the depths at which rooted submergent plants will be found. See also discussion of stem length and branching patterns in the section on growth habit above.

Emergent, floating-leaf, and free-floating plants have unimpeded access to sunlight. However, they can form a canopy that reduces the amount of light reaching lower layers of aquatic vegetation, just as in a forest.

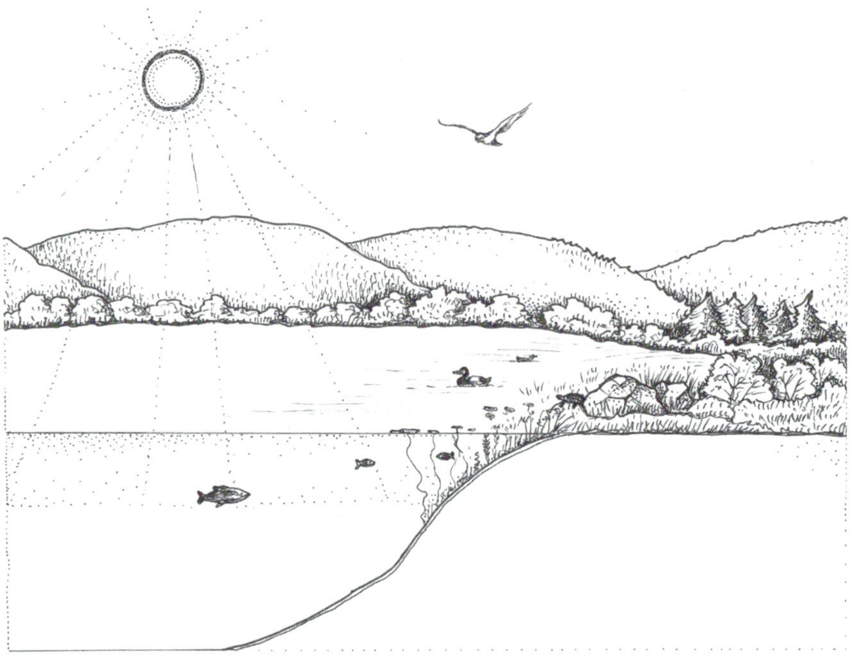

Figure 1.4. An aquatic food chain: powered by the sun, plants and algae in and around the lake (primary producers) photosynthesize, producing sugar and starch. Plants are food for many small fish, insects, and other invertebrates such as snails (herbivores), which are in turn consumed by carnivores. Larger fish and birds of prey are often the top carnivores in freshwater systems.

Figure 1.5. The Great Blue Heron and other fish-eating birds are top predators in aquatic ecosystems.

Carbon dioxide is the source of carbon for photosynthesis in terrestrial plants; in aquatic systems the carbon source can be either dissolved carbon dioxide (CO_2), bicarbonate (HCO_3^-), or carbonate (CO_3^{2-}). Sources include the atmosphere, geology, and carbon dioxide resulting from the respiration of aquatic organisms. The form in which carbon is available depends on the pH (acidity or alkalinity) of the water. Bicarbonate and carbonate are dominant at pH 6.4 and above. Dissolved CO_2 is more abundant at pH values below 6.4.

Dissolved CO_2 is used by most aquatic macrophytes and algae. Some algae and submersed macrophytes, including mosses and quillworts, can use only CO_2. Submersed, rosette-type plants, including water lobelia (*Lobelia dortmanna*) and quillworts (*Isoetes* spp.), obtain up to 90 percent of their carbon dioxide directly from lakebed sediments through their roots. Interestingly, these plants have stiff leaves with a cuticle, which prevents loss of CO_2 through diffusion into the surrounding water. Other plants, such as waterweed (*Elodea* spp.), can switch to bicarbonate when free CO_2 is in low supply; however, photosynthesis under those conditions is less efficient.

Another modification seen in hydrilla, Brazilian waterweed, and common waterweed is similar to the C_4 photosynthesis seen in warm season grasses and other terrestrial plants in high light/high temperature conditions. This allows the plants to capture carbon dioxide in the dark, later converting it to typical C_3 sugars (Casati et al. 2000).

Decomposers

Detritus feeders and decomposers that feed on decaying plants and other organic matter form another important link in the system. These organisms release minerals to be utilized in a new cycle of growth. Decomposition may be slow in aquatic systems because of a lack of oxygen at the lake or stream bottom; very acidic conditions also inhibit decomposition, leading to a build-up of peat.

An excessive amount of organic matter, such as occurs when high nutrient levels stimulate an algae bloom, may create anoxic conditions as decomposers deplete the oxygen level, depriving other organisms of an adequate source.

Habitat Structure

The zone in which aquatic plants can grow is called the littoral zone (Figure 1.6). Greater water depth, where light cannot reach the bottom to support photosynthesis, is referred to as the benthic zone. The littoral zone often shows strong zonation with emergent plants such as pickerel-weed (*Pontederia cordata*), arrowhead (*Sagittaria* spp.), bur-reeds (*Sparganium* spp.), and spike-rushes (*Eleocharis* spp.) closest to the shore (Figure 1.7). Beyond them are the water-lilies and other rooted floating-leaf plants. Rooted submergents can be found at greater depths and throughout the littoral zone, although a dense stand of water-lilies (*Nymphaea odorata* and *Nuphar* spp.) or watershield (*Brasenia schreberi*) may limit the light too much to allow other plants to grow beneath them.

Plants provide cover for young fish, tadpoles, and salamander larvae. Many algae grow attached to the surfaces of larger plants as epiphytes. Insects and other invertebrates such as fresh water sponges also live on underwater plant surfaces. Emergent plants provide sites for dragonflies, and other insects with aquatic larval stages, to crawl up out of the water in order to emerge as aerial adults (Figure 1.8).

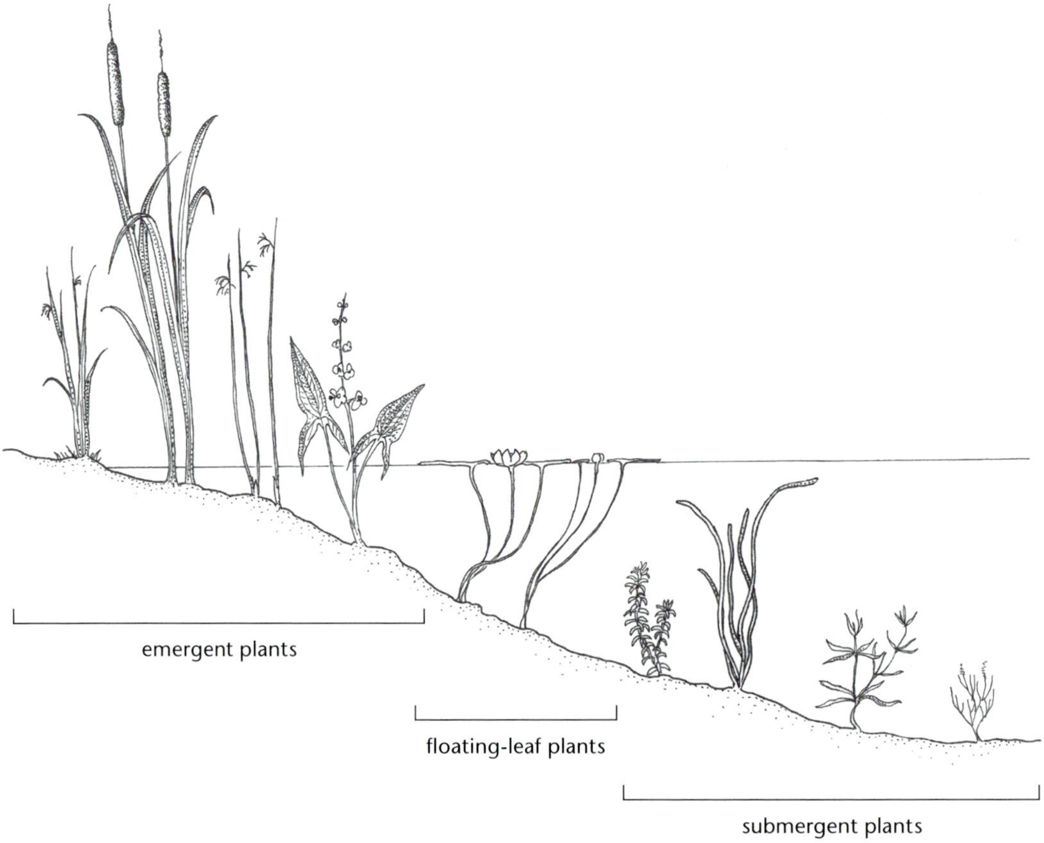

Figure 1.6. Diagram of the littoral zone showing bands of emergent, floating-leaf, and submergent plants.

Figure 1.7. Zonation in littoral zone vegetation at a lake in Wayne County, PA; note emergent plants closest to the shoreline with floating-leaf species farther out and open water beyond.

Figure 1.8. Shed skin (exuvia) left clinging to emergent vegetation following metamorphosis of a dragonfly from its aquatic larval stage.

TYPES OF AQUATIC ECOSYSTEMS

Lakes

All lakes, whether they are natural or created by damming, are on a trajectory that will result in filling and eventual conversion to swamp or marsh and then dry land. This process is accelerated by the production of abundant organic matter in the lake itself, but is slower in less productive systems.

Most of the natural lakes in Pennsylvania are found in the glaciated regions of the northeastern and northwestern corners of the state (Figure 1.9). Some are isolated kettle lakes that lack an outlet; depending on the depth and slope of the depression, they may be surrounded by a bog mat.

Bogs develop in shallow basins where acidic, low-nutrient conditions favor the growth of sphagnum mosses and associated bog vegetation around the lake margins. These lakes are termed dystrophic. Their water is typically stained brown by organic acids that leach out of the peat and other decaying vegetation. They will eventually become bogs, bog forests, and finally terrestrial forests as they fill with peat and dry out over time (Figures 1.10, 1.11).

Glacial lakes with low nutrient conditions, exceptionally clear water, and sparse growth of aquatic plants are termed oligotrophic. Some of these lakes drop off quickly to depths up to 30 m; as a result the littoral zone is narrow and sediment accumulation is low. Well developed bogs are not present. These lakes may be isolated kettle lakes, or water bodies with both inlets and outlets. Some appear to have been formed due to damming of streams by glacial deposits; but all are characterized by low within-lake biomass production. Many of the aquatic plants classified as endangered and threatened in Pennsylvania are found in lakes at the oligotrophic end of the spectrum (see below).

Other natural lakes in Pennsylvania would probably be classified as mesotrophic. These are lakes with moderate nutrient availability that support the growth of a variety of plants. Mesotrophic conditions may reflect natural inputs from the underlying geology or moderate enrichment due to water-

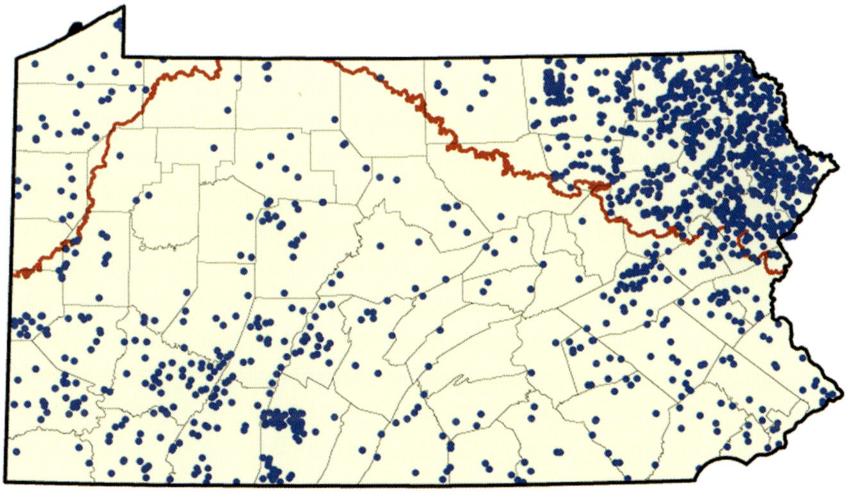

Figure 1.9. Distribution of lakes in Pennsylvania with respect to the southern boundary of the Wisconsinan (most recent) glaciation; lake location data provided by the Pennsylvania Department of Environmental Protection Clean Lakes Program.

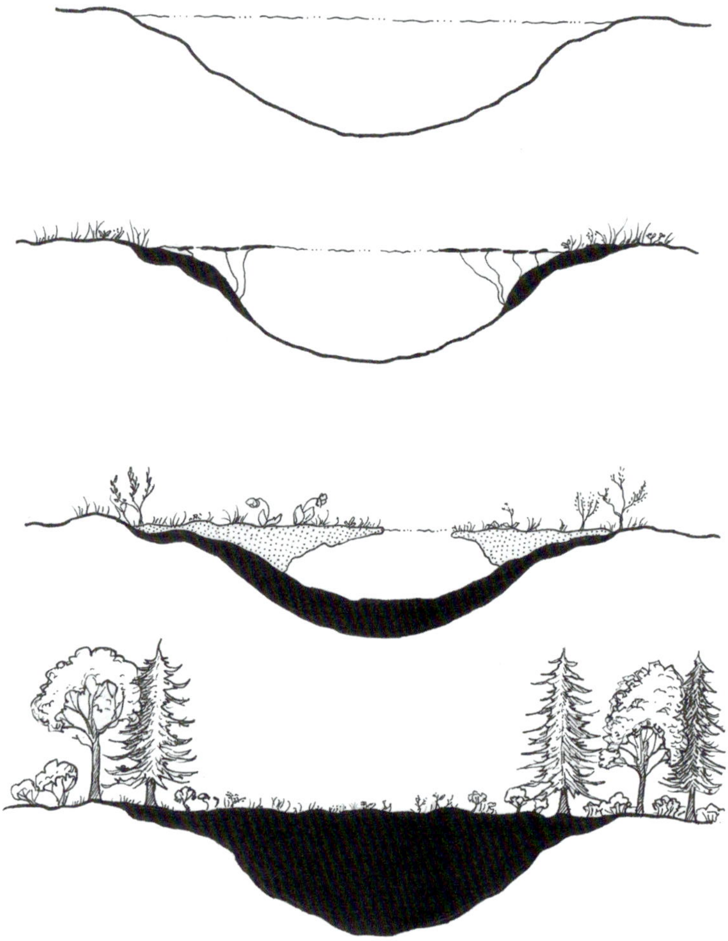

Figure 1.10. Succession from lake to bog and eventually dry land in a glacial kettle lake.

Figure 1.11. A glacial kettle lake surrounded by a floating bog mat in Sullivan County, PA.

shed conditions. Glacial lakes in northwestern Pennsylvania are in this category due to the limestone content of the underlying glacial deposits.

Eutrophic lakes, at the other end of the spectrum, are systems that have undergone nutrient enrichment with nitrogen and phosphorus derived from external sources such as agriculture, sewage systems, or fertilized lawns in the immediate watershed.

In addition to natural lakes, Pennsylvania has many lakes that were formed by humans damming a stream or wetland. These lakes are usually shallow (except perhaps in the former channel) and support abundant aquatic growth throughout. When coupled with nutrient enrichment, they become eutrophic. Abundant plant growth results in rapid build-up of sediments and accelerated successional development toward marsh and terrestrial conditions.

Rivers and Streams

Rivers and streams provide a variety of habitats for aquatic plants, from rapids, where riverweed (*Podostemum ceratophyllum*) and aquatic mosses cling to the rocks, to shoreline shallows and backwaters where a diversity of emergent and submergent species may be found. Factors such as water quality and rate of flow affect the occurrence of aquatic vegetation. Lack of light may limit the growth of some aquatic species in small streams in forested areas. In urban and suburban landscapes scouring due to heavy flow during and after rainstorms may limit the growth of rooted submergent species.

Delaware Estuary

The tidal influence of the Atlantic Ocean extends up the Delaware River to the fall line, where elevation of the river exceeds the elevation of the high tide line. The Pennsylvania portion of the estuary extends from the Pennsylvania/Delaware state line to Morrisville in Bucks County. Twice a day the water level in the estuary rises and falls as much as 2 m, creating tidal marshes along the shoreline of the river and its tributaries (Figure 1.12). The tidal influence also

Figure 1.12. Freshwater tidal marsh on the bank of the Delaware River in Bucks County, PA.

extends up tributary streams; however, in some cases, such as the Schuylkill River and Neshaminy Creek, dams have truncated its extent.

Classified as freshwater tidal marshes, despite the fact that the water in the estuary is slightly brackish, the marshes support a group of highly specialized aquatic plants. Many freshwater tidal marsh plants are classified as endangered, threatened, or rare by the Pennsylvania Natural Heritage Program, including water-hemp ragweed (*Amaranthus cannabinus*), swamp beggar-ticks (*Bidens bidentoides*), showy bur-marigold (*Bidens laevis*), American waterwort (*Elatine americana*), dwarf spike-rush (*Eleocharis parvula*), mud-plantain (*Heteranthera multiflora*), long-lobed arrowhead (*Sagittaria calycina*), subulate arrowhead (*Sagittaria subulata*), river bulrush (*Schoenoplectus fluviatilis*), Smith's bulrush (*Schoenoplectus smithii*), Walter's barnyard-grass (*Echinochloa walteri*), and wild-rice (*Zizania aquatica*).

Freshwater tidal marshes are an endangered habitat in Pennsylvania. They are limited to the narrow strip of Atlantic Coastal Plain in the state, and development of Philadelphia and the adjacent riverfront areas of Delaware and Bucks Counties has taken a heavy toll. Several plants that once grew along the tidal shores of the Delaware are believed to be extirpated. Parker's pipewort (*Eriocaulon parkeri*), awl-shaped mudwort (*Limosella australis*), and Nuttall's mud-flower (*Micranthemum micranthemoides*) have not been seen in Pennsylvania in over 50 years. Nuttall's mud-flower is believed to be extinct throughout its range.

Threats to freshwater tidal marshes include riverbank erosion, leading to loss of fine sediments (Figure 1.13) and colonization by non-native, invasive plants. Common reed (*Phragmites australis*) is the most troublesome of the invaders. In addition, Chinese lobelia (*Lobelia chinensis*), a low-growing plant from the Asia-Pacific region, has also become common along the tidal shores of the Delaware River.

Figure 1.13. Tidal riverbank showing absence of fine sediments and tidal marsh vegetation except in the area protected by a discarded section of dredge pipe.

Sea level rise, which is already occurring, is another threat. To survive, tidal marshes will have to accrete (grow upward) due to the gradual increase in elevation brought on by sediment deposition, or migrate inland. However, rates of sediment deposition are too slow to keep up with the current increase in sea level, which is expected to be approximately one meter by the year 2100. Unfortunately, a recent study has revealed that in very few cases is land available for freshwater tidal marshes in Pennsylvania to migrate inland (Titus et al. 2009).

MANAGEMENT OF AQUATIC ECOSYSTEMS

Endangered, Threatened, and Rare Species

About 60 species of aquatic plants are classified as endangered, threatened, rare, undetermined (candidate), or watch list by the Pennsylvania Natural Heritage Program at this time (PNHP 2010). One species, northeastern bulrush (*Scirpus ancistrochaetus*), is also listed as threatened under the Federal Endangered Species Act (Figure 1.14; Table 1.2).

Because of their rarity, these plants make an important contribution to the overall biological diversity of aquatic ecosystems. Many, such as water lobelia (*Lobelia dortmanna*), bayonet rush (*Juncus militaris*), floating-heart (*Nymphoides cordata*), slender water-milfoil (*Myriophyllum tenellum*), and horned bladderwort (*Utricularia cornuta*) are associated with oligotrophic glacial lakes; Pennsylvania populations represent the southern limit of range of these species.

An additional twelve species are believed to be extirpated in Pennsylvania. Of these, four are plants that were found in the freshwater intertidal zone.

PNHP-listed plants are protected under the Pennsylvania Code, Title 17, Chapter 45, Conservation of Pennsylvania Wild Plants and implementing

Figure 1.14. Northeastern bulrush (*Scirpus ancistrochaetus*), a federally threatened aquatic plant that occurs in Pennsylvania.

Table 1.2. Endangered, Threatened, and Rare Aquatic Plants of Pennsylvania

COMMON NAME	SCIENTIFIC NAME	GLOBAL (G) AND STATE (S) RANKS[a]	PNHP STATUS[b]
water-plantain	*Alisma triviale*	G5, S1	PE/PE
water-hemp ragweed	*Amaranthus cannabinus*	G5S3	PR/PR
water-marigold	*Bidens beckii*	G4G5, S1	PE/PE
swamp beggar-ticks	*Bidens bidentoides*	G3G4, S1	PT/PE
small beggar-ticks	*Bidens discoidea*	G5, S3	N/PR
showy bur-marigold	*Bidens laevis*	G5S3	N/TU
twig-rush	*Cladium mariscoides*	G5, S2	PE/PE
Walter's barnyard-grass	*Echinochloa walteri*	G5, S1	PE/PE
American waterwort	*Elatine americana*	G4, SU	PX/PE
dwarf spike-rush	*Eleocharis parvula*	G5, S1	PE/PE
spike-rush	*Eleocharis obtusa* var. *peasei*	G5, S1	PE/PE
four-angled spike-rush	*Eleocharis quadrangulata*	G4, S1	PE/PE
Robbins' spike-rush	*Eleocharis robbinsii*	G4G5, S2	PT/PT
ten-angle pipewort	*Eriocaulon decangulare*	G5, SX	PX/PX
Parker's pipewort	*Eriocaulon parkeri*	G3, SX	PX/PX
mud-plantain	*Heteranthera multiflora*	G4, S1	PE/PE
southern blue flag	*Iris virginica*	G5, S2	N/PE
quillwort	*Isoetes valida*	G4?, S1S3	N/PR
bayonet rush	*Juncus militaris*	G4, S1	PE/PE
little water duckweed	*Lemna obscura*	G5, SX	PX/PX
minute duckweed	*Lemna perpusilla*	G5, SU	N/TU
winter duckweed	*Lemna turionifera*	G5, SU	TU/TU
pale duckweed	*Lemna valdiviana*	G5, SH	PX/PX
awl-shaped mudwort	*Limosella australis*	G4G5, SX	PX/PX
water lobelia	*Lobelia dortmanna*	G4G5, S2	PT/PT
water-horehound	*Lycopus rubellus*	G5, S1	PE/PE
Nuttall's mud-flower	*Micranthemum micranthemoides*	SH/SX	PX/PX
Farwell's water-milfoil	*Myriophyllum farwellii*	G5, S3	PE/PR
broad-leaved water-milfoil	*Myriophyllum heterophyllum*	G5, S1	PE/SP
northern water-milfoil	*Myriophyllum sibiricum*	G5, S1	PE/PE
slender water-milfoil	*Myriophyllum tenellum*	G5, S2	PT/PT
whorled water-milfoil	*Myriophyllum verticillatum*	G5, S1	PE/PE
small yellow water-lily	*Nuphar microphylla*	G5, S1	TU/PE
floating-heart	*Nymphoides cordata*	G5, S2	PT/PT
goldenclub	*Orontium aquaticum*	G5, S4	PR/W
Tuckerman's pondweed	*Potamogeton confervoides*	G4, S2	PT/PT
threadleaf pondweed	*Potamogeton filiformis*	G5, SH	TU/PX
Fries' pondweed	*Potamogeton friesii*	G4, S1	PE/PE
grassy pondweed	*Potamogeton gramineus*	G5, S1	PE/PE
Hill's pondweed	*Potamogeton hillii*	G3, S1	PE/PE
Oakes' pondweed	*Potamogeton oakesianus*	G4, S1S2	TU/PE
blunt-leaved pondweed	*Potamogeton obtusifolius*	G5, S1	PE/PE
white-stem pondweed	*Potamogeton praelongus*	G5, S1	PX/PE
heartleaf pondweed	*Potamogeton pulcher*	G5, S1	PE/PE
redhead pondweed	*Potamogeton richardsonii*	G5, S3	PT/PR
narrow-leaved pondweed	*Potamogeton strictifolius*	G5, SH	PE/PE

Table 1.2 *(Continued)*

COMMON NAME	SCIENTIFIC NAME	GLOBAL (G) AND STATE (S) RANKS[a]	PNHP STATUS[b]
Tennessee pondweed	*Potamogeton tennesseensis*	G2, S1	PE/PE
Vasey's pondweed	*Potamogeton vaseyi*	G4, S1	PE/PE
flat-stemmed pondweed	*Potamogeton zosteriformis*	G5, S2S3	PR/PR
comb-leaved mermaid-weed	*Proserpinaca pectinata*	G5, SX	PX/PX
white water-crowfoot	*Ranunculus aquatilis* var. *diffusus*	G5, S3	N/PR
yellow water-crowfoot	*Ranunculus flabellaris*	G5, S2	N/PT
long-lobed arrowhead	*Sagittaria calycina*	G5T4, S1	PE/PE
arrowhead	*Sagittaria cuneata*	G5, S1	N/PE
arrowhead	*Sagittaria filiformis*	G4G5, SX	PX/PX
subulate arrowhead	*Sagittaria subulata*	G4, S3	PR/PR
great bulrush	*Schoenoplectus acutus*	G5, S2	PE/PE
river bulrush	*Schoenoplectus fluviatilis*	G5, S3	PR/PR
Smith's bulrush	*Schoenoplectus smithii*	G5?, S1	PE/PE
water bulrush	*Schoenoplectus subterminalis*	G4G5, S3	N/PR
Torrey's bulrush	*Schoenoplectus torreyi*	G5?, S1	PE/PE
northeastern bulrush	*Scirpus ancistrochaetus*	G3, S3, LT[c]	PE/ PT
branching bur-reed	*Sparganium androcladum*	G4G5, S1	PE/PE
narrow-leaf bur-reed	*Sparganium angustifolium*	G5, S2	N/PT
small bur-reed	*Sparganium natans*	G5, SX	PX/PX
horned bladderwort	*Utricularia cornuta*	G5, S2	N/PT
flat-leaved bladderwort	*Utricularia intermedia*	G5, S2	PT/PT
floating bladderwort	*Utricularia radiata*	G4, SX	PE/PX
northeastern bladderwort	*Utricularia resupinata*	G4, SX	PX/PX
slender bladderwort	*Utricularia subulata*	G5, SX	N/PX
bogmat	*Wolffiella gladiata*	G5, S2	PR/PR
wild-rice	*Zizania aquatica*	G5, S3	PR/PR

[a] Global ranks: G5 = secure, G4 = apparently secure, G3 = vulnerable; state ranks: S3 = vulnerable, S2 = imperiled, S1 = critically imperiled, SU = status uncertain, SX = apparently extirpated.

[b] PNHP status: PE = Pennsylvania endangered, PT = Pennsylvania threatened, PR = Pennsylvania rare, PX = extirpated in Pennsylvania, TU = tentatively undetermined, N = not listed, SP = special population, W = watch list.

[c] Federal rank: LT = listed threatened.

regulations (Commonwealth of Pennsylvania 1993). The program is administered by the Pennsylvania Natural Heritage Program and the Bureau of Forestry of the Department of Conservation and Natural Resources.

Problem Vegetation

Exotic invasive species—Non-native plants have invaded aquatic ecosystems as well as terrestrial habitats (Table 1.3). In low nutrient systems they are usually not a problem, as most serious invasives require high resource availability to grow vigorously. Where nutrients are not limiting, species like European water-chestnut (*Trapa natans*) (Figure 1.15), Eurasian water-milfoil (*Myriophyllum spicatum*), hydrilla (*Hydrilla verticillata*), fanwort (*Cabomba caroliniana*), or curly pondweed (*Potamogeton crispus*) can become dominant. Invasive spe-

Table 1.3. Non-native, Invasive Aquatic Plants in Pennsylvania

COMMON NAME	SCIENTIFIC NAME	% OF LAKES[a]
long-stem waterwort	*Elatine triandra*	15.7
curly pondweed	*Potamogeton crispus*	13.0
Eurasian water-milfoil	*Myriophyllum spicatum*	9.6
cultivated water-lilies	*Nymphaea* spp.	9.6
yellow iris	*Iris pseudacorus*	4.3
water-chestnut	*Trapa natans*	3.5
hydrilla	*Hydrilla verticillata*	3.0
fanwort	*Cabomba caroliniana*	2.6
waternymph	*Najas minor*	1.8
Brazilian waterweed	*Egeria densa*	<1.0
European water-clover	*Marsilea quadrifolia*	<1.0
flowering-rush	*Butomus umbellatus*	<1.0
mudmat	*Glossostigma cleistanthum*	<1.0
parrot-feather	*Myriophyllum aquaticum*	<1.0
watercress	*Nasturtium officinale*	streams and springs
water-starwort	*Callitriche stagnalis*	streams and springs

[a] Based on surveys of 115 lakes between 2000 and 2007.

cies are also more likely to become a problem in recently created impoundments where a native flora has not yet become established, or in lakes where excessive use of herbicides has eliminated the native plants. Dense growth of species like water-chestnut or Eurasian water-milfoil not only interferes with native plant growth, but also interferes with recreational uses such as boating and swimming.

Prevention is far more effective than attempts to control an invasive plant after it has become established. Prevention can take the form of limiting

Figure 1.15. An infestation of the non-native, invasive water-chestnut (*Trapa natans*) at a lake in Bucks County, PA.

access by boats from other areas that might carry seeds or plant fragments and educating lake users to prevent deliberate or accidental introductions of non-native plants. Another approach is to protect water quality; nutrient enrichment (eutrophication) will exacerbate problems with over-abundant vegetation.

Physical control efforts such as pulling, cutting, or raking may provide temporary relief; however, the ability of many aquatic plants to propagate themselves from detached fragments should be kept in mind.

Over-abundant native species—Not all problems arising from excessive growth of aquatic vegetation are caused by non-native species. Under favorable growing conditions some native species can also form dense stands that interfere with recreational uses of lakes. Shallow water and high nutrient availability are most often the causes.

Opportunistic species that have created problems include fragrant waterlily (*Nymphaea odorata*) (Figure 1.16), Farwell's water-milfoil (*Myriophyllum farwellii*), common water-milfoil (*M. humile*), broad-leaved water-milfoil (*M. heterophyllum*), waterweed (*Elodea nuttallii*), purple bladderwort (*Utricularia purpurea*), inflated bladderwort (*Utricularia inflata*), and coontail (*Ceratophyllum demersum*).

Prevention of problems caused by explosive growth of native species should focus on reducing nutrient inputs.

Figure 1.16. Water surface nearly covered by native fragrant water-lily (*Nymphaea odorata*) at a shallow pond in Lackawanna County, PA.

Chapter 2. Identification Keys for the Aquatic Plants of Pennsylvania

The keys in this book are designed to guide the user through a series of decisions leading to proper identification. At each step of the key, the user must examine a characteristic or characteristics of the plant and decide which of a pair of corresponding statements best fits the specimen in question. Each statement and its corresponding member are preceded by the same capital letter(s). There are a few simple rules that, if followed carefully, should ease the task of arriving at a correct identification.

1. Read each statement of a pair carefully and thoroughly before deciding which best fits the plant you are looking at.
2. Be sure you understand any botanical terminology used in the statements before moving on to the next step. Consult the glossary at the back of the book if you are not certain.
3. Look at more than one example of the characteristic in question. For example, if the key asks about the number of petals present in a flower, look at more than one flower.
4. Don't dwell on the exceptions! Most plant species are quite variable. Throughout the keys you will see qualifiers such as "often," "usually," or "mostly" to account for the lack of uniformity.
5. After reaching a tentative identification, read the descriptions and look at the illustrations to confirm that you are correct.
6. Make your observations slowly and carefully, especially if you are new to "keying out" plants.

For convenience the initial key below divides aquatic plants into six groups based on growth habit: emergent, floating-leaf, submergent, etc. The next step is to use the section keys to arrive at a genus and finally the keys to each genus, which are found in the succeeding chapters, to name a plant to species.

The keys assume the plant's appearance under normal water levels. Many species assume unusual growth habits and/or morphological forms when the water is unusually deep or shallow. Some species will appear in more than one habit category, since they assume more than one form under normal conditions. Whether a plant is considered emergent, submergent, or floating depends on whether the main body (leaves and/or primary axis) occurs above, below, or at the surface of the water. Species that produce only their flowers above the water surface are not considered emergent or floating-leaved. Be aware that exceptional variations may, and occasionally do, occur.

Key to Growth Habit Groups

A. plants without distinguishable leaves and stems, or tiny floating plants with minute leafy stems (leaves less than 1 mm long)..Thalloid and/or tiny floating plants, p. 163

A. plants with distinct leaves and/or stems; leaves, if present, more than 1 mm long

B. plants with at least some leaves and/or stems held above the water surface .... Emergent plants, p. 29

B. plants entirely under water or with at least some leaves floating at the surface

C. plants with at least some leaves floating on the water surface ....Floating-leaf plants, p. 143; or some pondweeds (*Potamogeton* spp.), p. 186

C. plants with leaves entirely or mostly under water, or buried in a sandy or peaty substrate; at most flowering above the surface or with only leaf tips floating at the surface; or stems sometimes leafless

D. underwater leaves broad to very narrow, but not finely divided

E. underwater leaves attached to an elongate stem, or submergent stems sometimes leafless, plants rooted or free-floating.. Submergent leafy-stemmed plants, p. 183

E. underwater leaves in a basal rosette, plant rooted on the bottom...Submergent rosette-forming plants, p. 241

D. underwater leaves very finely divided.. Submergent plants with finely divided leaves, p. 255

Key to Emergent Plants

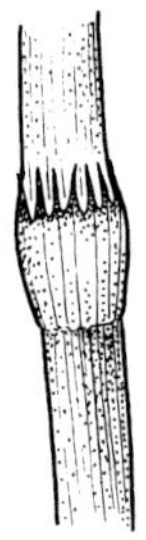

Figure 2.1. Stems jointed, tiny scale-like leaves in whorls at the node (×3/4).

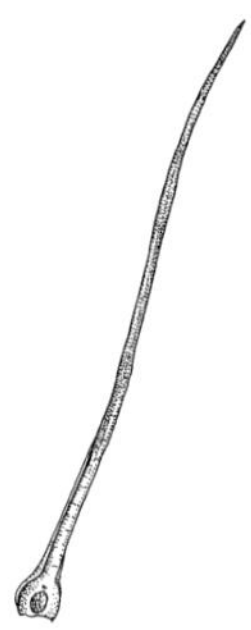

Figure 2.2. Leaf expanded at the base, containing spore sac (×3/8).

A. emergent leaves long and narrow, their blades more than 7 times longer than broad, or leaves absent

B. leaves absent

C. stems jointed; tiny, scale-like leaves in whorls at the nodes.. water horsetail (*Equisetum* spp.), p. 88

C. stems not jointed, not branching in whorls

D. inflorescence appearing lateral...... bulrush (*Schoenoplectus* spp.), p. 47

D. inflorescence terminal ....spike-rush (*Eleocharis* spp.), p. 111

B. leaves present

E. leaves all basal

F. leaves expanded at the base and containing spore sacs ....quillwort (*Isoetes* spp.), p. 245

F. leaves not expanded at the base, lacking spore sacs

G. leaves mostly not more than 10 cm long .... pipewort (*Eriocaulon* spp.), p. 243

G. leaves mostly more than 50 cm long.. flowering-rush (*Butomus umbellatus*), p. 76

E. at least some leaves cauline

H. perianth showy ....iris (*Iris* spp.), p. 90

H. perianth not showy, reduced to a few bristles, or absent

I. individual flowers of the inflorescence subtended by 1 or 2 bracts

J. stems mostly triangular at the base; each flower subtended by 1 bract

K. flowers in spikelets with 2-ranked scales................................ three-way sedge (*Dulichium arundinaceum*), p. 121

K. flowers in spikelets with spirally arranged scales
L. spikelets with 1 perfect flower and 1 achene ....twig-rush (*Cladium mariscoides*), p. 123
L. spikelets with many perfect flowers and many achenes .... bulrush (*Scirpus ancistrochaetus* and *Schoenoplectus* spp.), p. 47
J. stems round at base; each flower subtended by 2 bracts ....grasses (*Echinochloa, Glyceria* spp., *Leerzia oryzoides, Torreyochloa pallida, Zizania aquatica*), p. 79
I. individual flowers of the inflorescence not subtended by bracts
M. flowers perfect.................bayonet rush (*Juncus militaris*), p. 102
M. flowers unisexual
N. flowers in a dense, elongate spike; leaves flat... cat-tail (*Typha* spp.), p. 71
N. flowers in spherical heads; leaves keeled, at least at the base .... bur-reed (*Sparganium* spp.), p. 61
A. emergent leaves broad, their blades less than 7 times longer than broad
O. plants woody at the base; branches arching over the water's edge ....water-willow (*Decodon verticillatus*), p. 137
O. plants herbaceous throughout
P. leaves very large, roundish, often 30 cm or more across .... lotus (*Nelumbo* spp.), p. 147
P. leaves never more than 20 cm across
Q. leaves compound
R. leaves opposite..................beggar-ticks/bur-marigold (*Bidens*), p. 42
R. leaves alternate
S. plants to 2 m tall; leaflets toothed...water-parsnip (*Sium suave*), p. 128
S. plants seldom more than 0.5 m tall; leaflets entire ....watercress (*Nasturtium officinale*), p. 141
Q. leaves simple
T. leaves opposite..................beggar-ticks/bur-marigold (*Bidens*), p. 42
T. leaves alternate
U. leaf blades lobed or truncate at the base
V. leaves kidney-shaped, mostly broader than long...mud-plantain (*Heteranthera* spp.), p. 96
V. leaves longer than broad
W. leaf blades shallowly lobed, cordate, or truncate at the base
X. inflorescence a spike ....lizard's-tail (*Saururus cernuus*), p. 94
X. inflorescence a panicle.. water-plantain (*Alisma* spp.), p. 129
W. leaf blades deeply lobed at base
Y. inflorescence a dense fleshy spike enclosed in a sheathing bract .... arrow-arum (*Peltandra virginica*), p. 29
Y. inflorescence not enclosed in a sheathing bract
Z. flowers purple ....pickerel-weed (*Pontederia cordata*), p. 100
Z. flowers white......arrowhead (*Sagittaria* spp.), p. 31

U. leaf blades tapering at both ends

AA. leaves parallel-veined; inflorescence a thick spike of yellow flowers .... goldenclub (*Orontium aquaticum*), p. 77

AA. leaves net-veined; inflorescence various

BB. emergent leaves toothed or lobed......... mermaid-weed (*Proserpinaca* spp.), p. 273

BB. emergent leaves entire

CC. emergent leaves 2–5 mm long

DD. flowers unisexual, petals absent ....water-starwort (*Callitriche terrestris*), p. 229

DD. flowers bisexual, petals white .... Nuttall's mud-flower (*Micranthemum micranthemoides*), p. 14

CC. emergent leaves more than 5 mm long

EE. inflorescence spike-like .... smartweed (*Persicaria* spp.), p. 104

EE. inflorescence not spike-like

FF. flowers yellow; fruits 2 cm or more long ....water-primrose or primrose-willow (*Ludwigia* spp.), p. 132

FF. flowers pale violet to white or inconspicuous

GG. flowers small, inconspicuous, radially symmetrical; leaves ovate to obovate.... water-purslane (*Ludwigia palustris*), p. 136

GG. flowers pale violet to white, 2-lipped; leaves linear to lanceolate .... water-willow (*Justicia americana*), p. 137

Key to Floating-leaf Plants

A. leaves strap-like or ribbon-like, mostly submerged, floating only at the tips

B. midvein more prominent than the other veins of the leaf, cross veins not visible.. bur-reed (*Sparganium* spp.), p. 61

B. midvein not more prominent than the other veins of the leaf, cross veins conspicuous.................................. water-celery (*Vallisneria americana*), p. 253

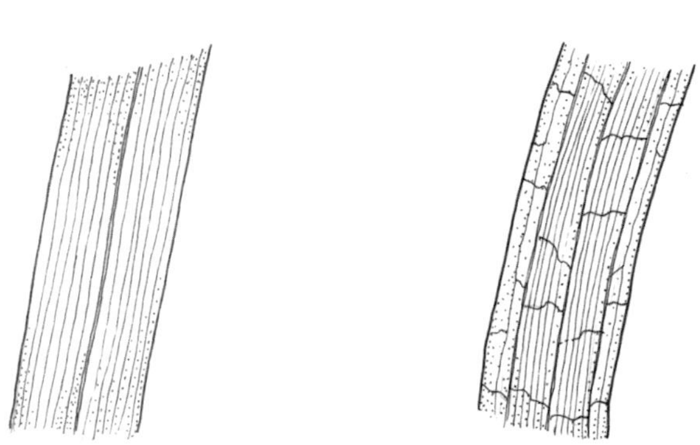

Figure 2.3. Leaf venation with and without prominent mid-vein.

Prominent midvein present
Bur-reed (*Sparganium*) ×1.4

Prominent midvein absent
Water-celery (*Vallisneria*) ×1.6

A. leaves not strap-like, entire blade floating
  C. floating leaves compound, shaped like a four-leaf clover .... European water-clover (*Marsilea quadrifolia*), p. 151
  C. floating leaves simple
    D. floating leaves deeply notched or lobed at base of blade, or petiole attached near center of blade
      E. petiole attached near the center of the blade
        F. leaves very large, roundish, often 30 cm or more across .... lotus (*Nelumbo* spp.), p. 147
        F. leaves elliptical to linear-elliptic, never more than 20 cm across
          G. floating leaves linear-elliptic, to about 2 cm long.. fanwort (*Cabomba caroliniana*), p. 272
          G. floating leaves elliptic, 10–15 cm long .... watershield (*Brasenia schreberi*), p. 161
      E. petiole attached at the summit of a deep notch or sinus
        H. flowers borne singly .... water-lily (*Nuphar* spp. and *Nymphaea* spp.), p. 153
        H. flowers in an umbel............floating-heart (*Nymphoides* spp.), p. 143
    D. floating leaves not deeply notched or lobed at base, petiole attached near base of blade
      I. floating leaves in a rosette
        J. leaf blades more than 2 cm long, petioles swollen in the middle.... .... European water-chestnut (*Trapa natans*), p. 149
        J. leaf blades less than 2 cm long, petioles not swollen .... water-starwort (*Callitriche* spp.), p. 229
      I. floating leaves borne singly
        K. veins of the leaf blade parallel .... pondweed (*Potamogeton* spp.), p. 186
        K. veins of the leaf blade netted
          L. floating or emersed leaves generally kidney-shaped, 3- to 5-lobed......................... water-crowfoot (*Ranunculus* spp.), p. 275
          L. floating or emersed leaves lance-shaped to elliptical, not lobed
            M. floating or emersed leaves toothed.......................... Beck's water-marigold (*Bidens beckii*), p. 278
            M. floating or emersed leaves entire .... smartweed (*Persicaria amphibia*), p. 104

Key to Thalloid or Tiny Floating Plants

A. plants without distinguishable leaves and stems
  B. plants appearing to consist entirely of repeatedly branching fine stems or stem-like structures ....Charophyte algae (*Chara* spp. or *Nitella* spp.), p. 164
  B. plants consisting entirely of flattened stems or segments
    C. plants segmented, individual segments elliptical or rounded, solitary or in small clusters, or attached by strap-like stalks...................................... duckweed, watermeal, and water flaxseed, p. 166
    C. plants forming branching colonies of flattened stems
      D. plants attached to rocks in fast-flowing water ....riverweed (*Podostemum ceratophyllum*), p. 180
      D. plants free-floating ....aquatic liverworts (*Riccia* spp. or *Ricciocarpus* spp.), p. 163
A. plants with minutely leafy stems..eastern mosquito fern (*Azolla caroliniana*), p. 178

Key to Submergent Leafy-stemmed Plants

A. leaves scale-like, or stems leafless
  B. stems stiffly erect, leafless or with scattered, scale-like leaves ....water-milfoil (*Myriophyllum tenellum*), p. 289
  B. stems flexuous, more regularly leafy .... aquatic mosses (*Fontanalis* spp. or *Sphagnum* spp.), p. 183
A. leaves not scale-like
  C. submersed leaves alternate
    D. leaves with a basal sheath or sheathing stipules variously fused to the leaf base or free; flowers in a terminal or axillary spike; tepals 4, inconspicuous .... pondweed (*Potamogeton* spp.), p. 186
    D. leaves without basal sheaths or stipules; flowers terminal; tepals 6, yellow and conspicuous..................... water star-grass (*Heteranthera dubia*), p. 221
  C. submersed leaves opposite or whorled
    E. leaves not more than 0.5 mm wide .... pondweed (*Zannichellia palustris*), p. 220
    E. leaves more than 0.5 mm wide
      F. leaves whorled
        G. leaves mostly in whorls of 3........... waterweed (*Elodea* spp.), p. 233
        G. leaves mostly in whorls of 4–8
          H. leaf margins very finely toothed; leaves crowded, overlapping... .... Brazilian waterweed (*Egeria densa*), p. 233
          H. leaf margins distinctly spine-toothed; leaves not overlapping ... . hydrilla (*Hydrilla verticillata*), p. 185
      F. leaves opposite
        I. leaves with a narrow blade and a broader, sheathing base............... .... waternymph (*Najas* spp.), p. 223
        I. leaves tapering or truncate at the base, broadest in the blade
          J. leaves linear, spatulate, or obovate, weakly 3-nerved................. .... water-starwort (*Callitriche* spp.), p. 229
          J. leaves oblong to narrowly obovate, 0–1-nerved.......................... .... waterwort (*Elatine* spp.), p. 237

Waterweed (*Elodea* spp.) Brazilian waterweed (*Egeria densa*) Hydrilla (*Hydrilla verticillata*)

Figure 2.4. Leaf arrangement in waterweeds (not to scale).

Key to Submergent Rosette-forming Plants

A. leaves strongly flattened
  B. leaves mostly more than 20 times longer than wide
    C. midvein not more prominent than the other veins of the leaf; cross veins conspicuous
      D. cross veins in the central portion of leaves more densely reticulate..... .... water-celery (*Vallisneria americana*), p. 253

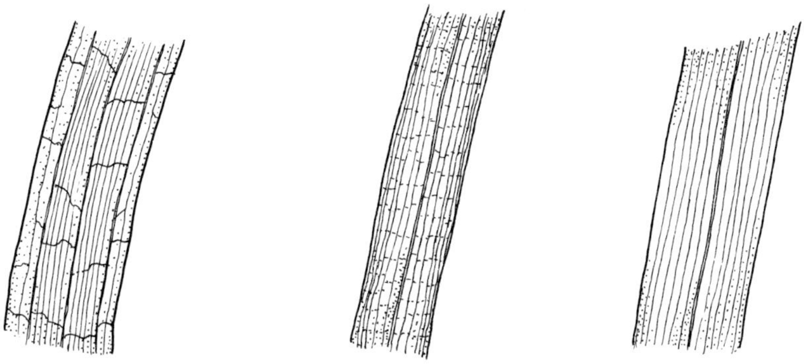

Figure 2.5. Vein reticulation (not to scale).

D. reticulate network of cross veins uniform throughout the leaf .... arrowhead (*Sagittaria* spp.), p. 31

C. midvein more prominent than the other veins of the leaf; cross veins not readily visible........ bur-reed (*Sparganium* spp.), p. 61

B. leaves less than 20 times longer than wide

E. plants mat-forming

F. leaves acute at the tips........ pipewort (*Eriocaulon* spp.), p. 243

F. leaves rounded at the tips

G. leaves to about 2 cm long, in pairs along the stolons........ mudmat (*Glossostigma cleistanthum*), p. 241

G. leaves mostly 2–5 cm long, in tufts of 5–10 along the stolons........ mudwort (*Limosella australis*), p. 14

E. plants not mat-forming

H. leaves rounded at the tips..... water lobelia (*Lobelia dortmanna*), p. 251

H. leaves acute at the tips........ pipewort (*Eriocaulon* spp.), p. 243

A. leaves generally round in cross section, not strongly flattened, or leaves absent, only stems apparent

I. leaves or stems stiff and firm, holding their shape out of water

J. leaf bases expanded and containing spore sacs........ quillwort (*Isoetes* spp.), p. 245

J. leaf or stem bases not containing spore sacs........ spike-rush (*Eleocharis* spp.), p. 111

I. leaves or stems limp, not holding their shape out of water

K. leaves reduced to bladeless sheaths at the base of the stems; flowering stems with a single terminal spikelet.... spike-rush (*Eleocharis* spp.), p. 111

K. basal leaves filamentous; flowering stems lacking a basal sheath, each bearing a single lateral spikelet.... bulrush (*Schoenoplectus subterminalis*). p. 53

Figure 2.6. Quillwort leaf with spore sac in the base (×3/8).

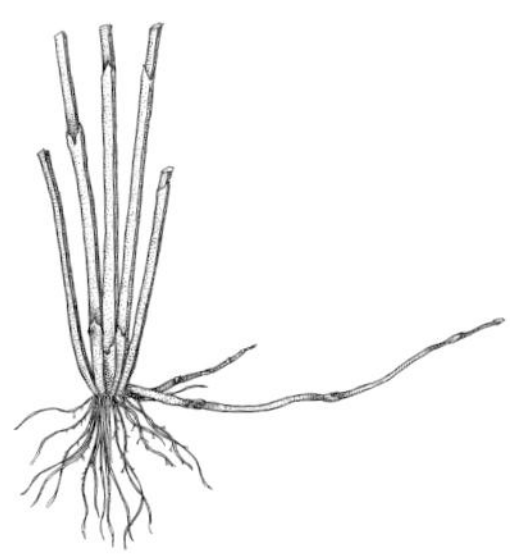

Figure 2.7. Bladeless leaf sheaths in a spike-rush (*Eleocharis* sp., ×1/4).

Key to Submergent Plants with Finely Divided Leaves

A. leaves and/or stems bearing small bladders.. bladderwort (*Utricularia* spp.), p. 255
A. leaves and/or stems not bearing bladders
- B. leaves opposite or whorled
  - C. leaves pinnately divided............ water-milfoil (*Myriophyllum* spp.), p. 280
  - C. leaves palmately divided, fan-like; the segments repeatedly forking
    - D. leaves remaining rigid out of water, often crowded toward the stem tips .... coontail (*Ceratophyllum* spp.), p. 269
    - D. leaves limp out of water, not crowded toward the stem tips
      - E. submersed leaves with a distinct petiole; flowers white .... fanwort (*Cabomba caroliniana*), p. 272
      - E. submersed leaves without a distinct petiole; flowers yellow .... Beck's water-marigold (*Bidens beckii*), p. 278
- B. leaves alternate
  - F. underwater leaves pinnately divided; simple emergent leaves often present .... mermaid-weed (*Proserpinaca* spp.), p. 273
  - F. underwater leaves palmately divided, fan-like; the segments repeatedly forking; emergent leaves never present .... water-crowfoot (*Ranunculus* spp.), p. 275

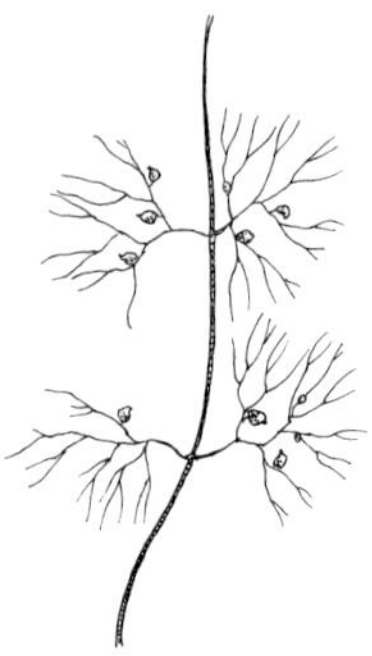

Palmately divided,
opposite leaves bearing traps
Bladderwort (*Utricularia*)

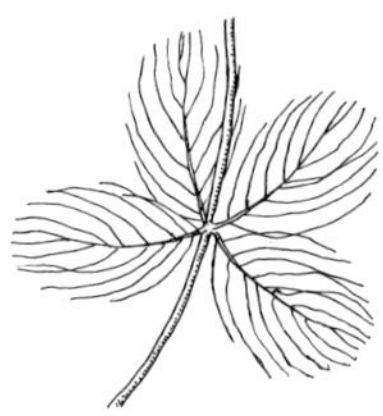

Pinnately divided,
whorled leaves
Water-milfoil (*Myriophyllum*)

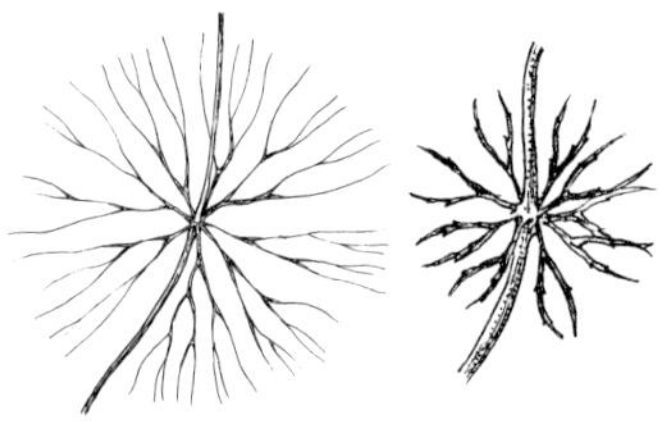

Palmately divided,
whorled leaves
Coontail (*Ceratophyllum*)

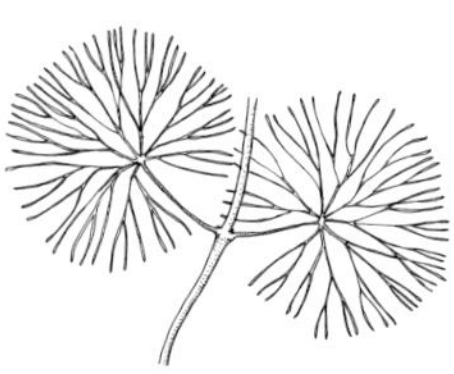

Palmately divided,
opposite leaves with petioles
Fanwort (*Cabomba*)

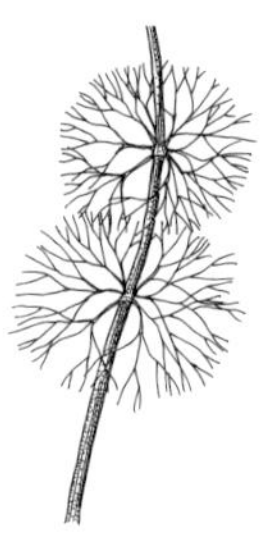

Palmately divided,
whorled leaves
Water-marigold (*Bidens beckii*)

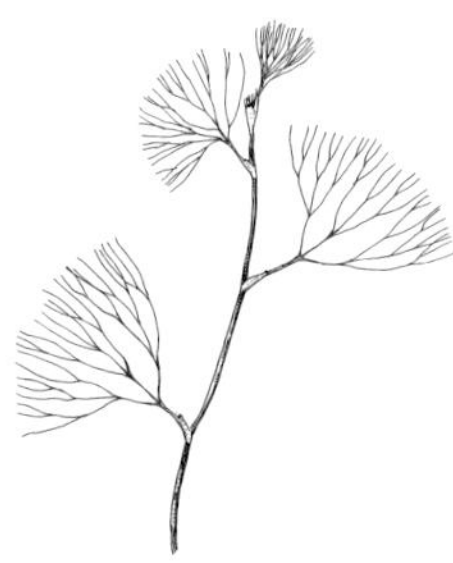

Palmately divided,
alternate leaves
Water-crowfoot (*Ranunculus aquatilis*)

Figure 2.8. Leaf form and arrangement (not to scale).

Chapter 3. Emergent Plants

Emergent species grow in shallow water at the margins of lakes or streams. These species are rooted below the water surface, but the stems, leaves, and flowers are mostly above the surface. Emergent plants often form a continuous band at the edge of the water at depths to about 1 m; the width of the emergent zone depends on the slope of the lake bottom.

ARROW-ARUM, TUCKAHOE
PELTANDRA RAF.
Arum Family—Araceae

Arrow-arum, tuckahoe — ***Peltandra virginica* (L.) Schott & Endl.**

Native
Rooted, emergent perennial

Arrow-arum is one of several emergent aquatic plants with large, arrow-shaped leaves. The plants consist of a cluster of long-petioled leaves that grow from buried vertical rhizomes. The strongly veined leaf blades are 5–20 cm long. Separate male and female flowers are packed closely on a fleshy stalk (spadix) that is enclosed in a tightly rolled green spathe. After pollination the flower stalk bends downward, pulling the developing fruits under the water surface, where the spathe eventually rots away, releasing the pea-sized fruits. The height of mature plants is up to 1 m.

Other emergent aquatic plants of similar stature and leaf shape include pickerel-weed (*Pontederia cordata*) and duck-potato (*Sagittaria latifolia*). Neither has a flower and fruit structure at all like arrow-arum. In addition, pickerel-weed has more rounded, heart-shaped leaves, and those of arrowhead are much thinner in texture and more variable in shape than arrow-arum.

The range of arrow-arum extends from southern Maine to Florida and inland to Michigan, Missouri, eastern Kansas, and Texas. In Pennsylvania it is found in the eastern and western portions of the state. Arrow-arum typically grows in water to 0.3 m deep in swamps and along the shorelines of rivers, lakes, or ponds. Flowering occurs from late spring through summer.

Native Americans used grated roots of arrow-arum in milk as a tonic for babies. Leaves, seeds, and fruits were also used for food. The fruits of arrow-arum are also eaten by wildlife, including waterfowl.

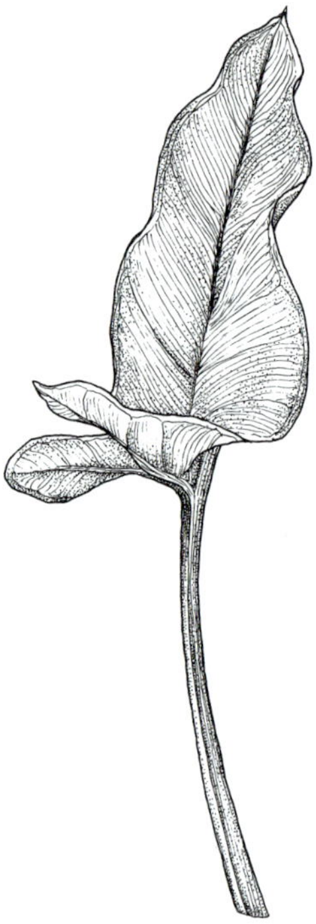

Peltandra virginica—
leaf ×1/4

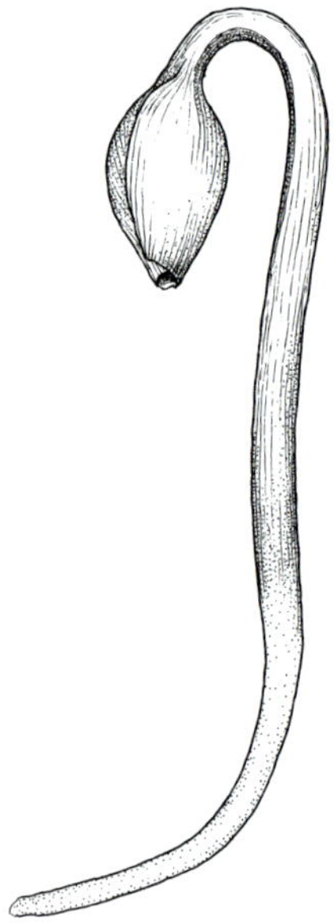

Peltandra virginica—
inflorescence ×1/4

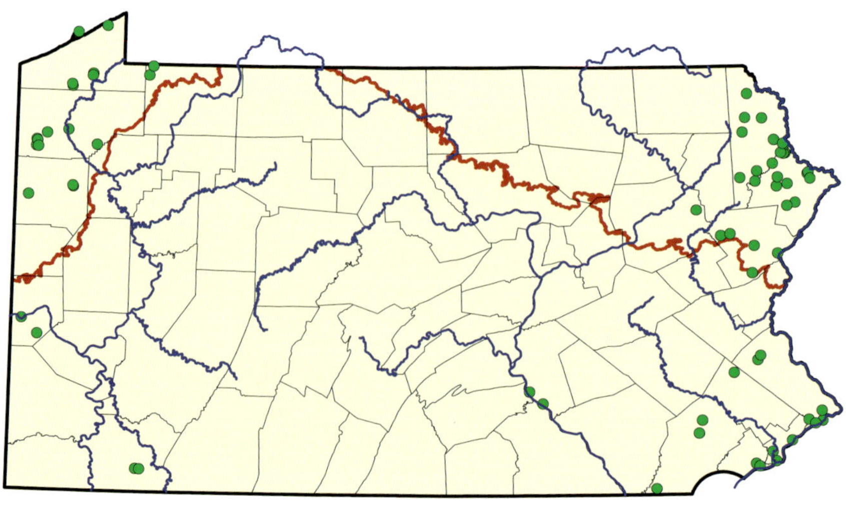

Peltandra virginica

ARROWHEAD

Sagittaria L.

Water-plantain Family—Alismataceae

The arrowheads are named for the shape of the leaf blade of the more prominent members. However, leaf shape is highly variable and influenced by water depth. Most species are emergent perennials; however, several grow completely submersed, and one species (*S. calycina*) is an annual. Individual flowers are unisexual, but both male and female flowers are borne on the same plant, making the species monoecious. The inflorescence is typically a raceme with female flowers in several whorls, each containing 3 flowers, at the bottom and male flowers in additional whorls above. The flowers are insect-pollinated. The fruit is a cluster of achenes. Seven species of arrowhead, all native, are present in Pennsylvania. An eighth, *S. filiformis*, known only from a few early collections in southern Bucks County, is believed to be extirpated.

Key to the Species of Arrowhead

A. sepals closely appressed in fruit .... *Sagittaria calycina*
A. sepals spreading or recurved in fruit
  B. leaf blades rarely, or never, sagittate
    C. stamens 7–15; plants low-growing, stoloniferous, mat-forming; leaf tips rounded; found only in freshwater intertidal marshes along the Delaware River .... *Sagittaria subulata*
    C. stamens 12–many; plants not mat-forming; leaf tips pointed in the rosette stage; of more widespread occurrence
      D. female flowers sessile .... *Sagittaria rigida*
      D. female flowers with a distinct pedicel .... *Sagittaria graminea*
  B. sagittate leaf blades usually present
    E. beak of the achene spreading to nearly horizontal .... *Sagittaria latifolia*
    E. beak of the achene ascending
      F. beak to 0.5 mm; achene without facial wings; petioles not ridged .... *Sagittaria cuneata*
      F. beak 0.7–1.7 mm; achene with 1 wing on each face; petioles ridged .... *Sagittaria australis*

Appalachian arrowhead — *Sagittaria australis* (J.G. Smith) Small

Native
Rooted, emergent perennial

Appalachian arrowhead is as much as 1 m tall with broad sagittate leaf blades. In overall appearance Appalachian arrowhead is very similar to the common *Sagittaria latifolia*, from which it is most easily distinguished by the presence of 5 longitudinal wings on the petioles. Like *S. latifolia* it produces stolons and corms. The inflorescence bears 5–12 whorls of 2.5 cm-wide flowers; the achenes have a strongly recurved beak.

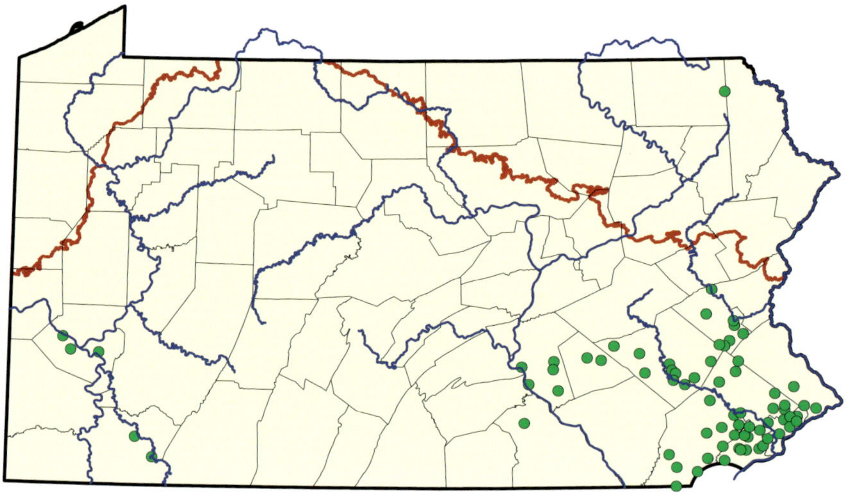

Sagittaria australis

Appalachian arrowhead grows from New York south to Mississippi and Louisiana. In Pennsylvania it is found in river backwaters, wet meadows, and wet woods in the southern half of the state. Flowering occurs from late July to early September.

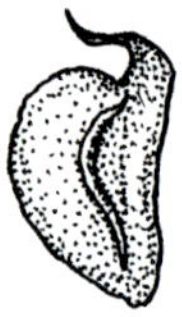

Sagittaria australis—achene ×5

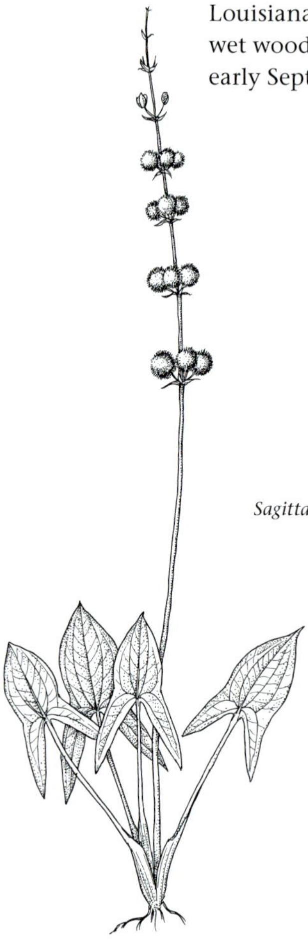

Sagittaria australis ×1/4

Long-lobed arrowhead

Sagittaria calycina Engelm. [synonym: _Sagittaria montevidensis_ Chamisso & Schlechtendal spp. _calycina_ (Engelm.) Bogin]

PNHP
Native
Rooted, emergent or submergent annual

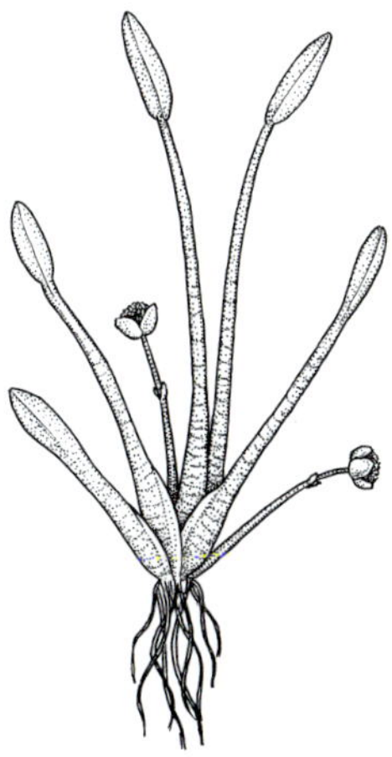

Sagittaria calycina ×1/2

In Pennsylvania long-lobed arrowhead is a rare plant of the freshwater tidal zone of the Delaware River and its tidal tributaries. Submersed plants have strap-shaped leaves with a blunt tip; under emergent conditions sagittate leaves with long, spongy petioles are produced. Inflorescences contain several whorls of flowers; sepals remain closely appressed to the fruiting head. Unlike subulate arrowhead, another freshwater tidal species, long-lobed arrowhead is not turf forming, but rather occurs singly or in small groups of 2–3 plants. Flowering occurs July to early September.

Along the east coast long-lobed arrowhead is limited to tidal mudflats in Delaware, Pennsylvania, and New York. It also has a large interior range in the Mississippi River basin, but the main range of this plant is in South America. It has apparently expanded its range northward since it was first collected near Wilmington, Delaware, in 1860.

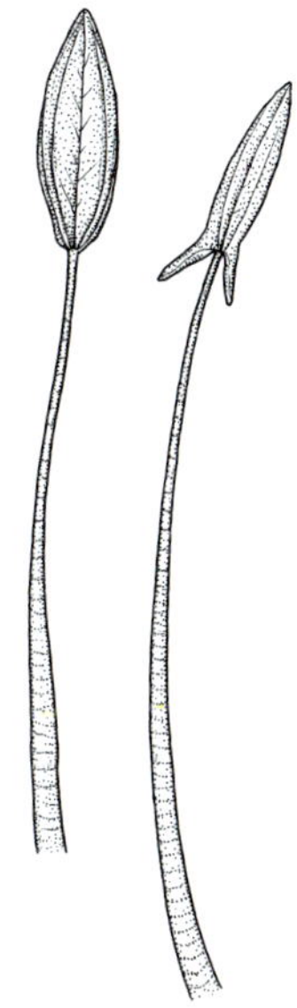

Sagittaria calycina—leaf variation ×1/2

Slightly wider leaves and appressed sepals help to differentiate long-lobed arrowhead from *S. subulata*. *Sagittaria rigida* and *S. graminea* also occur in the tidal zone, but these species have pointed leaves in the rosette stage.

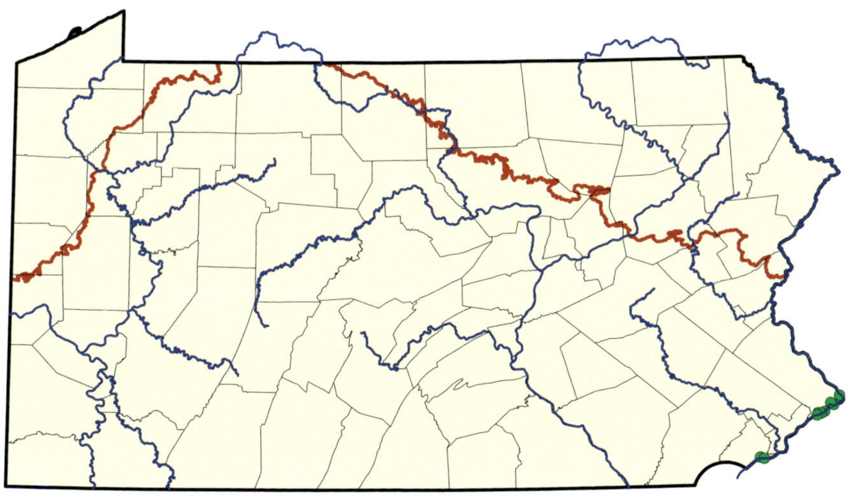

Sagittaria calycina

Sagittaria calycina

Northern arrowhead — *Sagittaria cuneata* Sheldon

PNHP
Native
Rooted, emergent or submergent perennial

The leaves of northern arrowhead are sometimes completely submersed, or they may extend above the water surface on long petioles. The leaf blades are sagittate when emersed, with a central lobe that is larger than the basal lobes. When submersed the blade is elliptical (lacking basal lobes) or long and ribbon-like. Floating leaves may also be produced in deep water. The plants form large, edible corms.

The flowering stalk contains 2–10 whorls of white flowers, the upper staminate and the lower pistillate; the petals are 7–10 mm long. Stamen number ranges from 5–25. The achenes have a short, erect beak.

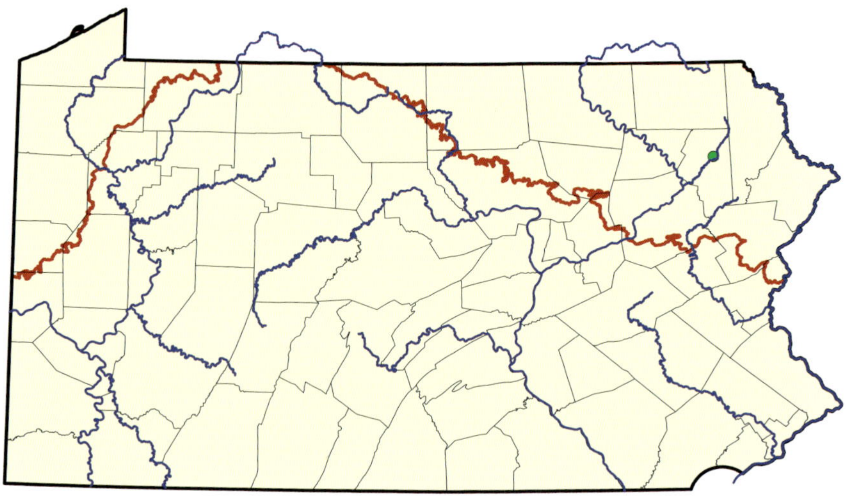

Sagittaria cuneata

This species is rare in pond margins and swamps in northwestern Pennsylvania. It is also known from one site in northeastern Pennsylvania where it was likely planted in a restored wetland. The main range of northern arrowhead extends from Nova Scotia and Quebec to Alaska and south to New York, Ohio, Texas, and California.

The corms of northern arrowhead were harvested by Native Americans; they are also an important source of food for wildlife. The achenes are eaten by ducks.

Submersed plants with ribbon-like leaves can be confused with water-celery (*Vallisneria americana*) or some of the bur-reeds (*Sparganium* spp.), but the venation is different and the leaves are not as strictly parallel-sided. See additional discussion under water-celery.

Sagittaria cuneata—achene ×5

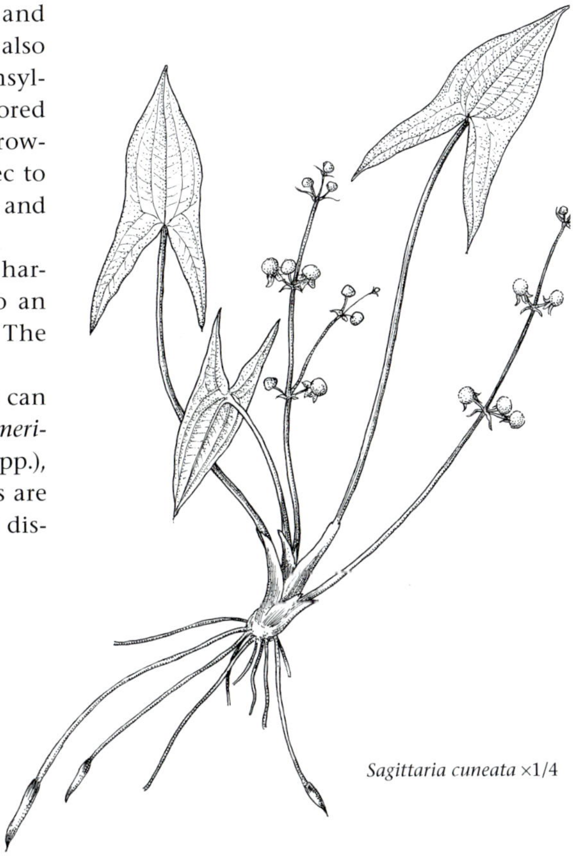

Sagittaria cuneata ×1/4

Grass-leaved arrowhead *Sagittaria graminea* Michx.

Native
Rooted, emergent or submergent perennial

Grass-leaved arrowhead is a rhizomatous perennial; it grows as an emergent or completely submersed where it sometimes forms a coarse turf. Underwater, the leaves are stiff, linear, and tapered to an acute tip; length and width can be highly variable. Emergent leaves typically develop an elliptical blade; lobed leaves are rare. The flowering stem contains 2–12 whorls of flowers; both male and female flowers are borne on individual pedicels. Flowering occurs from June to early August.

Grass-leaved arrowhead has a range covering the eastern half of the United States and adjacent areas of southern Canada. In Pennsylvania it occurs in the eastern half of the state in shallow water of lakes, ponds, and streams and on tidal mudflats.

Stalked female flowers distinguish this species from sessile-fruited arrowhead. In the absence of an inflorescence it is impossible to differentiate grass-leaved arrowhead from the sterile rosettes of other species based on appearance alone.

Sagittaria graminea ×1/2

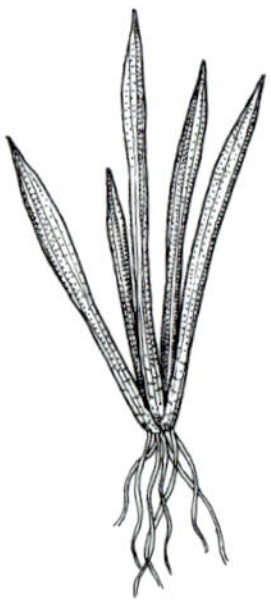

Sagittaria graminea—juvenile rosette ×1/2

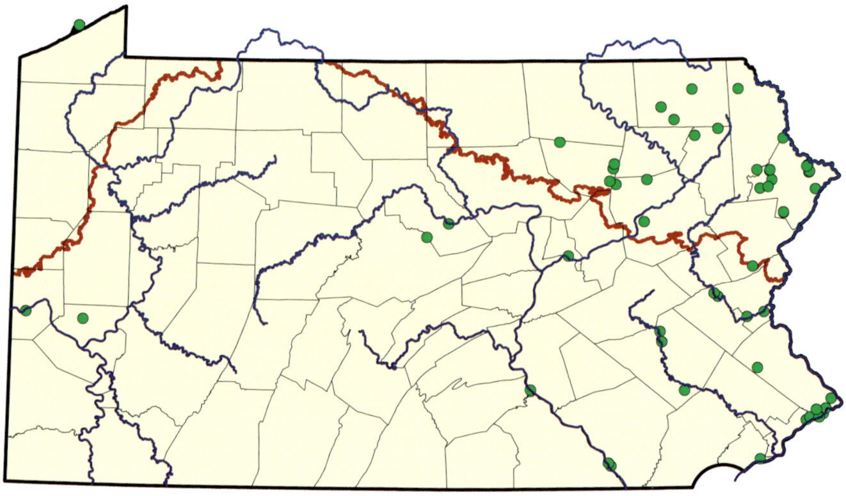

Sagittaria graminea

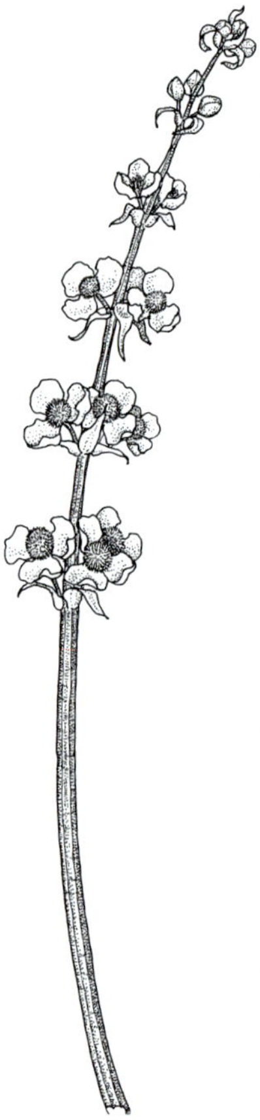

Sagittaria latifolia—inflorescence ×1/2

Duck-potato, wapato *Sagittaria latifolia* Willd.

Native

Rooted, emergent perennial

Duck-potato is one of the most common emergent plants of lake and pond margins, swamps, marshes, and stream edges throughout Pennsylvania. The leaves are arrow-shaped with 2 backward-pointing basal lobes; the width of the main part of the blade and the lobes varies from plant to plant, ranging from very narrow to broad. Both leaves and flowering stems of this native perennial, which are up to 0.6 m tall, arise directly from a large, starchy, edible corm; stolons may also be present.

The flowering stalks bear 2–15 whorls of 3 flowers each; the lower whorls contain female flowers; male flowers are borne above. The flowers are white and 0.5–2 cm across; flowering occurs from July to early September. Fruiting heads are round and globe-like to somewhat flattened, up to 2.5 cm across, and the tiny, 6 mm-long achenes have a beak extending at a 90-degree angle to the body of the fruit.

The total range of arrowhead extends from Newfoundland and New Brunswick west to British Columbia, and south to northern Florida, Alabama, Louisiana, Texas, California, and Mexico.

The starchy corms of arrowhead were an important food for many Native American tribes. Known as "wappatoo" or "katniss," the corms were prepared by baking, roasting, or boiling; they were also stored for winter use. Ducks, geese, and muskrats also eat the corms and seeds of this plant.

Sagittaria latifolia—leaf variation ×1/2

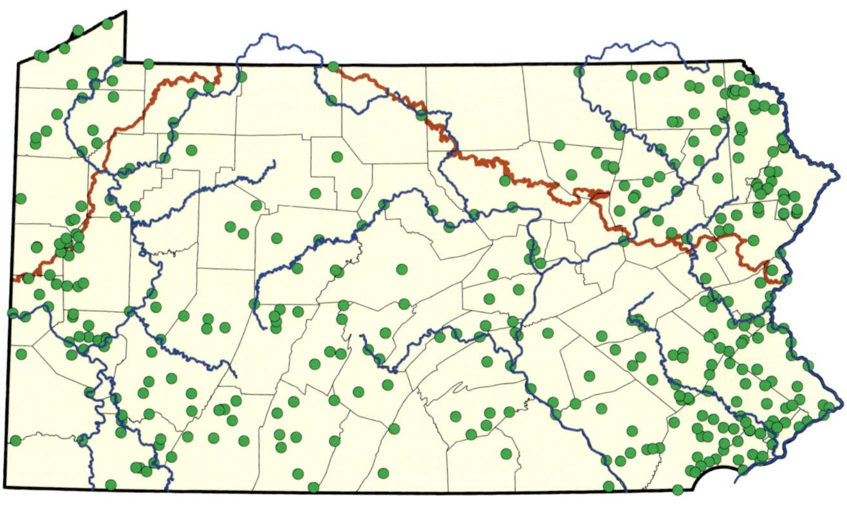

Sagittaria latifolia

Sagittaria latifolia

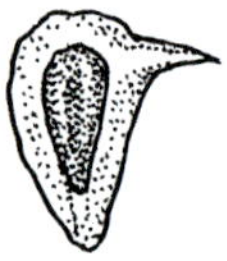

Sagittaria latifolia—achene ×5

Sessile-fruited arrowhead *Sagittaria rigida* Pursh

Native

Rooted, emergent or submergent perennial

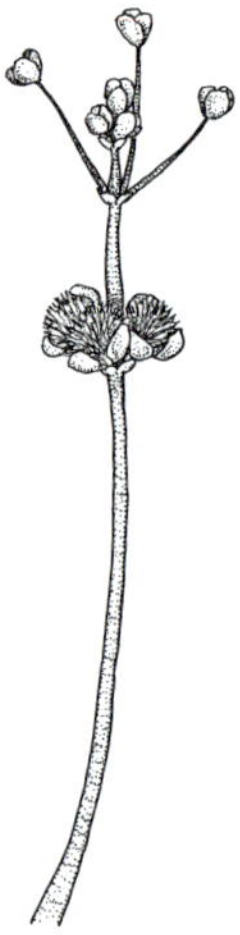

Sagittaria rigida—
inflorescence ×1/2

Sessile-fruited arrowhead is a stoloniferous perennial with emergent or submergent leaves; corms also are produced. Underwater leaves are stiff, flattened, about 1 cm wide, and narrowed to an acute tip. Emergent leaves are typically elliptical or sometimes even develop a sagittate blade. Flowering can occur on emergent plants or those that are submersed in water 1–2 m deep. The inflorescence is characterized by female flowers that are attached directly to the main axis (sessile). In addition, the flowering stems are typically bent at the lowermost whorl of flowers.

Sessile-fruited arrowhead grows from Prince Edward Island to Wisconsin and south to Virginia and Missouri. Disjunct populations are known in Washington and California. In Pennsylvania, it is a plant of lake and pond margins, mudflats, and stream edges at scattered locations throughout. Flowering occurs from late July to late August.

If inflorescences are present, sessile-fruited arrowhead can be distinguished from the very similar grass-leaved arrowhead by the sessile female flowers. The two species are almost impossible to differentiate based on sterile rosettes alone.

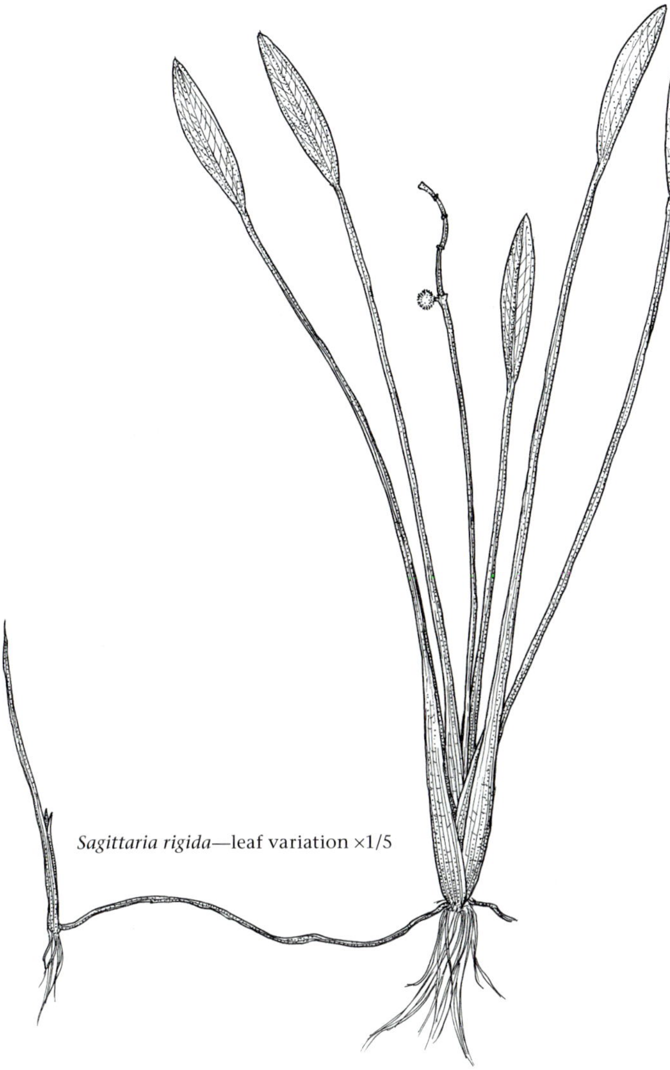

Sagittaria rigida—leaf variation ×1/5

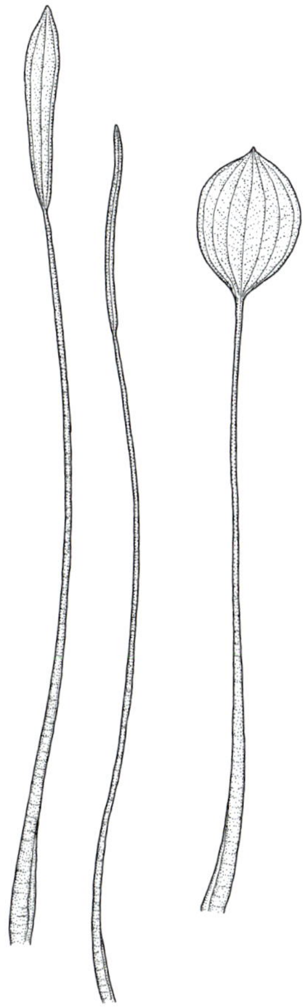

Sagittaria rigida ×1/4

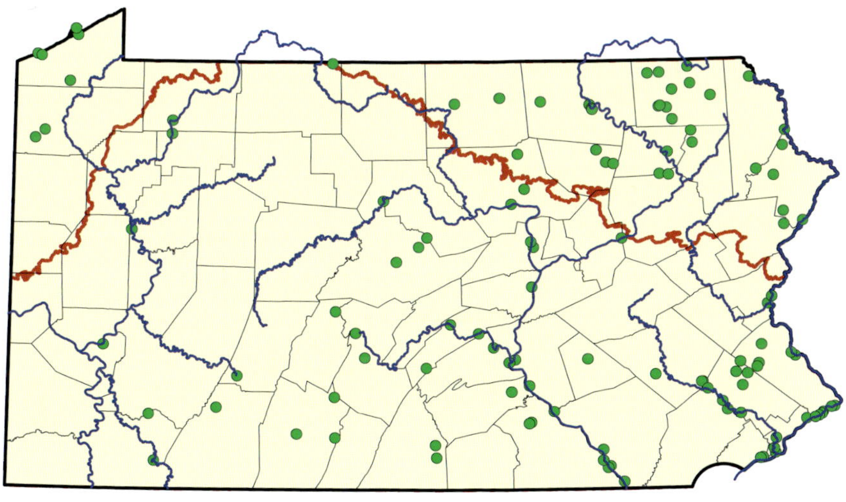

Sagittaria rigida

Subulate arrowhead *Sagittaria subulata* (L.) L. Buch

PNHP
Native
Rooted, emergent or submergent perennial

This perennial species spreads by stolons to form a turf in the freshwater intertidal zone along the Delaware River and some of its tributaries. The plants are completely exposed at low tide and inundated at high tide. The 2.5–5 cm-long leaves are linear, sometimes with just a suggestion of an expanded blade at the tip. Flowering stems are slightly taller than the leaves and bear 2–3 whorls of flowers. Sepals of the fruiting heads are spreading or recurved. Flowering occurs from June to early September.

Subulate arrowhead has a limited global range in coastal areas of North America from Massachusetts to Mississippi. It is classified as a rare species in Pennsylvania.

Subulate arrowhead often occurs with long-lobed arrowhead, another species of freshwater tidal flats. See discussion above under *Sagittaria calycina* on how to distinguish these two species.

Sagittaria subulata ×2/3

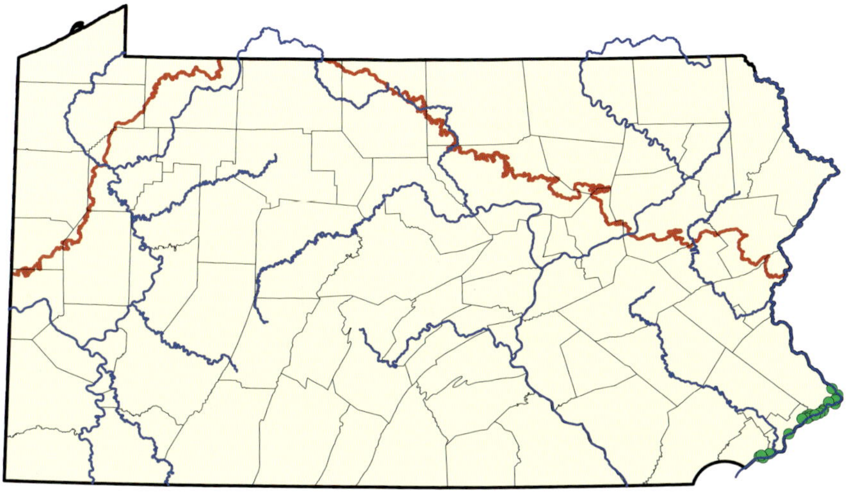

Sagittaria subulata

Sagittaria subulata

BEGGAR-TICKS/BUR-MARIGOLD

Bidens L.

Aster Family—Asteraceae

Bidens species are herbaceous annuals with opposite leaves and small flowers clustered in heads surrounded by an involucre which consists of large, leafy outer bracts and smaller membranous inner bracts. Additional (receptacular) bracts are associated with each individual floret.

Of the species treated here, only *Bidens laevis* produces ray flowers; all, however, are insect-pollinated. The fruits are achenes with 2–4 awns. These species are plants of shallow water of stream or lake margins or freshwater tidal marshes. The seeds are eaten by waterfowl.

Note: *Bidens frondosa* and *Bidens cernua* are included in the key due to their similarity to *Bidens discoidea* and *Bidens laevis* respectively. These species are not described below as they do not fit the criteria for aquatic plants treated in this work, despite the fact that they may occasionally occur in marginal habitat areas.

Key to Aquatic Species of Beggar-ticks and Bur-marigold

A. leaves compound
- B. outer involucral bracts 3–5, not ciliate .... *Bidens discoidea*
- B. outer involucral bracts 5–10, ciliate .... *Bidens frondosa*

A. leaves simple
- C. leaves sessile; showy ray florets present
  - D. rays 1.5–3 cm long, receptacular bracts red-tinged .... *Bidens laevis*
  - D. rays less than 1.5 cm long, receptacular bracts yellowish .... *Bidens cernua*
- C. leaves with distinct petioles; ray florets not present .... *Bidens bidentoides*

Swamp Beggar-ticks — *Bidens bidentoides* (Nutt.) Britton

PNHP
Native
Rooted, emergent annual

Bidens bidentoides—achene ×2, note upward-pointing barbs on awns

Swamp beggar-ticks is a plant of freshwater tidal marshes. It has opposite leaves that are simple and coarsely toothed; the heads contain only disk flowers. A characteristic that distinguishes it from all the other *Bidens* species in our range is the upward pointing barbs on the awns of the achenes (visible with a 10X lens).

Swamp beggar-ticks typically grows in the upper zone of the tidal marsh where its roots are inundated at high tide but the upper stem remains above water. Its range includes freshwater tidal marshes of the estuaries of coastal rivers in New York, New Jersey, Pennsylvania, Delaware, and Maryland.

Bidens bidentoides ×1/3

Bidens bidentoides

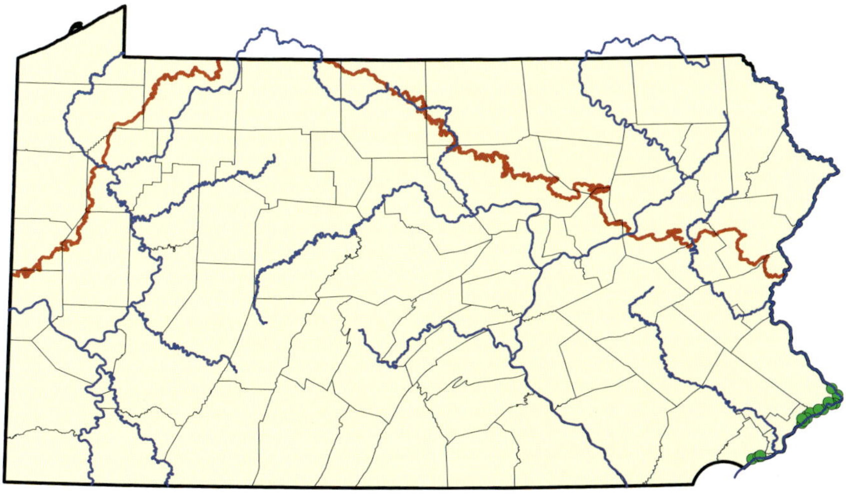

Bidens bidentoides

Small beggar-ticks

Bidens discoidea (Torr. & A.Gray) Britton

PNHP
Native
Rooted, emergent annual

Bidens discoidea ×1/2

Small beggar-ticks is primarily a plant of lake margins, where it grows with its roots in water but the upper stem and leaves exposed. It is known from glacial lakes in northeastern Pennsylvania and scattered sites elsewhere in the state.

The range of small beggar-ticks includes the eastern United States and adjacent areas of eastern Canada from Nova Scotia, southern Quebec, and Minnesota south to Florida and Texas.

With compound leaves and numerous small cylindrical heads that contain only disk flowers, this species is very similar to the common and widespread *Bidens frondosa*. Distinguishing characters include the smaller number of outer involucral bracts and the lack of cilia in the margins of the bracts (see key above).

Bidens discoidea

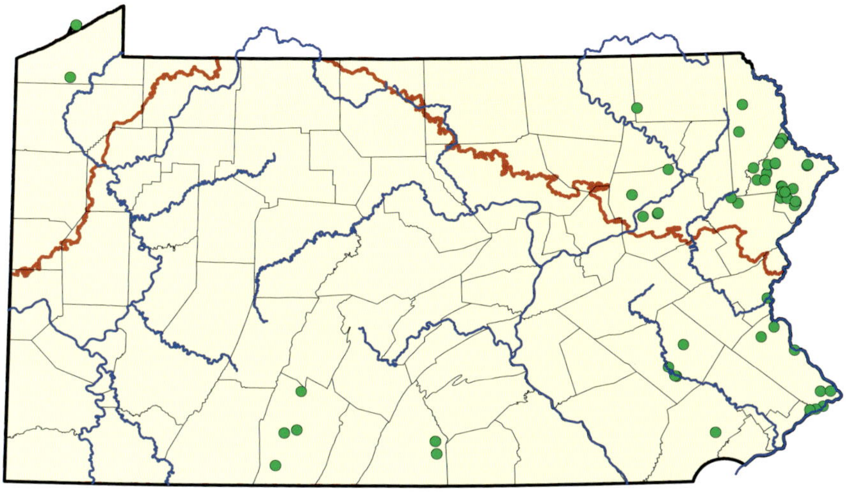

Bidens discoidea

Showy bur-marigold

Bidens laevis (L.) Britton, Stearns & Poggen

PNHP
Native
Rooted, emergent annual

Showy bur-marigold, as the name suggests, is the most conspicuous of our native *Bidens* species due to the 1.5–3 cm-long yellow rays that surround each flowering head. This species grows in the upper portion of freshwater tidal marshes where the roots and lower stem are inundated at high tide, often with swamp beggar-ticks (*Bidens bidentoides*). It is also found in non-tidal habitats growing in shallow water at the margins of lakes or slow-moving streams, especially in areas of diabase geology.

The plants are up to 1 m tall and have simple, but coarsely toothed, opposite leaves.

The range of showy bur-marigold includes coastal and inland areas from New Hampshire and Massachusetts south to Florida and Louisi-

Bidens laevis

ana. It also grows in scattered locations from Texas to California and south to South America. Showy bur-marigold was used as a food plant by some Native American tribes.

Showy bur-marigold is similar to *Bidens cernua*. The rays of *B. cernua* rarely exceed 1.5 cm in length and the flowering heads become nodding with age. Another distinguishing characteristic is the red tinge on the receptacular bracts of showy bur-marigold. The floral bracts of *Bidens cernua* are yellowish (see key above).

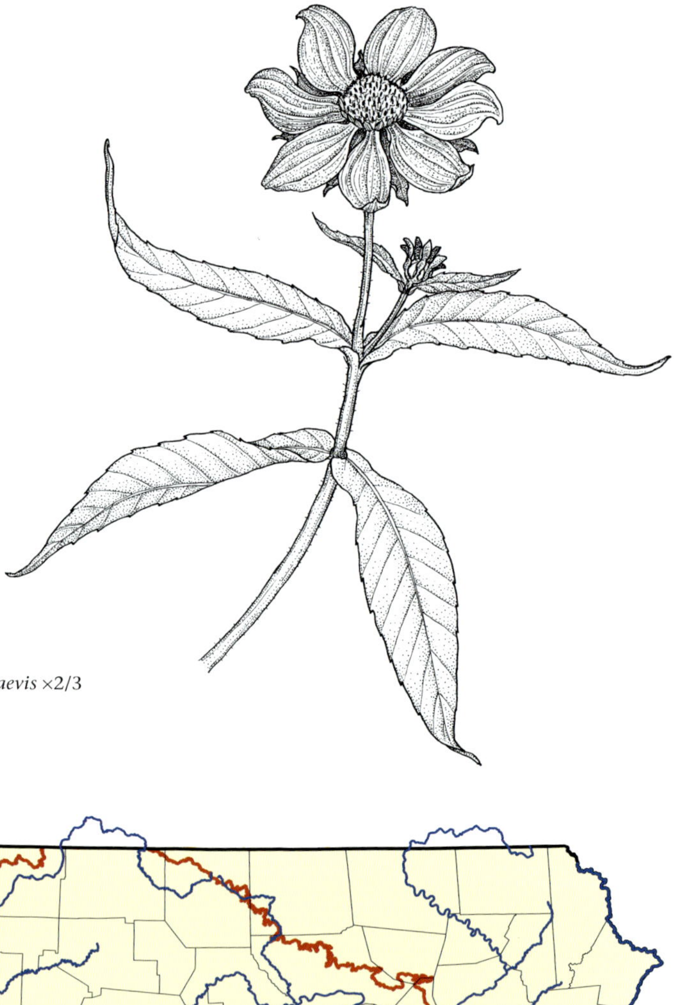

Bidens laevis ×2/3

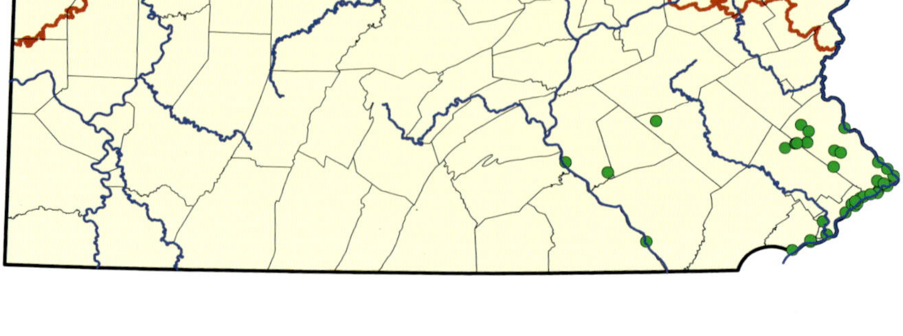

Bidens laevis

BULRUSH
Schoenoplectus (Rchb.) Palla and *Scirpus* L.
Sedge Family—Cyperaceae

The bulrushes are herbaceous plants with 3-angled or rounded (terete) stems. Leaves of some species are represented by bladeless sheaths; others have narrow, parallel-veined blades with a tubular basal sheath. Flowers are produced in inflorescences at or near the tips of upright stems. As in other members of the sedge family, the flowers are aggregated in spikelets and each flower is subtended by a scale. Spikelets may be sessile and clustered or arranged in a branched inflorescence. Bulrush flowers are perfect and wind-pollinated. The perianth is reduced to bristles (or completely absent), and the fruit is an achene.

Fruits of the bulrushes are eaten by a wide range of marsh and water birds; in addition, the rhizomes and stems are food for muskrats. Bulrushes also provide nest sites and cover for birds and mammals that use the emergent zone habitat.

Although most bulrushes are restricted to wetland habitats, we have included only those species that typically grow with at least their roots under water.

Key to the Species of Aquatic Bulrushes

A. spikelets produced in a branching terminal inflorescence
  B. spikelets 3–5 mm long; leaves spaced along the stem .... *Scirpus ancistrochaetus*
  B. spikelets 10–40 mm long; stem leafy to about the middle .... *Schoenoplectus fluviatilis*
A. spikelets in a single, apparently lateral, cluster arising in the upper part of the stem
  C. spikelets in small stalked clusters
    D. peduncles relatively stiff; scales of the spikelets dull gray-brown with brown spots; awn tips of the scales bent .... *Schoenoplectus acutus*
    D. peduncles arching; scales of the spikelets pale gray to orange-brown; awn tips of the scales straight .... *Schoenoplectus tabernaemontani*
  C. spikelets not stalked (sessile)
    E. rhizomes not produced, stems tufted
      F. perianth bristles absent; plants of freshwater tidal marshes .... *Schoenoplectus smithii*
      F. perianth bristles present; plants of lake and pond margins and freshwater tidal marshes .... *Schoenoplectus purshianus*
    E. plants rhizomatous with scattered stems
      G. plants with capillary underwater leaves .... *Schoenoplectus subterminalis*
      G. capillary underwater leaves not produced
        H. rhizomes flaccid; leaf blades triangular-channeled; scales scarcely notched at the tip, awns lacking or inconspicuous .... *Schoenoplectus torreyi*
        H. rhizomes firm; leaf blades v-shaped; scale tips with conspicuous notches and awns .... *Schoenoplectus pungens*

Great or hard-stemmed bulrush — *Schoenoplectus acutus* (Muhl. Ex Bigel.) Löve & Löve var. *acutus*

Schoenoplectus acutus ×1/20

PNHP
Native
Rooted, emergent perennial

Schoenoplectus acutus—inflorescence ×1/2

Hard-stemmed bulrush is a rare plant of calcareous lake and pond margins in western and central Pennsylvania. It is similar to *Schoenoplectus tabernaemontani*, but the stems are 1–3 m tall and harder and firmer with the central air cavities only 0.5 mm in diameter. Sheaths are often dark reddish at the base. Branches of the inflorescence are mostly stiffly ascending, and the mature achenes are black and lustrous.

The total range of hard-stemmed bulrush includes the northeastern, central, and northwestern states and adjacent areas of Canada.

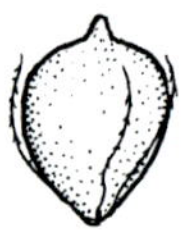

Schoenoplectus acutus—achene ×5

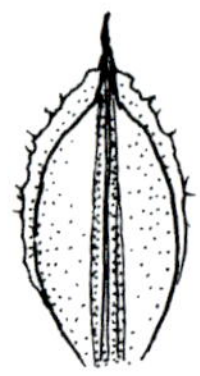

Schoenoplectus acutus—scale ×5

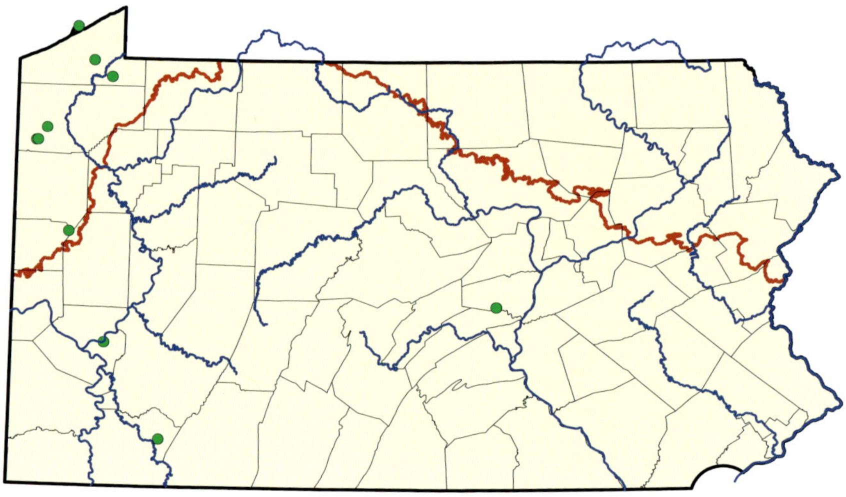

Schoenoplectus acutus

River bulrush *Schoenoplectus fluviatilis* (Torr.) Strong

PNHP
Native
Rooted, submergent or emergent perennial

River bulrush is leafy-stemmed plant 1–2 m tall with the inflorescence located at the tip of the stem. It is rhizomatous and often forms dense stands, which may be entirely vegetative.

Stems are triangular in cross-section and bear leaves to about the middle. The inflorescence is umbel-like with spikelets solitary or in clusters at the ends of short spreading or drooping branches. Three to six spreading, leaf-like bracts surround the inflorescence. Achenes are dark brown, 3-sided or somewhat compressed, and surrounded by bristles that remain attached when the fruits are shed.

River bulrush occurs throughout the United States and adjacent areas of Canada, with the exception of Florida, Louisiana, Texas, and Oklahoma. It also grows in Japan, Australia, and New Zealand. In Pennsylvania it is known only from a few scattered sites across the state, including several freshwater tidal marshes along the Delaware River and sites along Presque Isle Bay in Erie County.

Schoenoplectus fluviatilis—inflorescence ×1/2

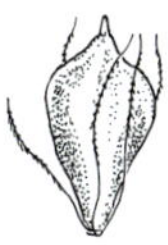

Schoenoplectus fluviatilis—achene ×2.5

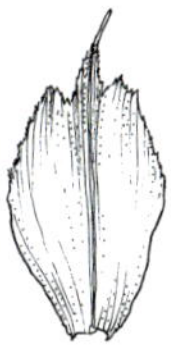

Schoenoplectus fluviatilis—scale ×2.5

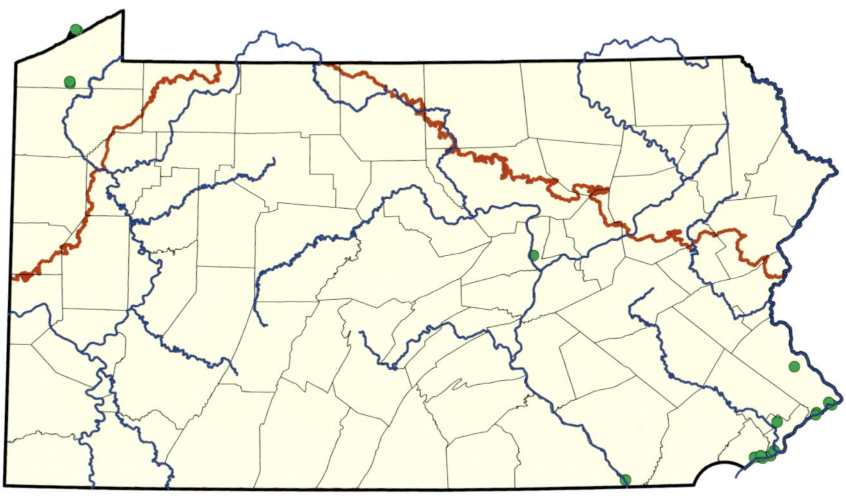

Schoenoplectus fluviatilis

Schoenoplectus fluviatilis

Three-square, chairmaker's-rush ***Schoenoplectus pungens* (Vahl) Palla**

Native
Rooted, emergent perennial

Three-square forms colonies in shallow water of freshwater tidal riverbanks; lake, pond, and stream margins; and marshes and fens. Its sharply triangular stems are produced from a rhizome; each is surrounded at the base by a few v-shaped linear leaves.

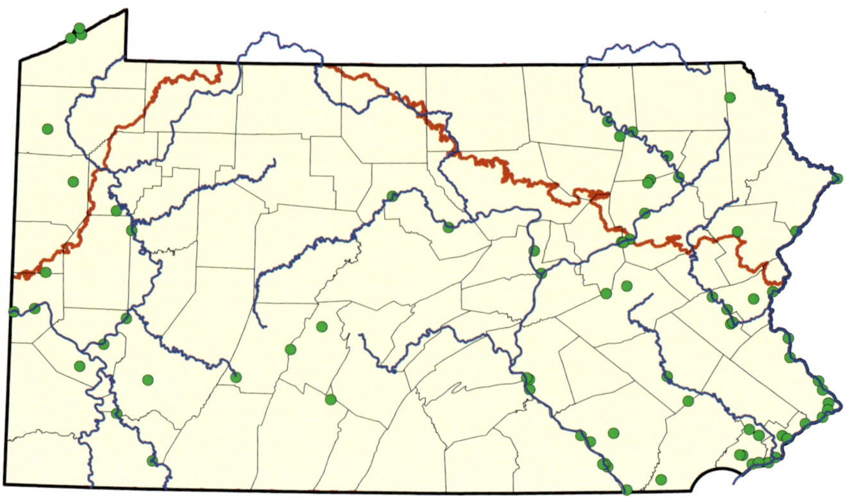

Schoenoplectus pungens

Flowers and fruits are produced in a small cluster of sessile spikelets located on one side just below the stem tip. Achenes are brown, and compressed or somewhat 3-sided.

Globally, three-square is a wide ranging species occurring in both the northern and southern hemispheres. In Pennsylvania it is found at scattered locations, mostly along major river systems.

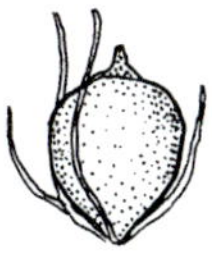

Schoenoplectus pungens—
achene ×5

Schoenoplectus pungens ×3/16

Pursh's bulrush — *Schoenoplectus purshianus* (Fernald) Strong

Native
Rooted, emergent annual

Pursh's bulrush has densely tufted stems 20–60 cm tall with 2–5 spikelets in a sessile cluster about one-third of the way down from the tip. Leaves are reduced to bladeless sheaths. The achenes are biconvex with a brown or blackish pitted surface and are surrounded by bristles. This species grows in shallow water or on muddy shores, including the freshwater tidal marshes along the Delaware River.

Pursh's bulrush occurs in eastern North America from southern Canada to Mississippi and Alabama; in Pennsylvania it is most common in the southeastern region, but also occurs at scattered locations elsewhere.

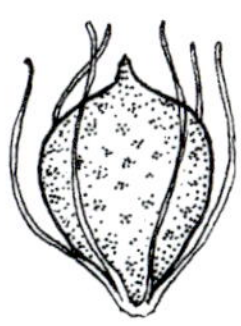

Schoenoplectus purshianus—
achene with bristles ×10

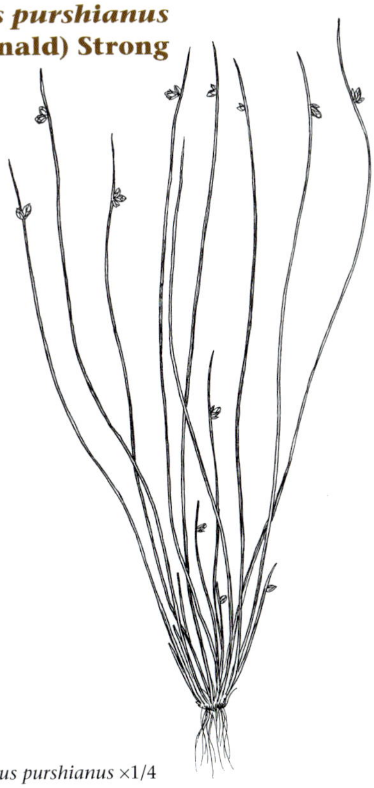

Schoenoplectus purshianus ×1/4

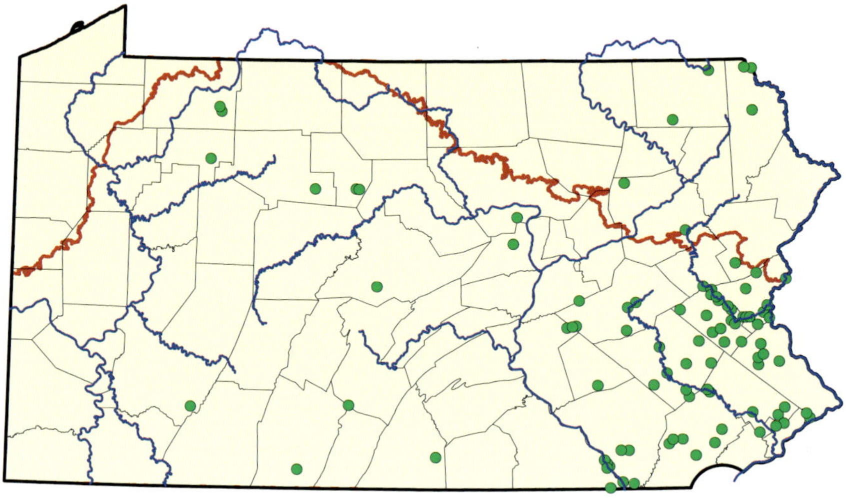

Schoenoplectus purshianus

Smith's bulrush — *Schoenoplectus smithii* (A. Gray) Sojak

PNHP
Native
Rooted, emergent perennial

Smith's bulrush is very similar to Pursh's bulrush (*S. purshianus*), but in our area lacks bristles surrounding the achenes. In Pennsylvania it is a plant of freshwater intertidal mudflats along the Delaware River and its tributaries, and the shores of Lake Erie. Its global range includes the Atlantic coast from New England to Virginia and the Great Lakes region.

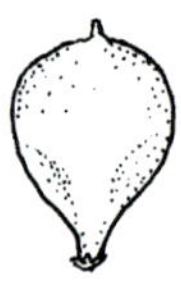

Schoenoplectus smithii—achene ×10, note lack of bristles

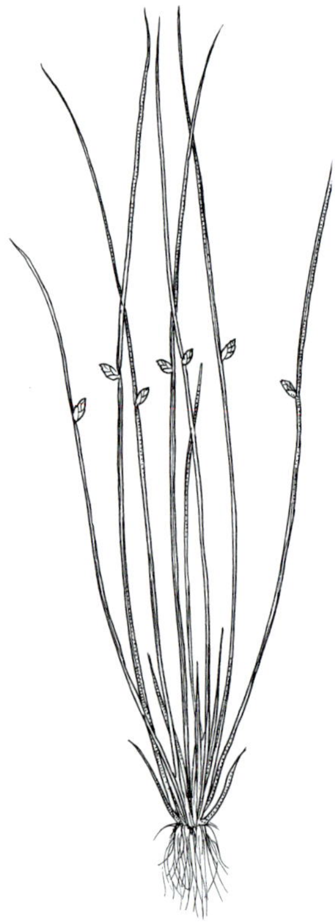

Schoenoplectus smithii ×1/2

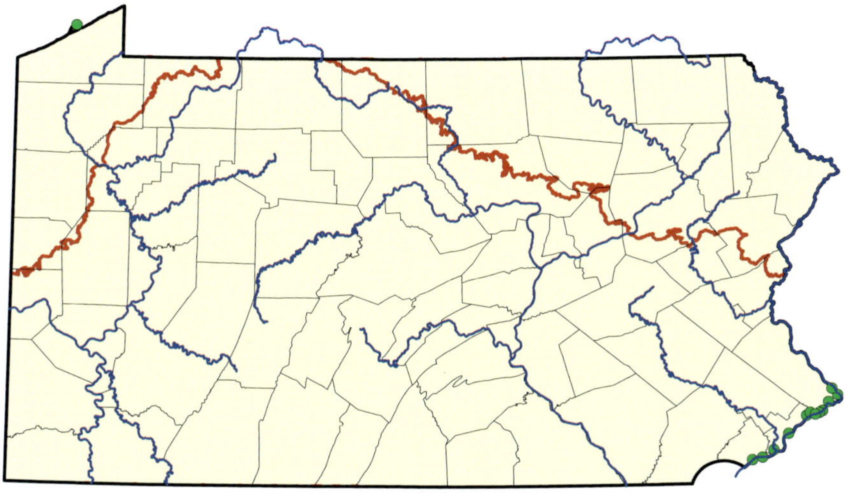

Schoenoplectus smithii

Schoenoplectus smithii growing with *Sagittaria subulata*

Water bulrush — *Schoenoplectus subterminalis* (Torr.) Sojak

PNHP
Native
Rooted, emergent or submergent perennial

Water bulrush is a native, emergent perennial with only the tips of the flowering stems extending above the water surface. The erect flowering stems, which can be 1–1.5 m long, are surrounded by numerous, flexuous capillary leaves that move with the water currents.

Rooted in the lake bottom, water bulrush spreads by means of rhizomes to form a loose turf. Leaves are usually entirely submergent and have elongate,

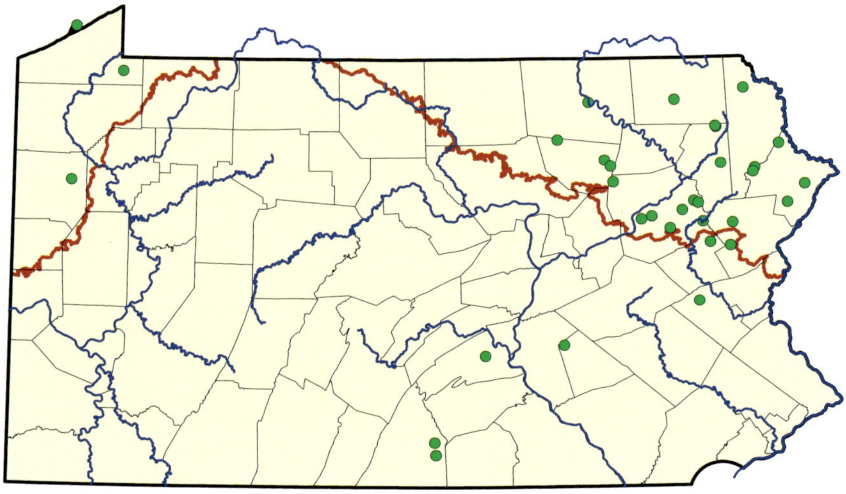

Schoenoplectus subterminalis

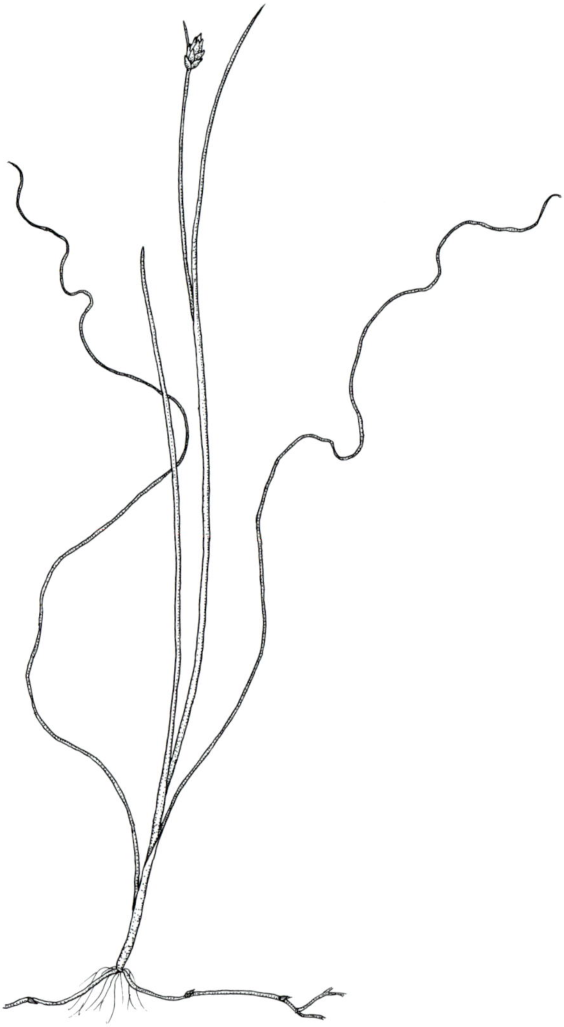

Schoenoplectus subterminalis ×1/2

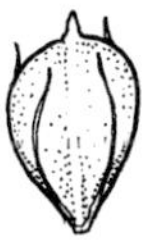

Schoenoplectus subterminalis—achene ×5

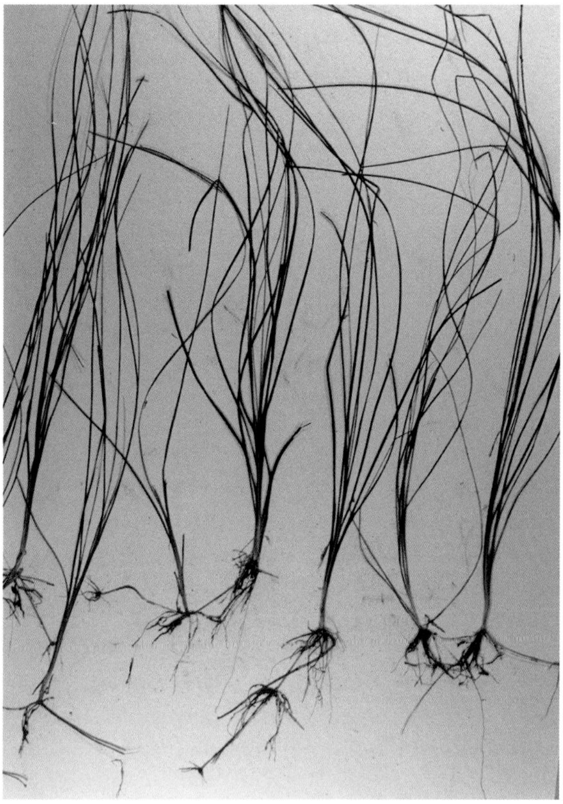

Schoenoplectus subterminalis

very slender, flaccid blades and sheathing bases; they are attached basally in clusters along the rhizomes. Flowering and fruiting is more common in shallow water or on seasonally exposed mud flats. In deeper water fertile stems may be very sparse or even absent.

Flowers and fruits are produced in a single lateral spikelet a few centimeters below the emergent stem tip. Flowers are perfect; the fruit, a three-sided achene, is surrounded by bristles. Fruiting occurs in the summer.

The range of water bulrush includes the northeastern United States and adjacent provinces of Canada. It is also present in the Pacific Northwest, British Columbia, and Alaska. In Pennsylvania it is present in lakes and slow-moving streams in northeastern and northwestern counties as well as scattered locations in the south central region.

Water bulrush is similar to several other aquatic plants that form capillary underwater leaves or stems, including Robbins' spike-rush and bayonet rush. Careful examination will reveal that the leaves of water bulrush have a sheathing base, unlike the capillary stems of the other two species. If flowering stems are present, the species can easily be distinguished by the nature and position of the inflorescence.

Water bulrush might also be confused with robust, non-flowering plants of *Eleocharis acicularis*; to distinguish them, note the basal leaf sheath of water bulrush. *Eleocharis* does not have leaves; the flaccid capillary structures are stems. In addition, *Eleocharis acicularis* rhizomes are much more slender than the 1 mm-diameter rhizomes of water bulrush.

Great or soft-stemmed bulrush

Schoenoplectus tabernaemontani (Gmel.) Palla [synonym: _Scirpus validus_ Vahl.]

Native
Rooted, emergent perennial

Soft-stemmed bulrush is a commanding plant along lake and pond shorelines due to its upright, unbranched stems that are 1–2.5 m tall. The stems are round in cross-section, hollow, and easily compressed; leaves are represented by bladeless sheaths at the base of the stem. Soft-stemmed bulrush grows in small to large colonies in shallow water, the stems arising from stout, deeply buried rhizomes.

Flowers are borne in a small, branched, usually drooping inflorescence produced on one side of the stem just below the tip. Fruiting occurs from late spring through fall. The achenes are dark gray-brown and flattened. They are eaten by waterfowl.

Soft-stemmed bulrush is a cosmopolitan species that is distributed throughout the northern and southern hemispheres. It is present throughout Pennsylvania. In addition it is frequently planted in ornamental water gardens and ponds.

Hard-stemmed bulrush (*Schoenoplectus acutus*) is very similar in overall size and appearance, but has firmer stems with a very small (0.5 mm) hollow space in the center. In addition the bladeless sheaths at the base of the stem of hard-stemmed bulrush are often red-tinged at the base.

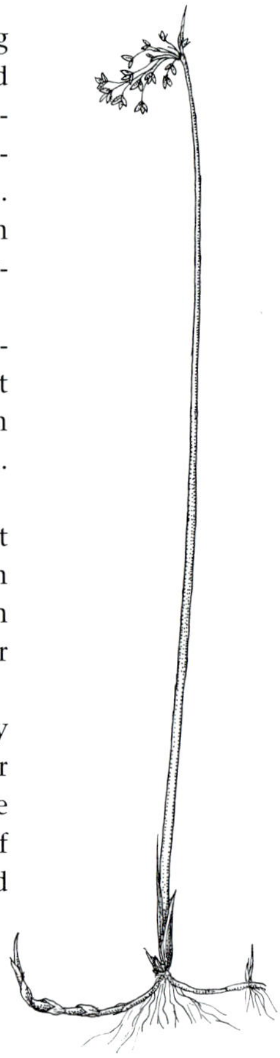

Schoenoplectus tabernaemontani ×1/10

Schoenoplectus tabernaemontani—achene ×5

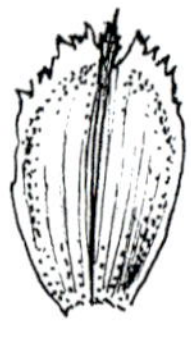

Schoenoplectus tabernaemontani—scale ×5

Schoenoplectus tabernaemontani—inflorescence

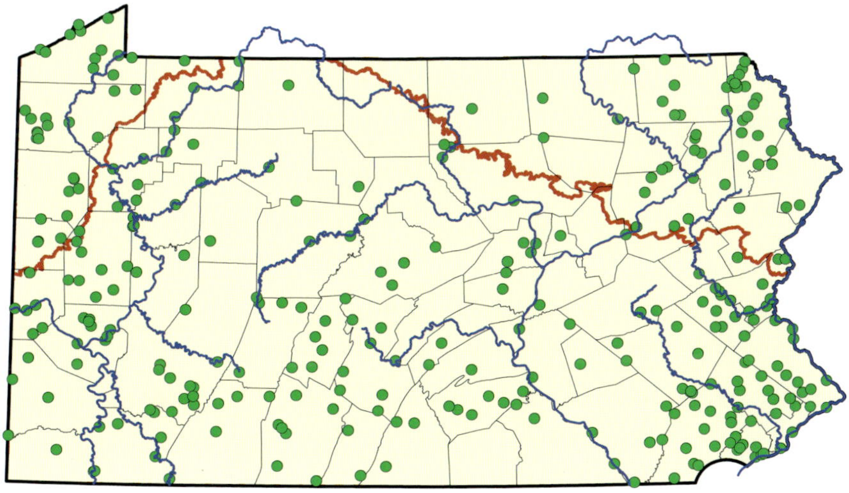

Schoenoplectus tabernaemontani

Torrey's bulrush *Schoenoplectus torreyi* (Olney) Palla [synonym: *Scirpus torreyi* Olney]

PNHP
Native
Rooted, emergent perennial

Torrey's bulrush is a rare plant in Pennsylvania that forms colonies in shallow water at the margins of lakes. It is rhizomatous and mat-forming, with the sharply triangular stems bearing 4–7 erect leaves that are nearly as long as the stem. Torrey's bulrush bears a cluster of 1–4 sessile spikelets on the side of the stem just below the tip. Flowering and fruiting occur in the summer; achenes are brown and 3-angled but somewhat compressed.

The range of Torrey's bulrush extends from southern Canada through the northeastern United States. In Pennsylvania it is known mainly from the northeast, with a few scattered locations in the central and northwestern regions.

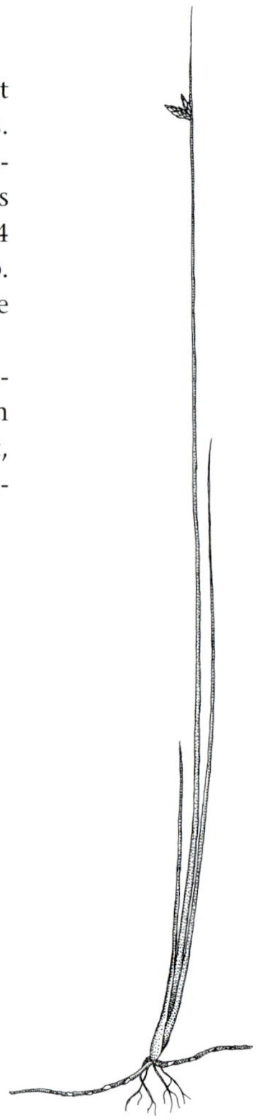

Schoenoplectus torreyi ×1/8

Schoenoplectus torreyi—inflorescence

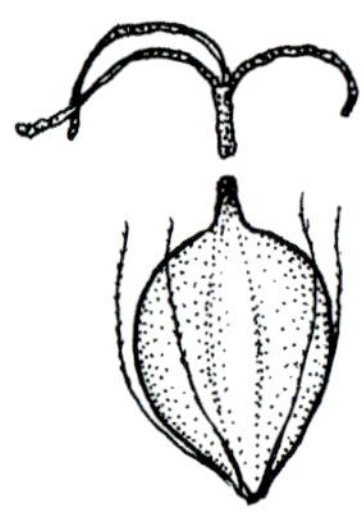

Schoenoplectus torreyi—achene ×5

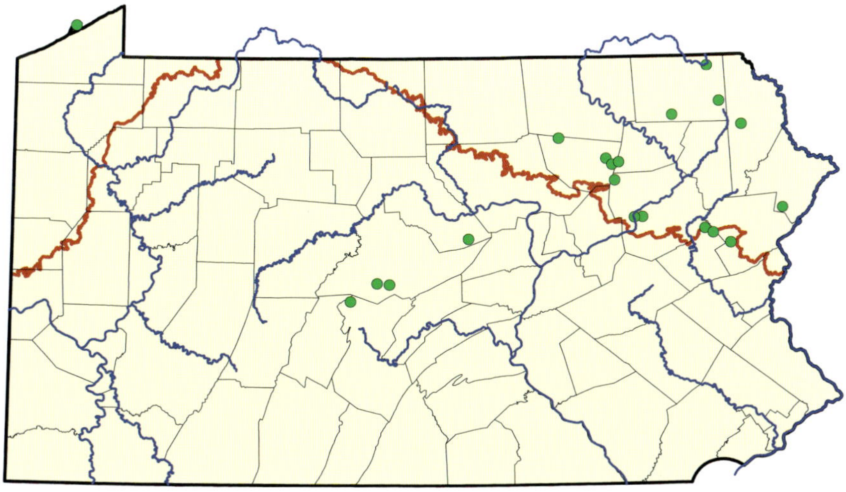

Schoenoplectus torreyi

Northeastern bulrush *Scirpus ancistrochaetus* Schuyler

PNHP, federal endangered
Native
Rooted, emergent perennial

Northeastern bulrush is one of 11 species of *Scirpus* in Pennsylvania. Most are wetland plants, but not true aquatics; northeastern bulrush is included here due to its occurrence in areas with fluctuating water levels, ranging from inundated to merely moist.

Stems are clustered from short, tough rhizomes; bulblets often form at the nodes late in the season. Five to nine leaves are spaced along the stem. The

Scirpus ancistrochaetus—inflorescence

inflorescence is terminal and consists of arching branches bearing clusters of spikelets. The scales subtending each floret are brown to blackish-brown with a green midrib. Achenes are pale brown and lens-shaped; the 6 stout bristles are slightly shorter than the achene and bear downward-pointing, thick-walled pointed teeth.

Northeastern bulrush is globally rare; its range extends from New England to Virginia. In Pennsylvania it is known from scattered sites in the central and eastern parts of the state.

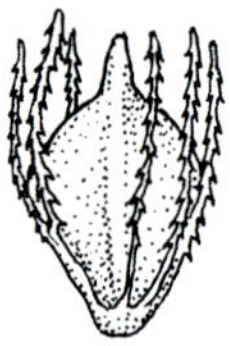

Scirpus ancistrochaetus—achene with bristles ×15

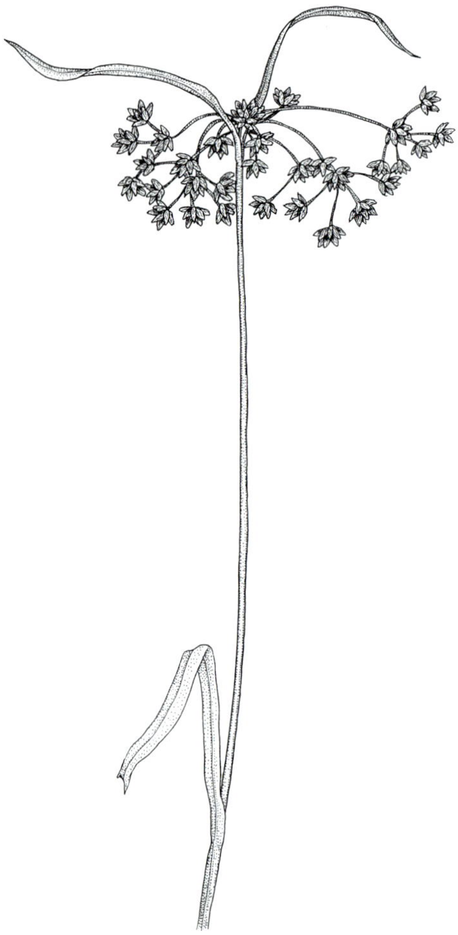

Scirpus ancistrochaetus ×1/2

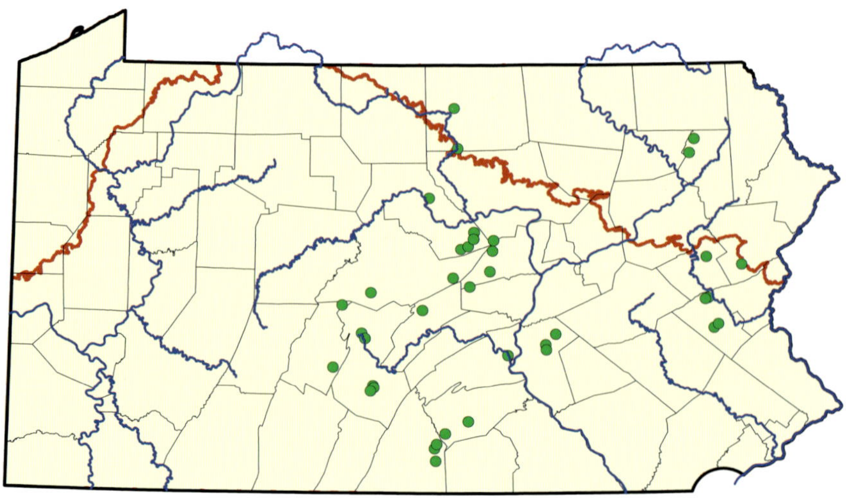

Scirpus ancistrochaetus

BUR-REED

SPARGANIUM L.

Bur-reed Family—Sparganiaceae

The spherical, bur-like heads of flowers and fruits give the bur-reeds their common name. The bur-reeds are rooted, rhizomatous perennials with emergent inflorescences and emergent or floating leaves. Seven species of bur-reed are native to Pennsylvania; however, *S. natans* L. [synonym: *S. minimum* (Hartman) Fries], a northern species, is known from only two locations and has not been collected since the late 1800s. It is believed to be extirpated.

Leaves are linear, basal, or alternate on the erect flowering stem, and stiffly erect and keeled in most species, but sometimes lax and floating at the tips. Leaf form varies greatly with water depth; consequently, an inflorescence, preferably containing mature fruits, is necessary for positive identification of most species.

The inflorescence is typically shorter than the leaves. Flowers are small and aggregated in spherical heads that are located in the axils of reduced leaves or bracts (axillary) or above the bracts (supra-axillary). Both male and female flowers occur on each plant, making the species monoecious. Female flower heads occupy the lower positions on the inflorescence, and the male heads are toward the tips. Pollination is by wind; male flowers fall off soon after pollen is released. Fruits consist of a prickly sphere of hard, green achene-like drupes, each with a persistent style that forms a beak (diameter measurements for the heads given below are from tip of beak to tip of beak). The heads eventually shatter, allowing for dispersal of individual fruits.

Native Americans used an infusion of *Sparganium* to treat chills; the bulbous stem base and rootstock were used as food. Wildlife, including waterfowl, marsh birds, and muskrats, eat the fruits.

Key to the Species of Bur-reed

A. stigmas 2; mature fruits broadest at the top .... *Sparganium eurycarpum*
A. stigma 1; mature fruits tapered at both ends
  B. leaves floating at the tips, not keeled or keeled only toward the base
    C. inflorescence branched .... *Sparganium fluctuans*
    C. inflorescence unbranched
      D. leaves 4–10 mm wide, flat, keeled toward the base .... *Sparganium emersum*
      D. leaves 2–5 mm wide, rounded on the back, not keeled .... *Sparganium angustifolium*
  B. leaves and inflorescences erect and emergent, leaves distinctly keeled
    E. inflorescence unbranched .... *Sparganium emersum*
    E. inflorescence branched
      F. inflorescence branches bearing some female heads .... *Sparganium americanum*
      F. inflorescence branches bearing male flower heads only .... *Sparganium androcladum*

Bur-reed *Sparganium americanum* Nutt.

Native
Rooted, emergent perennial

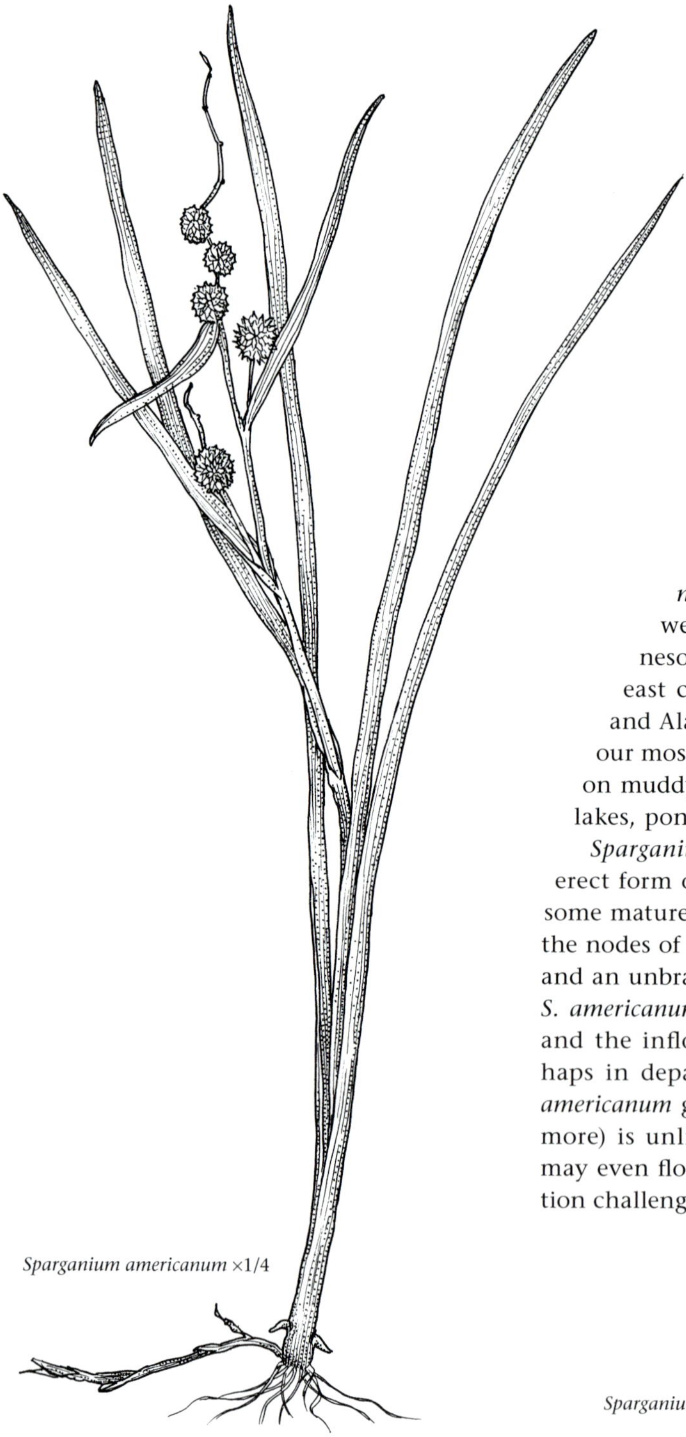

Sparganium americanum ×1/4

Sparganium americanum is a very common native emergent, rhizomatous aquatic plant that occurs throughout Pennsylvania. This 0.5–1 m-tall perennial herb produces stiffly erect, keeled leaves up to 1 m long and 1.5 cm wide. Sterile shoots formed late in the season may have even wider leaves.

The inflorescence is typically branched. Female flower heads are borne directly in the axils of reduced leaves (bracts), with the male flower heads above the female heads. Fruiting takes place from July to September.

The total range of *S. americanum* extends from Newfoundland west to Ontario, Wisconsin, Minnesota, and North Dakota. Along the east coast it extends south to Florida and Alabama, *Sparganium americanum* is our most common bur-reed, and is found on muddy shores and in shallow water of lakes, ponds, rivers, streams, and swamps. *Sparganium americanum* resembles the erect form of *S. emersum*, which usually has some mature fruiting heads positioned above the nodes of the inflorescence (supra-axillary) and an unbranched inflorescence, whereas in *S. americanum* the heads are strictly axillary and the inflorescence branched, except perhaps in depauperate specimens. *Sparganium americanum* growing in deep water (0.6 m or more) is unlikely to flower and some leaves may even float at the tips, making identification challenging.

Sparganium americanum—fruit ×3

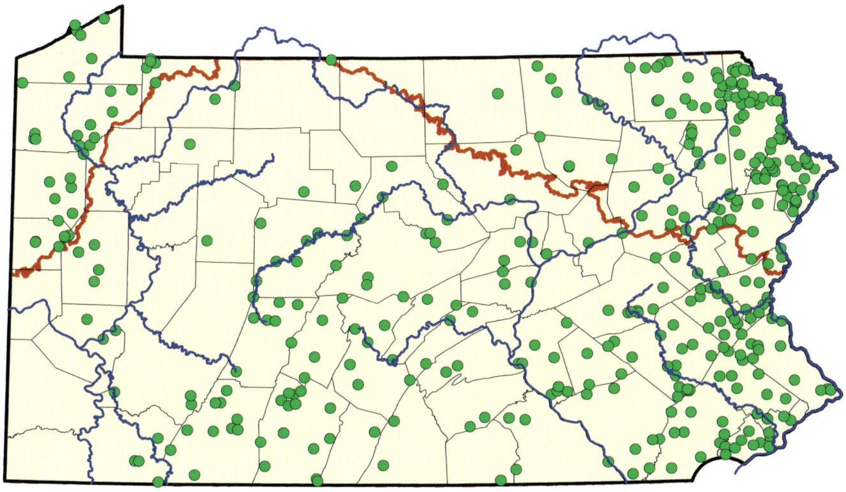

Sparganium americanum

Sparganium americanum—in flower

Branching bur-reed

***Sparganium androcladum* (Engelmann) Morong**

PNHP
Native
Rooted, emergent perennial

Sparganium androcladum—fruit ×2

Branching bur-reed is an erect emergent, rhizomatous plant to 1 m tall with stiff leaves that are up to 1.5 cm wide and keeled from the base to the middle or more. The inflorescence is branched and bears female heads on the main axis and male heads on the branches as well as the tip of the main stem. Fruiting heads are 2–3.5 cm in diameter; individual fruits are usually constricted at the middle and have a 4–7 mm-long beak.

The range of *Sparganium androcladum* extends from Maine, Quebec, and Ontario south to New Jersey and Pennsylvania, and south in the Mississippi Valley to Missouri. In Pennsylvania it is found in wet meadows, swales, and stream banks at widely scattered locations.

Sparganium androcladum is similar to *S. americanum* but generally more robust. Immature fruiting heads closely resemble mature heads of *S. americanum*, but the lack of female flower heads on the branches of the inflorescence is a good distinguishing character.

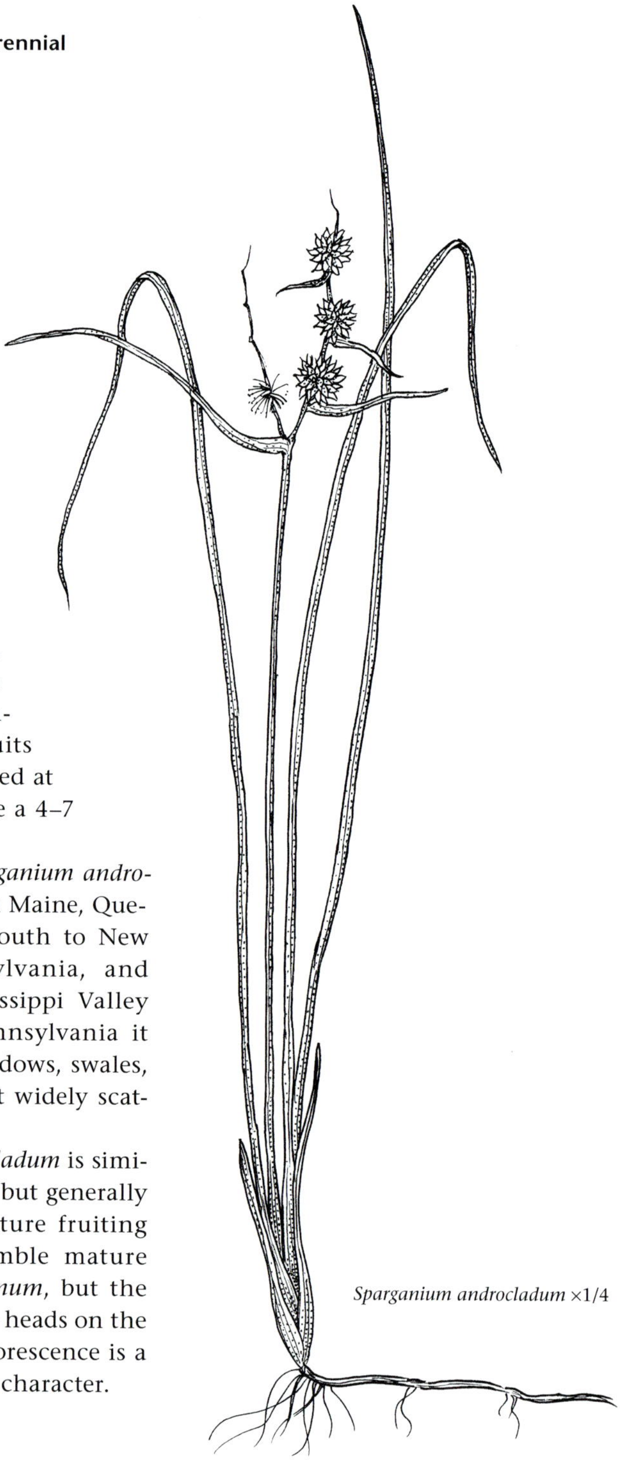

Sparganium androcladum ×1/4

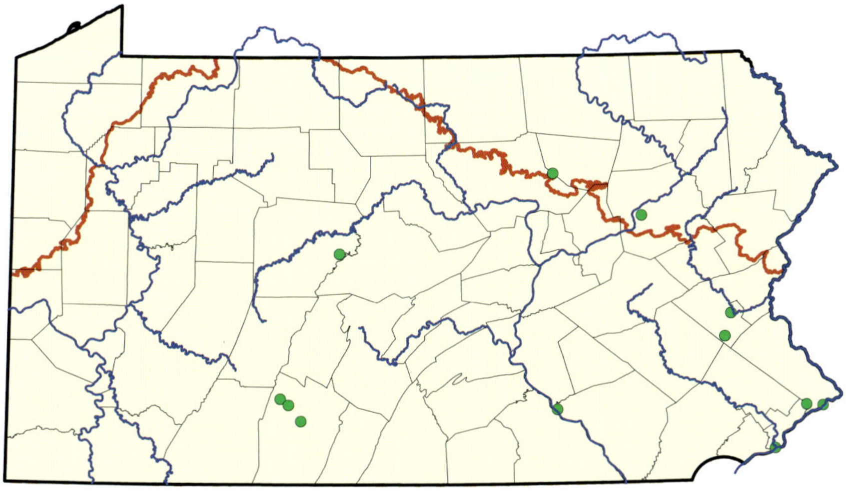

Sparganium androcladum

Sparganium androcladum

Narrow-leaf bur-reed *Sparganium angustifolium* Michx.

PNHP
Native
Rooted, floating-leaf perennial

Sparganium angustifolium—inflorescence ×1/2

Narrow-leaf bur-reed is a mostly submergent aquatic plant with long, narrow, ribbon-like leaves that extend upward to the surface and float at the ends. The plants spread by rhizomes and usually form colonies several meters in diameter. When the plants are not in flower, their floating leaf tips, which tend to align themselves in one direction due to wind or water currents, are the most conspicuous indication of their presence.

Leaves can be 1–1.2 m long, usually less than 5 mm wide, and slightly convex on the upper surface. Flowering stems extend above the water surface and bear several spherical heads of flowers in an unbranched inflorescence. The lowest female flowering head is on a peduncle arising from the axil of a leafy bract, the upper are sessile; at least some of the female heads are supra-axillary. Mature fruits are dull reddish or brown with a 1.5–2 mm-long beak.

Narrow-leaf bur-reed is a northern species; its range extends from Newfoundland to Alaska and south to New Jersey, Pennsylvania, Michigan, and northern California. In Pennsylvania it is a rare plant of natural glacial lakes in the northeast, where it grows in water about 1 m deep.

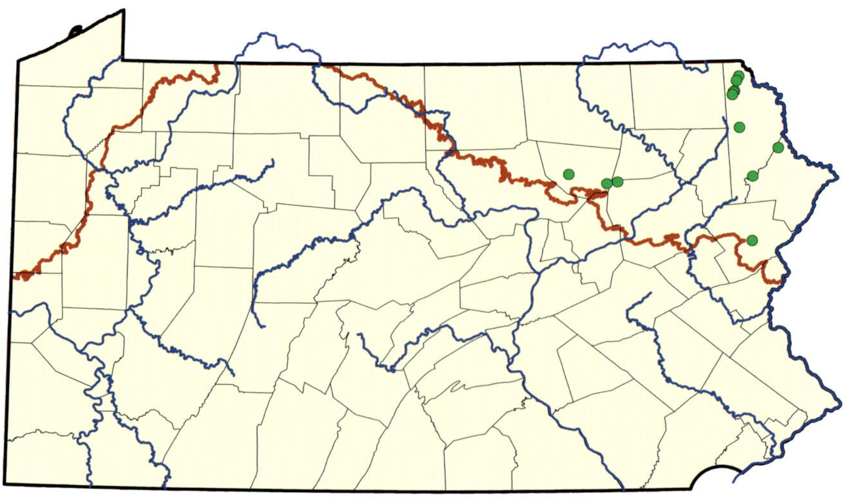

Sparganium angustifolium

Narrow-leaf bur-reed is most likely to be confused with *S. fluctuans* or *S. emersum*, both of which can also have floating leaves. Leaf width is the best guide for distinguishing the species in the absence of flowering or fruiting stems; in addition, the leaves of narrow-leaf bur-reed are not keeled (Table 3.1).

Because the leaves and inflorescences extend to the surface of the water, narrow-leaf bur-reed could be damaged by outboard motors. In lakes that permit motors, a no-entry zone should be established to protect colonies of this rare native plant.

Sparganium angustifolium—emergent inflorescence with floating leaf tips

Table 3.1. Comparison of Key Characters of Bur-reed Species

SPECIES	PLANT HEIGHT (M)	POSITION OF FRUITING HEADS	DIAMETER OF FRUITING HEADS	BEAK LENGTH (MM)	INFLORESCENCE	LEAF	LEAF WIDTH (MM)
S. americanum	1	axillary	1.5–2.5	3–5	branched[b]	erect, keeled	6–12
S. androcladum	1.2	axillary	2–3.5	4–7	branched[b]	erect, keeled	5–15
S. angustifolium	2+	supra-axillary[a]	1–3	1.5–2	unbranched	floating, (plano-convex)	2–5
S. emersum	to 2	supra-axillary[a]	1.6–3.5	2–4.5	unbranched	erect or floating, keeled	4–10
S. eurycarpum	2.5	axillary	2–3.5	2–4	branched	erect, keeled	6–20
S. fluctuans	1+	supra-axillary[a]	1.5–2.3	2–3.5	branched[b]	floating, flat, keeled at base	4–10

[a] At least some female heads supra-axillary.

[b] Inflorescence may be unbranched in depauperate plants.

Bur-reed *Sparganium emersum* Rehmann [synonym: *S. chlorocarpum* Rydberg]

Native
Rooted, emergent perennial

Leaves of *Sparganium emersum* are stiff and erect, up to 0.8 m tall and 4–10 mm wide, or limp and floating, in which case they may be as much as 2 m long. They are keeled at least at the base. The inflorescence is unbranched; at least some of the female heads are supra-axillary. Mature fruits are shiny light green with ribbed beaks 2–4.5 mm long.

Plants with floating leaves are easily confused with *S. angustifolium*; they are best differentiated by their leaves, which are keeled, at least at the base, and generally wider than those of *S. angustifolium* (Table 3.1). In addition, the fruits are green, and have longer beaks.

Sparganium emersum—inflorescence ×1/2

Sparganium emersum—fruit ×3

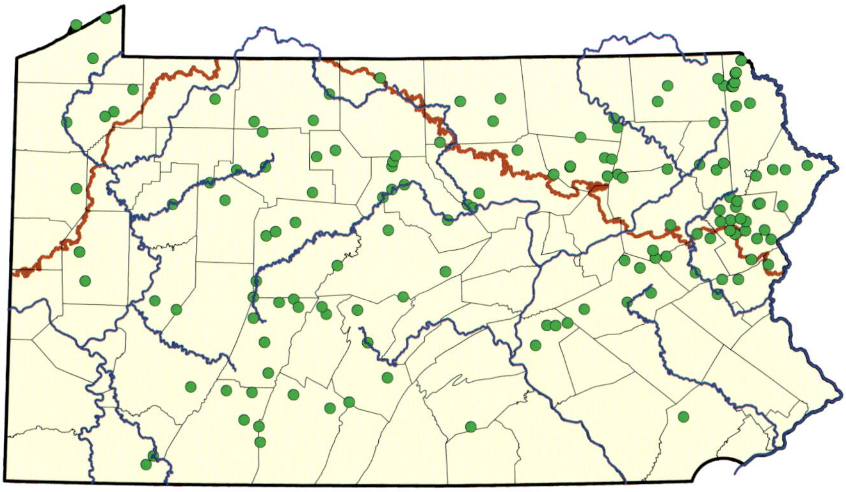

Sparganium emersum

Bur-reed — *Sparganium eurycarpum* Engelmann in A. Gray

Native
Rooted, emergent perennial

Sparganium eurycarpum is a robust plant up to 2.5 m tall, with stiffly erect, strongly keeled leaves. The inflorescence is erect, branched, and has prominent ascending bracts. Female heads are in the axils of bracts; those on the main axis are stalked and heads on the branches are sessile. Male heads are crowded at the tips of the inflorescence branches. Fruiting heads are 3–4 cm in diameter; achenes are brown at maturity and wedge-shaped, being broadest at the flattened to somewhat rounded top.

Sparganium eurycarpum is the only bur-reed in our flora with 2 stigmas; however, late in the season the tips of the stigmas may be broken off, making this character hard to see. The size of the fruiting heads and the lack of a constriction in the achenes are diagnostic.

The range of *Sparganium eurycarpum* extends across North America from Newfoundland and Labrador south to West Virginia and west to California and British Columbia. In Pennsylvania it is found along lake and river margins and in bogs and swamps throughout.

Sparganium eurycarpum—fruiting inflorescence ×1/4

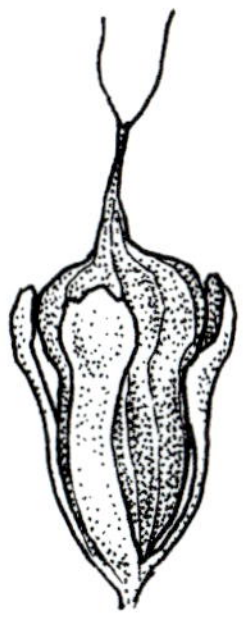

Sparganium eurycarpum—fruit ×3

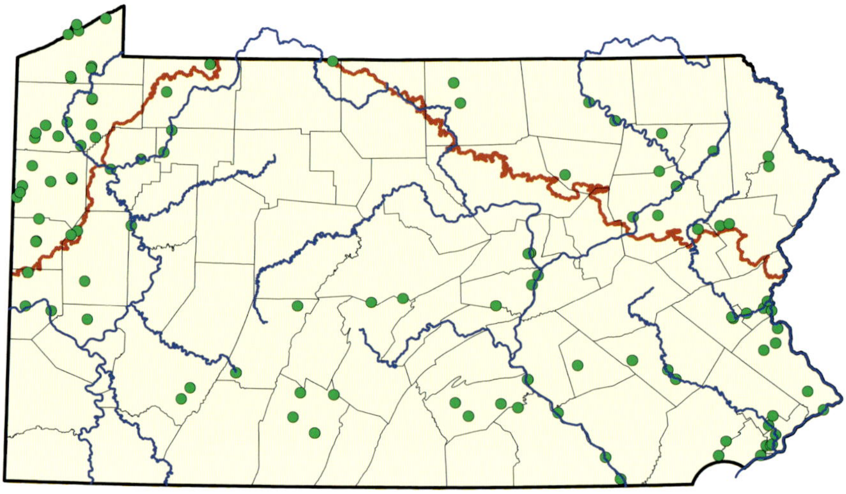

Sparganium eurycarpum

Bur-reed ***Sparganium fluctuans* (Engelmann ex. Morong) B. L. Robinson**

Native

Rooted, floating-leaf perennial

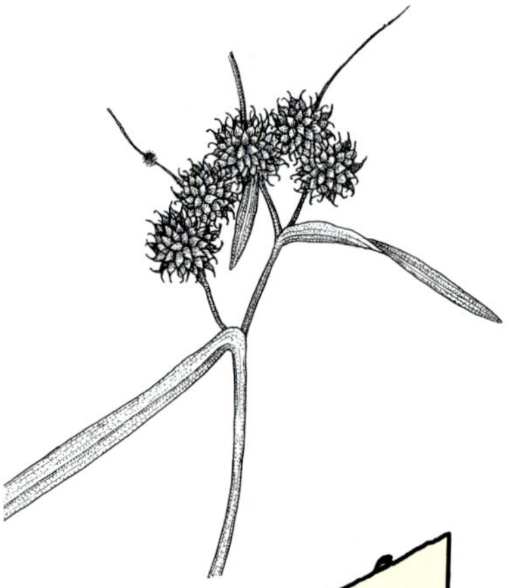

Sparganium fluctuans—fruiting inflorescence ×1/2

Sparganium fluctuans has slender, flexuous stems and limp floating leaves that can be 1.3–1.6 m long and 4–10 mm wide. The inflorescence, which extends above the water surface, is branched. Flower heads are axillary or supra-axillary, 1–2.5 cm in diameter.

The range of *Sparganium fluctuans* extends from Newfoundland to New Jersey and west to Wisconsin. In Pennsylvania it grows in lakes and slow-moving streams in the glaciated northeast, in water to 1.3 m deep. Non-flowering plants can be confused with *S. angustifolium*, but the leaves are consistently wider and keeled at the base (see Table 3.1).

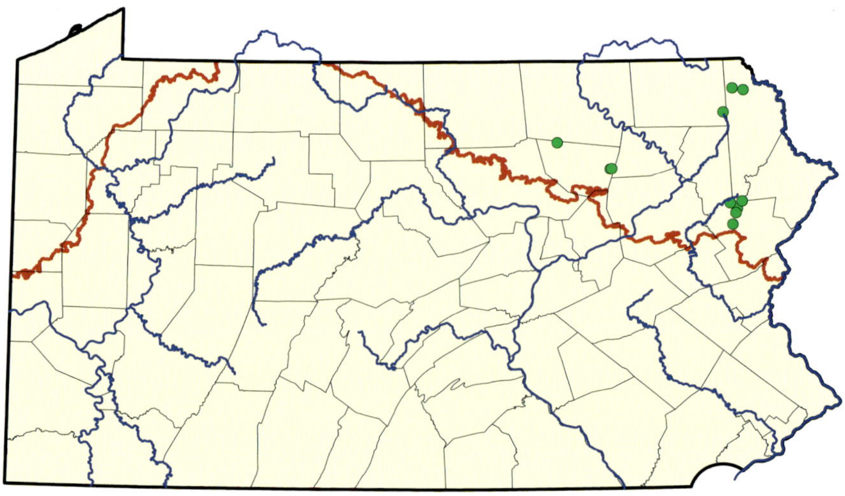

Sparganium fluctuans

CAT-TAIL
TYPHA L.
Cat-tail Family—Typhaceae

Cat-tails are erect plants of wet soils and shallow water; their cylindrical, brown spikes of flowers or fruits are readily recognized. Cat-tails are rhizomatous, usually forming dense stands. Leaves are narrow, elongate, and somewhat blue-green. Flowers are unisexual and crowded in dense cylindrical spikes. Male flowers are located at the top of the stem; wind is the agent of pollination. After shedding pollen the male flowers shrivel. Female flowers mature into a dense, cylindrical mass of fruits, which are eventually wind-dispersed.

Key to the Species of Cat-tail

A. male and female portions of the inflorescence continuous; leaves 10–23 mm wide, flat .... *Typha latifolia*
A. male and female portions of the inflorescence separated by 2–12 cm of bare stem; leaves 5–11 mm wide, rounded on the back........ *Typha angustifolia*

Common cat-tail — *Typha latifolia* Pursh

Native
Rooted, emergent perennial

Common cat-tail is a plant of swamps, marshes, wet shores, and ditches throughout Pennsylvania. Its abundant, wind-dispersed seeds spread it far and wide. The total range of common cat-tail extends throughout the continental United States and southern Canada; it is also known from Central and South America, Eurasia, and Africa.

The stems of common cat-tail are 1–3 m tall; the leaves are 10–23 mm wide. Male and female portions of the inflorescence are continuous or separated by a gap of less than 4 mm.

Common cat-tail is most abundant where nutrient levels are elevated, such as the inlets of lakes or other areas that receive runoff from agriculture or other sources of nutrients.

The abundant pollen of cat-tails was collected by Native Americans and others and used as flour to make bread, cake, and cereal; rhizomes and roots were also eaten. Medicinal uses of the plant included poultices to treat sores, and remedies for kidney stones and whooping cough. The leaves were used to weave baskets and mats.

The rhizomes of cat-tail are an important food source for muskrats and various waterfowl. Sandpipers eat the seeds. Marsh wrens and red-wing blackbirds nest in stands of cat-tail.

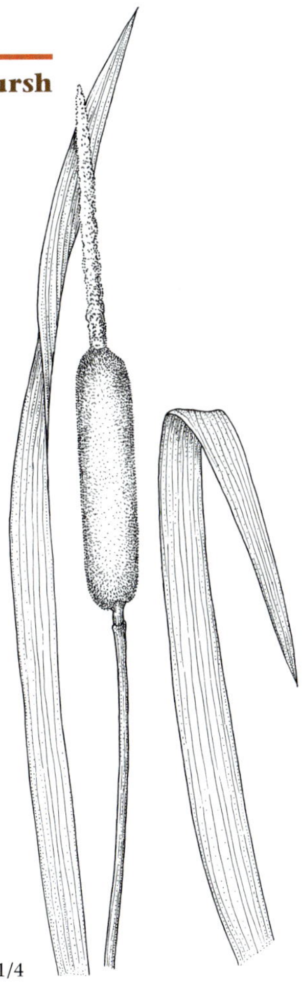

Typha latifolia ×1/4

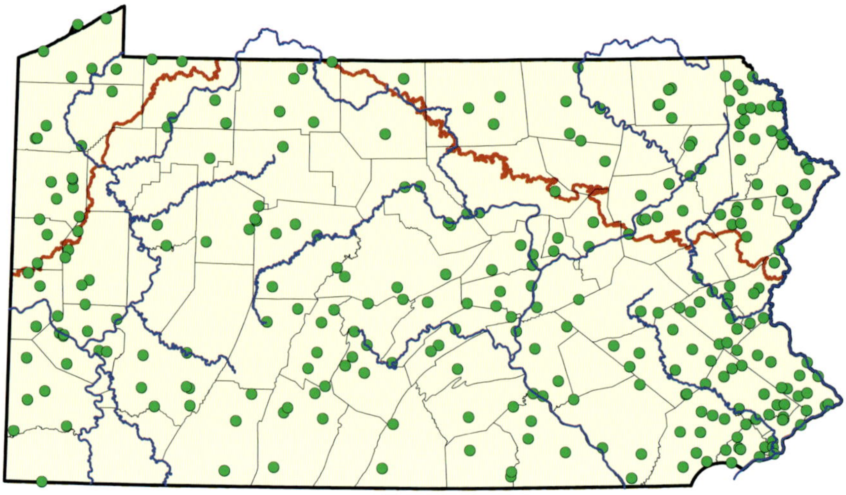

Typha latifolia

Narrow-leaf cat-tail *Typha angustifolia* Raf.

Introduced

Rooted, emergent perennial

Narrow-leaf cat-tail has stems 1–1.5 m tall; its leaves are 5–11 mm wide at the base of the blade. The staminate portion of the inflorescence is separated from the lower pistillate portion by a distinct sterile zone 2–12 cm long. Narrow-leaf cat-tail is a plant of roadside ditches, shores and marshes, often in calcareous or saline habitats across Pennsylvania. It is increasingly common in roadside ditches where deicing salts are used.

There is some debate as to whether narrow-leaf cat-tail is native to North America; it is considered by some to be an early European introduction (Flora of North America Editorial Committee 2000). Currently its range extends throughout the northeastern and mid-western states with scattered locations along the Pacific coast from Washington to southern California.

A hybrid between narrow-leaf cat-tail and common cat-tail, known as *Typha* x *glauca,* is intermediate between the two parent species in leaf width and separation of the male and female portions of the inflorescence. *Typha* x *glauca* is extremely vigorous and capable of out-competing native vegetation. It is considered a noxious weed on Presque Isle and in other areas of the state.

Typha angustifolia—inflorescence ×1/4

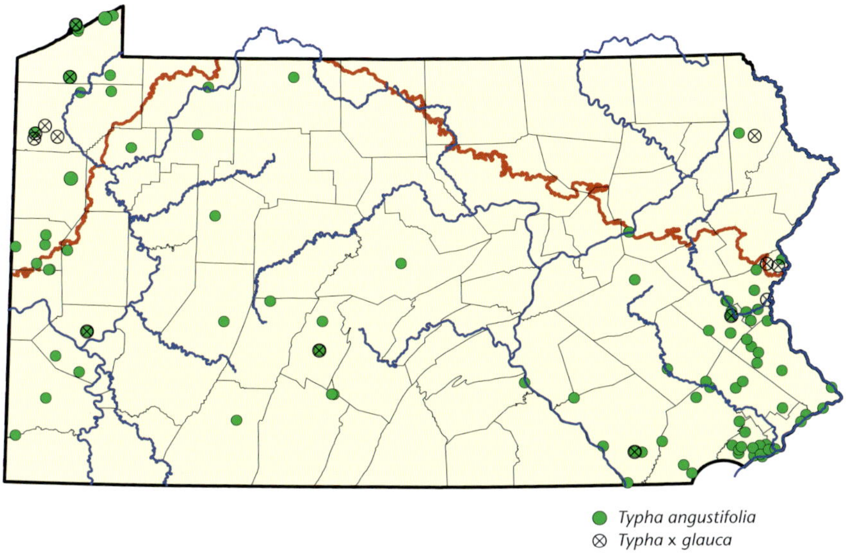

Typha angustifolia and *Typha* x *glauca*

Typha x *glauca*

CHINESE LOBELIA

LOBELIA L.

Bellflower Family—Campanulaceae

Chinese lobelia *Lobelia chinensis* Lour.

Introduced
Rooted, emergent perennial

Chinese lobelia is a low-growing, creeping, mat-forming plant of tidal shores of the Delaware River. It has narrow, coarsely toothed, alternate leaves and stems that root at the nodes.

The flowers are produced singly in the leaf axils; the corolla is pale lavender and approximately 1.5 cm wide. Flowering is more frequent in the upper portion of the tidal zone where the plants are exposed to the sun for longer periods; it may not flower at all if it is in a submersed position most of the time.

Chinese lobelia was first observed in our area in 1985; it is assumed that it reached the Delaware Estuary in ballast water of a ship from the Asia-Pacific region, where the plant is native. It has subsequently become well established at numerous locations along the banks of the Delaware River in Bucks and Philadelphia Counties.

Lobelia chinensis ×2/3

Lobelia chinensis on the tidal shore of the Delaware River

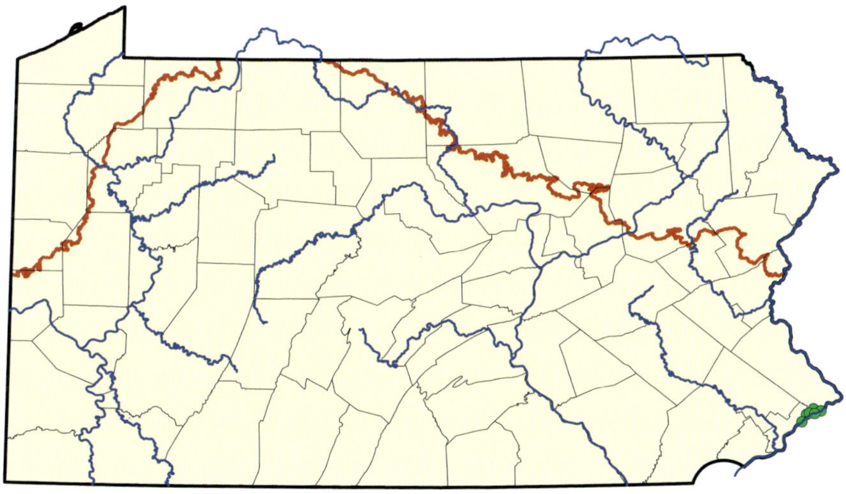

Lobelia chinensis

FLOWERING-RUSH
BUTOMUS L.
Flowering-rush Family—Butomaceae

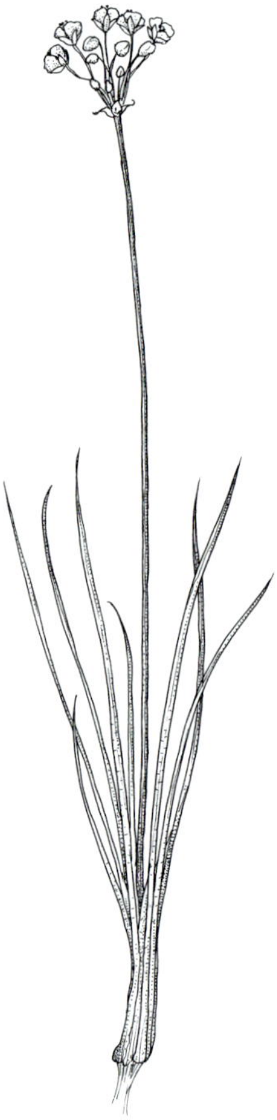

Butomus umbellatus ×1/6

Flowering-rush ***Butomus umbellatus* L.**

Introduced
Rooted, emergent perennial

Butomus umbellatus—flower ×1

Flowering-rush is an emergent aquatic with an erect stem bearing a terminal umbel of pinkish-purple, insect-pollinated flowers. It is a rhizomatous perennial with linear, parallel-veined leaves that can be submersed, floating, or emergent. The flowers have 3 petals, 3 sepals, 9 stamens, and 6 carpels, each of which becomes a beaked achene.

Native to Eurasia, flowering-rush is naturalized in North America, where it is known from Nova Scotia to Alberta and south to Pennsylvania, Ohio, and Idaho. In Pennsylvania it is a rare plant of pond margins and marshes known only from the northwestern corner of the state.

Flowering-rush has been used to treat worms in cows and horses.

It is the only emergent aquatic plant in our flora with a simple terminal umbel of showy flowers, thus difficult to confuse with anything else.

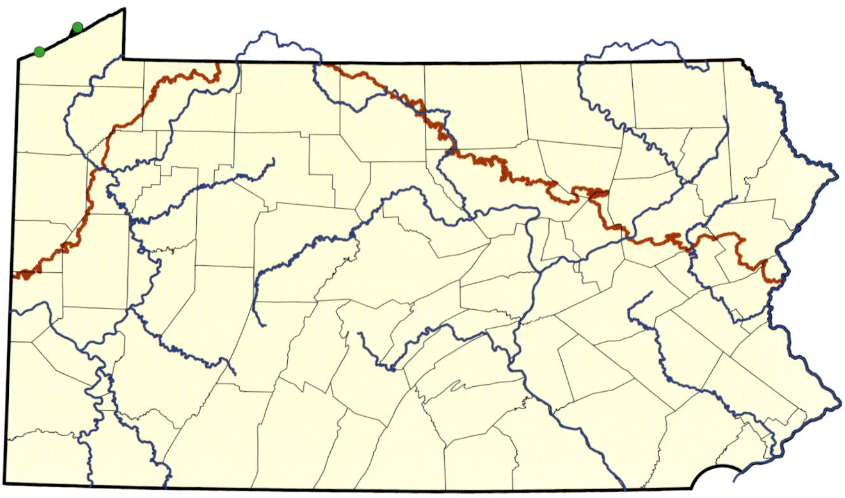

Butomus umbellatus

GOLDENCLUB
Orontium L.
Arum Family—Araceae

Goldenclub *Orontium aquaticum* L.

PNHP
Native
Rooted, emergent perennial

It is the inflorescence that gives goldenclub its name; small bright yellow flowers are packed together on a thick white stalk (spadix) that extends above the water surface. Flowering occurs in May; the flowers are insect-pollinated. Unlike most members of the Arum Family, goldenclub inflorescences are not surrounded by a spathe. Each fertilized flower matures into a spherical green fruit the size of a large pea. The seeds may germinate while floating; upon coming to rest on a muddy substrate the roots grow into the mud and contract, pulling the base of the young plant into the mud.

Goldenclub has dark blue-green, narrowly elliptical leaf blades that extend above the water surface, or sometimes float. The stout petioles may be 0.5 m long depending on the depth of the water. The surfaces of the 20–30 cm-long leaf blades are strongly hydrophobic: drops of water roll right off them. The leaves arise in clusters from rhizomes that are rooted in lake-bottom sediments.

Goldenclub is a plant of lakes, ponds, streams, and vernal ponds; it often forms a band along the shoreline extending out to depths of about 1 m. Deer and/or geese graze on the leaves; by mid to late summer it can be hard to find an intact leaf in some locations.

Orontium aquaticum, Photograph by Roger E. Latham

Orontium aquaticum ×1/4

Ranging from southern New England to Texas, goldenclub is primarily a plant of the Atlantic and Gulf Coastal Plains and adjacent portions of the Appalachians. It has disappeared from many sites along the Delaware River in southeastern Pennsylvania, but remains locally abundant at more inland sites from the Pocono Plateau to south central and southwestern counties.

Native Americans and early European settlers ate the dried or boiled seeds of goldenclub. This plant is also sometimes cultivated in water gardens; it is available from nurseries specializing in aquatic plants.

The leaves and flowering stems of goldenclub are very distinctive: there is no other aquatic plant with which it is likely to be confused. Flowering occurs in late spring; the leaves remain conspicuous throughout the growing season.

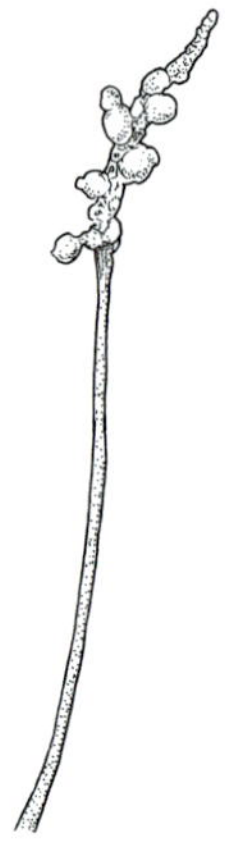

Orontium aquaticum—
fruiting inflorescence ×1/4

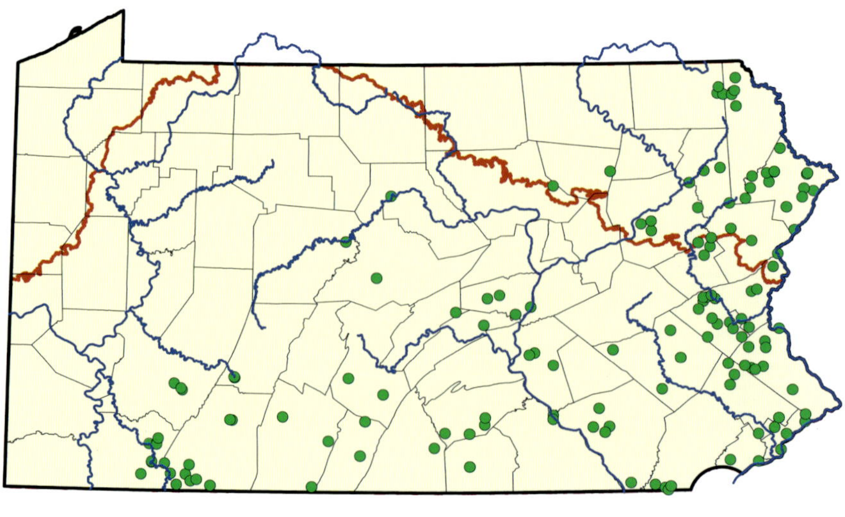

Orontium aquaticum

GRASSES
ECHINOCHLOA P. BEAUV.
GLYCERIA R.BR.
LEERSIA SW.
TORREYOCHLOA CHURCH
ZIZANIA L.
Grass Family—Poaceae

Grasses are herbaceous plants with a reduced flower structure that is adapted for wind pollination. Inflorescences consist of individual florets arranged in spikelets subtended by 2 bracts (glumes). Each flower is enclosed by a pair of additional bracts, the lemma and the palea.

The grass family is a huge family with worldwide distribution. In Pennsylvania, only a handful of species in five genera are considered truly aquatic.

Key to the Aquatic Grasses of Pennsylvania

A. flowers unisexual, the male flowers on the lower branches of the inflorescence .... *Zizania aquatica*
A. flowers perfect, each containing both male and female structures
  B. spikelets lacking glumes .... *Leersia oryzoides*
  B. glumes present
    C. leaf sheaths free to the base
      D. ligules absent; spikelets awned, containing a single fertile floret .... *Echinochloa walteri*
      D. ligules present; spikelets lacking awns, containing 2–8 florets .... *Torreyochloa pallida*
    C. leaf sheaths united for at least half their length .... *Glyceria* spp.

Walter's barnyard-grass — *Echinochloa walteri* (Pursh) A. Heller

PNHP
Native
Rooted, emergent annual

Walter's barnyard-grass is a coarse grass 1–2 m tall that grows in the freshwater tidal marshes and mud flats of the Delaware Estuary in Pennsylvania.

The spikelets are arranged in thick racemes that are aggregated into panicles. Each spikelet contains a single floret and is subtended by a sterile lemma bearing an awn up to 3 cm long.

The range of Walter's barnyard-grass includes the east coast of North America, the Gulf coast, and the Great Lakes region. It also grows in Mexico and Guatemala. In Pennsylvania it is limited to the southeast corner of the state, where it grows with other freshwater tidal marsh species such as wild rice (*Zizania aquatica*).

Echinochloa walteri—spikelet ×3

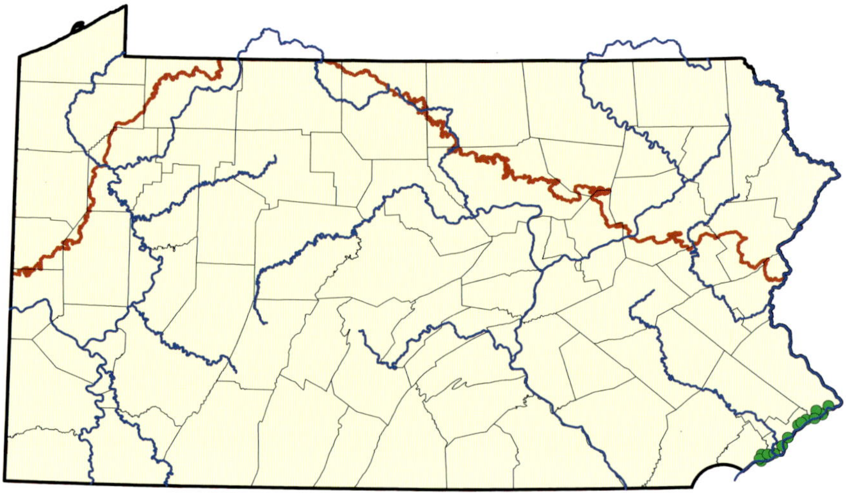

Echinochloa walteri

Echinochloa walteri—inflorescence ×1/4

Mannagrass *Glyceria* R.Br.

The mannagrasses are perennials with 5–12 florets in each spikelet. The glumes are shorter than the lowest lemma; lemmas are rounded on the back, with 5–9 conspicuous nerves.

Key to the Species of Mannagrass

A. lemmas acute; palea much longer than the lemma ....*Glyceria acutiflora*
A. lemmas obtuse; palea only slightly longer than the lemma........*Glyceria borealis*

Creeping mannagrass *Glyceria acutiflora* Torr.

Native
Rooted, emergent perennial

Creeping mannagrass grows up to 1 m tall from a decumbent base, and often produces roots at the lower nodes. The inflorescence is a panicle with a few erect branches. Lemmas are 7–8.5 mm long with 7 veins; the palea is up to 3 mm longer than the lemma.

Creeping mannagrass is a plant of the northeastern United States from Massachusetts to Tennessee and eastern Missouri. It also occurs in eastern Asia. In Pennsylvania it is found mainly in the central and northeastern regions of the state in shallow water of ponds and swamps. The fruits are an important food for wildlife.

Glyceria acutiflora —spikelet ×2

Glyceria acutiflora—inflorescence ×1/2

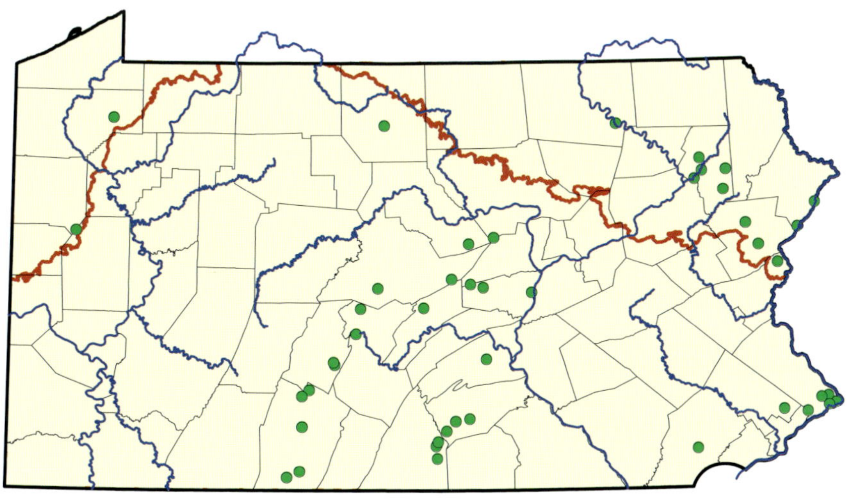
Glyceria acutiflora

Northern mannagrass *Glyceria borealis* (Nash) Batch.

Native

Rooted, emergent perennial

Northern mannagrass has stems to 1.2 m tall that are decumbent and rooting at the base; it grows in water to 1 m deep. The leaves are typically floating, aided by densely papillose leaf surfaces, which are strongly hydrophobic. The inflorescence is a slender panicle with several branches. The 1–2 cm-long spikelets contain 8–12 florets. Lemmas are 3–4 mm long.

Northern mannagrass is widespread across the northern United States and Canada. It extends farther south in the western mountains. In Pennsylvania it is concentrated in the glaciated northeast, with a few scattered sites in the center of the state.

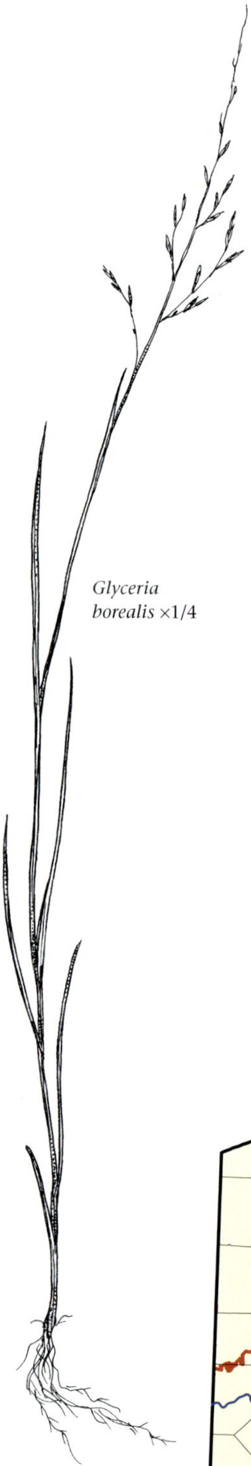

Glyceria borealis ×1/4

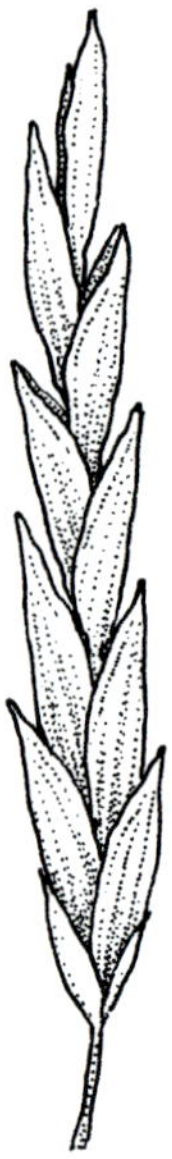

Glyceria borealis—spikelet ×5

Glyceria borealis (foreground) growing in shallow water in Wayne County, Pennsylvania

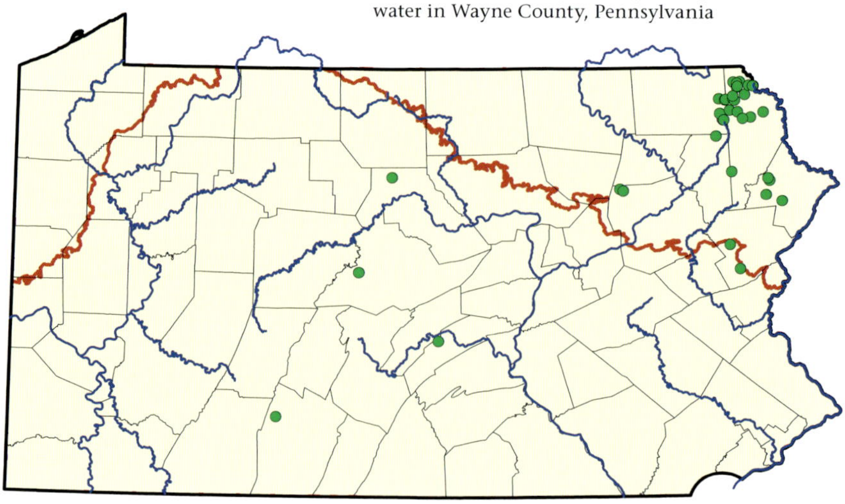

Glyceria borealis

Rice cutgrass *Leersia oryzoides* (L.) Sw.

Native
Rooted, emergent perennial

This grass is best identified by running your hand upward along the whole plant to feel the rough edges of the stems and leaves. The stems are up to 1.5 m tall and decumbent at the base, especially when growing in standing water.

The inflorescence is a panicle with the spikelets clustered on the on upper half of each panicle branch. The spikelets lack glumes, and contain a single floret.

Rice cutgrass is a plant of shallow water, tidal shores, marshes, and bogs throughout Pennsylvania. Its total range extends throughout the eastern half of North America and along the Pacific coast. It has become established in Europe and Asia. Ducks feed on the seeds and rootstock.

Leersia oryzoides—spikelet ×3

Leersia oryzoides—inflorescence ×1/2

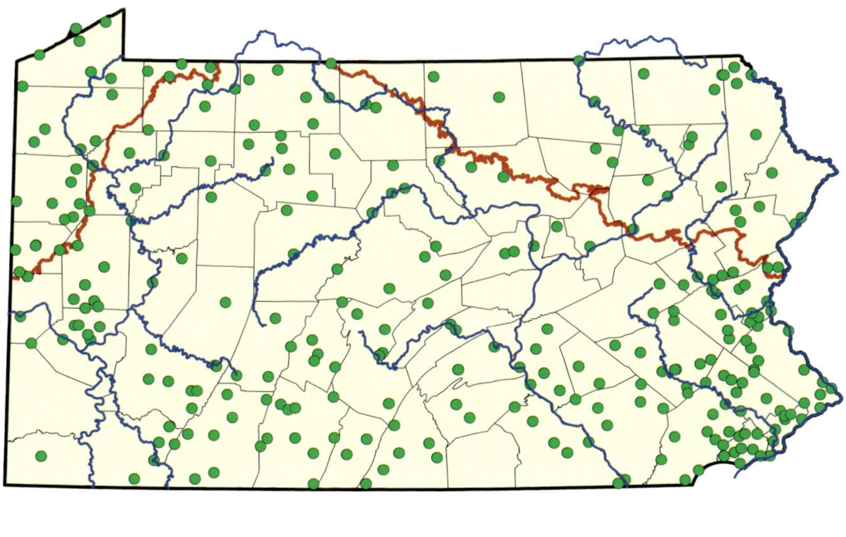

Leersia oryzoides

Pale meadowgrass *Torreyochloa pallida* (Torr.) Church

Native

Rooted, emergent perennial

Pale meadowgrass is a perennial with stems to 1 m tall; it usually grows in standing water with the upper leaves floating. The inflorescence is an open panicle with few branches. Spikelets contain 4–7 florets subtended by 2 short glumes; the lemmas are 2–3.6 mm long and strongly 5–7-veined.

Pale meadowgrass occurs throughout the northeastern and western United States and Canada, including Alaska. In Pennsylvania it is scattered throughout the state in lakes, stream margins, marshes, and swamps.

Two varieties are recognized:

A. leaf blades 2–3 mm wide; flowering May-June .... *Torreyochloa pallida* var. *fernaldii*

A. leaf blades to 8 mm wide; flowering July-August .... *Torreyochloa pallida* var. *pallida*

Torreyochloa pallida var. *fernaldii*—spikelet ×3

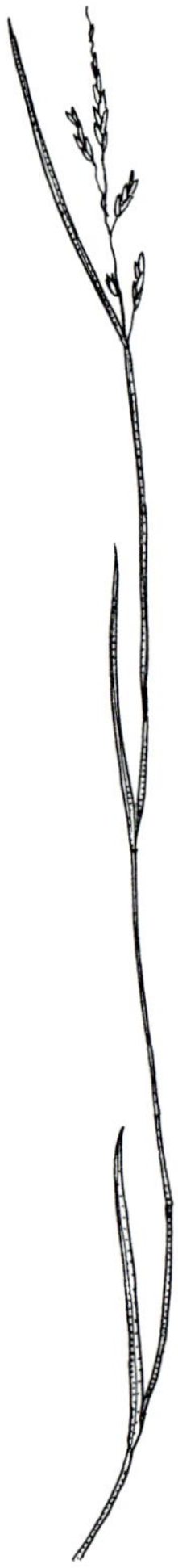

Torreyochloa pallida var. *fernaldii* ×2/3

Torreyochloa pallida var. *pallida*—spikelet ×5

Torreyochloa pallida var. *pallida* ×1/8

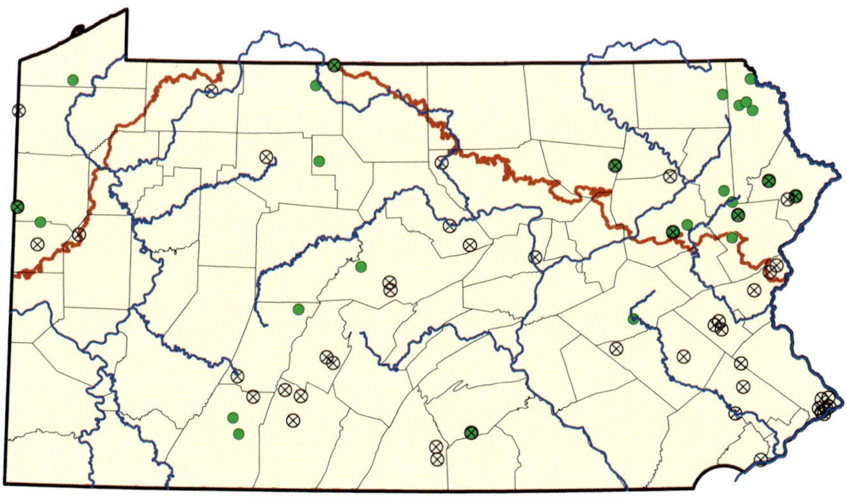

Torreyochloa pallida with varieties

Wild-rice *Zizania aquatica* L.

PNHP
Native
Rooted, emergent annual

Wild-rice is an erect annual grass that can reach 3 m in height; the large terminal inflorescence contains female flowers on erect upper branches and male flowers on spreading lower branches. Each spikelet contains a single flower. Mature spikelets are about 2 cm long with a 1–5 cm-long awn.

Wild-rice is native from the Dakotas and the Great Lakes region to the eastern seaboard. In Pennsylvania it grows in the upper portions of freshwater intertidal marshes along the Delaware River, on Presque Isle in the northwestern corner of the state, and in a few scattered locations between. The grains are an important food for both waterfowl and songbirds and were also harvested by Native Americans. Commercial production of wild-rice utilizes cultivated strains that do not shatter as quickly as the wild type.

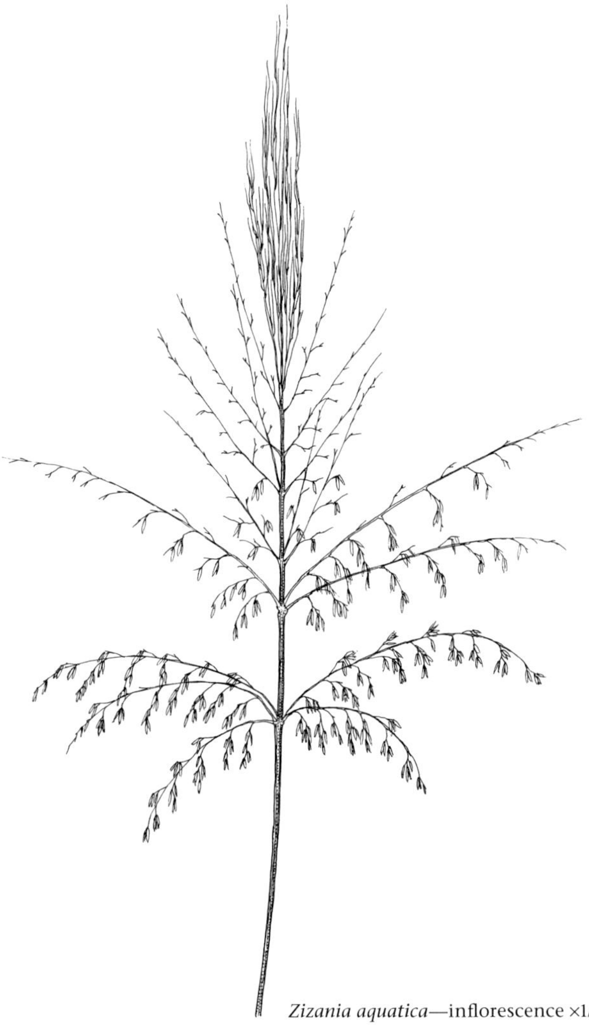

Zizania aquatica—inflorescence ×1/6

Zizania aquatica—female flower ×2

Zizania aquatica—male flower ×2

Stand of *Zizania aquatica* along the tidal bank of the Delaware River in Bucks County, Pennsylvania

Zizania aquatica—inflorescence

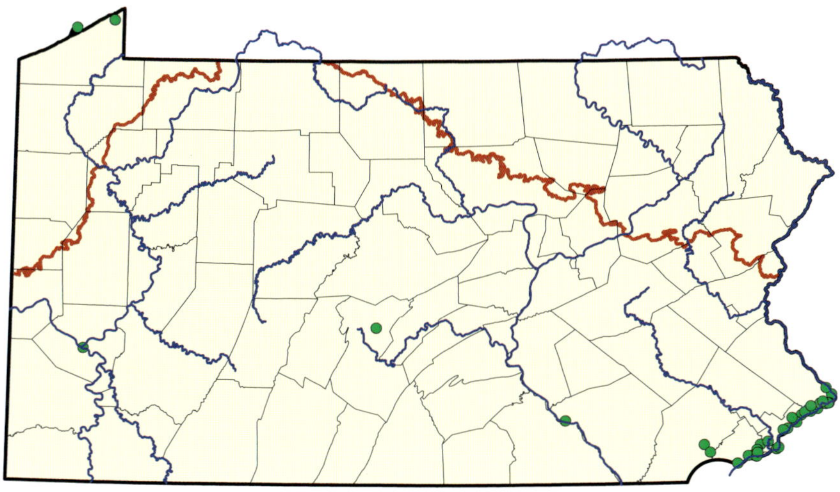

Zizania aquatica

WATER HORSETAIL
EQUISETUM L.
Horsetail Family—Equisetaceae

Water horsetail — *Equisetum fluviatile* L.

Native
Rooted, emergent perennial

Water horsetail is an emergent species of lake, pond, and stream margins. Its dark green, ribbed and segmented stems grow from black rhizomes buried in the lake or stream bottom sediments.

The horsetails are spore-bearing plants; they do not produce flowers or seeds. Their growth habit consists of dark green, hollow, finely ribbed stems that are jointed at regular intervals. The joints (or nodes) often bear whorls of slender branches. Leaves are much reduced and appear as a sheath around the stem. Spores are produced in a cone-like strobilus at the stem tip.

Water horsetail occurs throughout Canada, Alaska, and the northern third of the continental United States. It also grows in Europe and Asia. In Pennsylvania, which is near the southern limit of its range, it is scattered throughout the state with the exception of the north central and southwestern regions.

Waterfowl, bear, muskrat, and farther north, moose, eat the stems and rhizomes of water horsetail.

Shore horsetail (*E.* x *litorale*) is a hybrid between water horsetail (*E. fluviatile*) and field horsetail (*E. arvense*) that is known from scattered locations throughout the state. It has few if any branches; they are solid rather than hollow as in *E. fluviatile*. In addition, the teeth on the sheath of shore horsetail have a white edge; in water horsetail the teeth are uniformly dark. Shore horsetail grows along riverbanks and lakeshores, usually in the vicinity of the parent species. Although strobili are produced, the spores are misshapen and abortive.

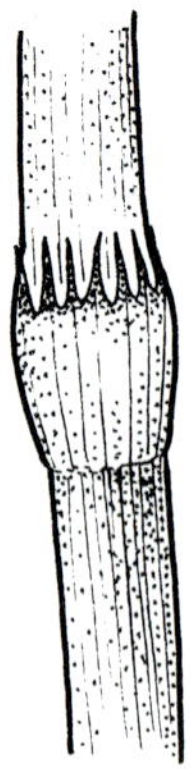

Equisetum fluviatile—
node with scale-like leaves ×2

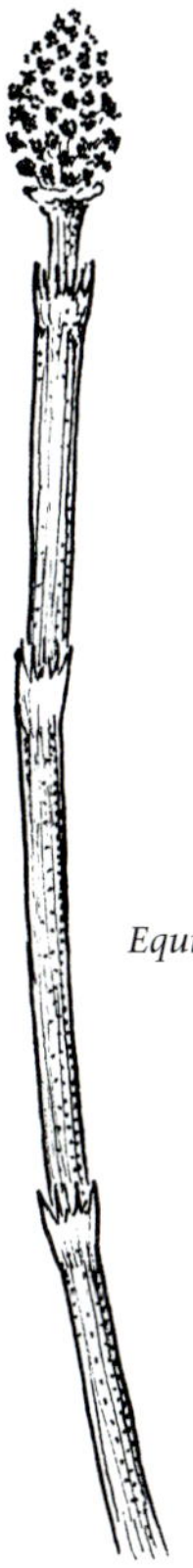

Equisetum fluviatile—strobilus ×1

Equisetum fluviatile ×1/6

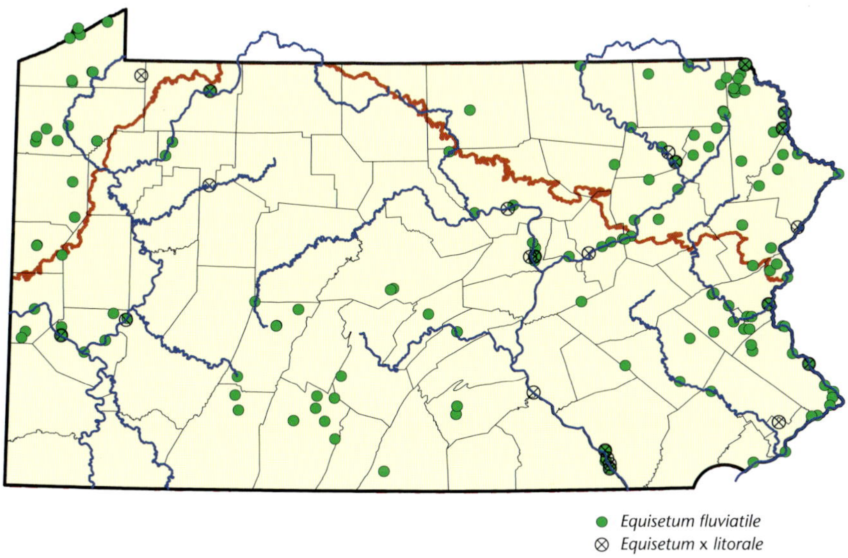

Equisetum fluviatile and *Equisetum* x *litorale*

IRIS
Iris L.
Iris Family—Iridaceae

The irises are herbaceous perennials with a prominent fleshy rhizome. The aquatic species grow in distinct clumps. Leaves are narrow, pointed and enfold the younger shoots at the base. The flowers are showy, 3-merous, with 3 spreading sepals, 3 upright petals and 3 arching petal-like stigma lobes. The 3 stamens are hidden between the sepals and the stigma lobes. The inferior ovary matures into a 3- or 6-angled capsule.

Key to the Aquatic Species of Iris

A. flowers yellow .... *Iris pseudacorus*
A. flowers blue or violet
  B. leaves stiff and erect; sepals with a yellow patch at the base of the blade .... *Iris virginica*
  B. leaves lax and arching; sepals with green or greenish-yellow patch surrounded by a purple and white veined zone .... *Iris versicolor*

Yellow Iris — *Iris pseudacorus* L.

Introduced
Rooted, emergent perennial

Yellow iris is a robust plant with stems to 1.5 m tall in dense clumps; it is by far the largest of the aquatic irises in our area. The leaves are up to 2 cm wide, erect and dark green or glaucous at the base. The flowers, which are produced from late May to early July, are bright yellow and very showy, in part because of the nearly orbicular blade of the sepals.

Yellow iris is native to Eurasia and northern Africa; it is widely planted and frequently naturalized throughout the northeastern United States and adjacent areas of Canada. In Pennsylvania it is present in swamps, marshes, wet shores, and shallow water, mainly in the east and northwest.

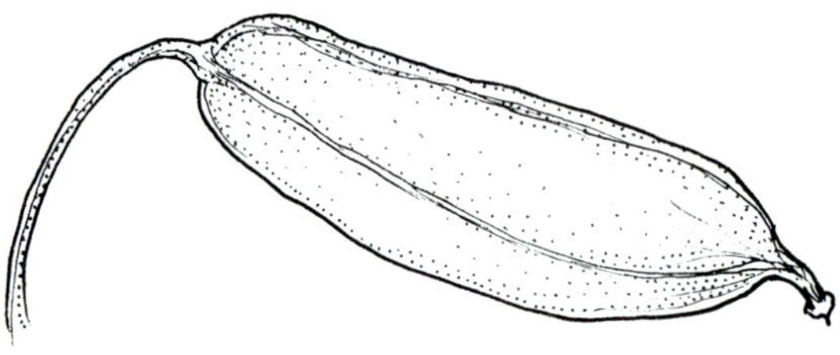

Iris pseudacorus—capsule ×1

Iris pseudacorus

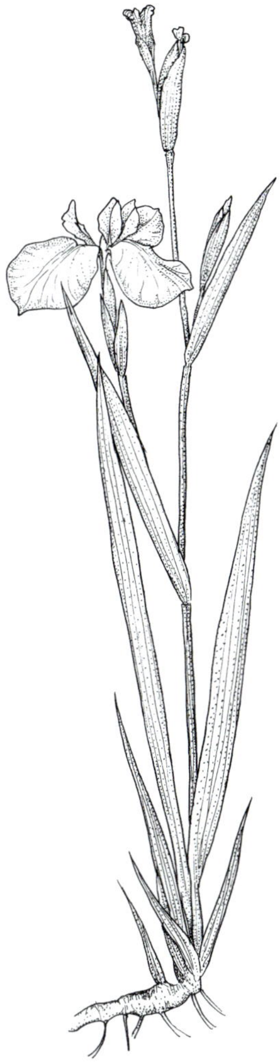

Iris pseudacorus ×3/16

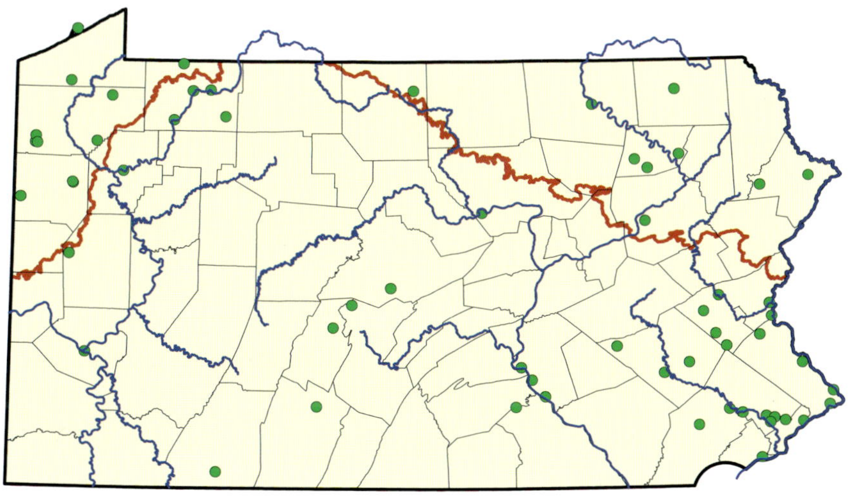

Iris pseudacorus

Northern blue flag *Iris versicolor* L.

Native
Rooted, emergent perennial

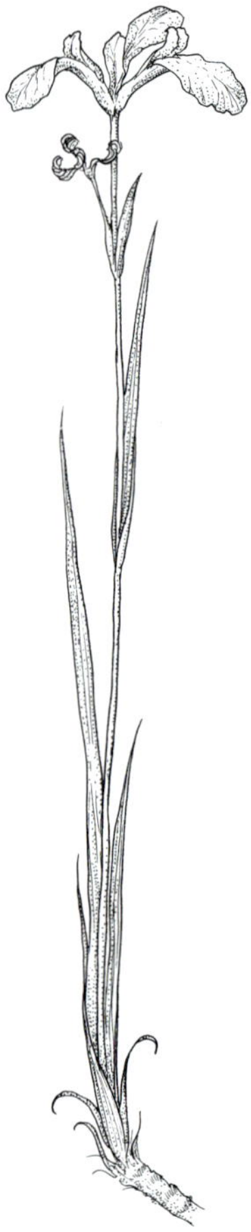

Iris versicolor ×1/4

Northern blue flag forms dense clumps to 0.6 m tall; the stems arise from branching rhizomes. Leaves are grayish-green, often with a purple tinge at the base. The showy flowers are violet-blue with a greenish-yellow patch surrounded by a veiny purple and white zone near the base of each sepal. Flowering occurs from May to July.

A plant of the northeast, northern blue flag occurs from Nova Scotia to Minnesota and south to West Virginia. In Pennsylvania it grows primarily in the eastern and western portions of the state with scattered sites in the central area. It is a plant of bogs, swamps, and shallow water of lake and pond margins.

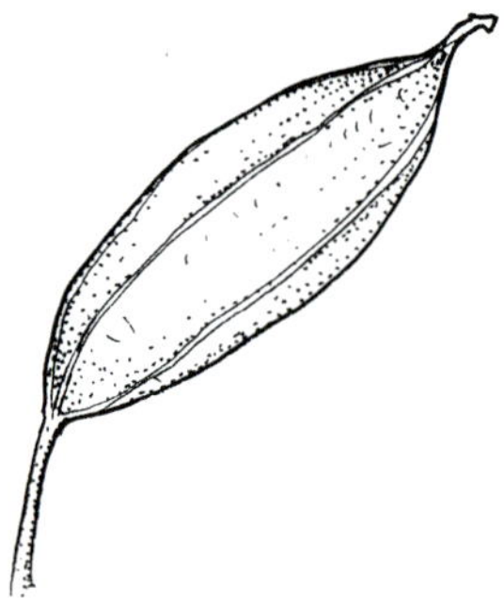

Iris versicolor—capsule ×1

Iris versicolor

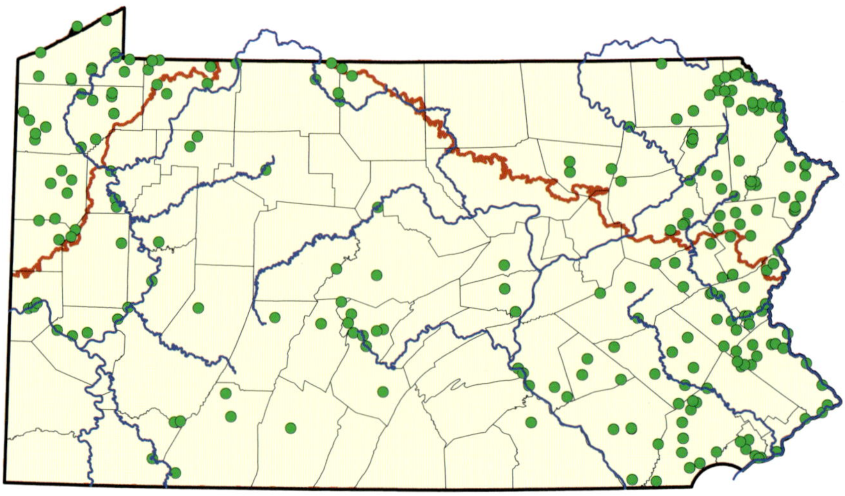

Iris versicolor

Southern blue flag — *Iris virginica* L.

PNHP
Native
Rooted, emergent perennial

Southern blue flag is a rare plant in Pennsylvania, known only from a few sites along Lake Erie in the northwestern corner of the state, where it grows in shallow water of pond margins. Flowering occurs from May to July.

Southern blue flag is very similar to northern blue flag, from which it can be distinguished by the bright yellow, hairy blotch at the base of the sepals and its more lax, arching leaves.

Pennsylvania is at the northeastern limit of the range of southern blue flag, which extends to Florida and west to Iowa and Louisiana.

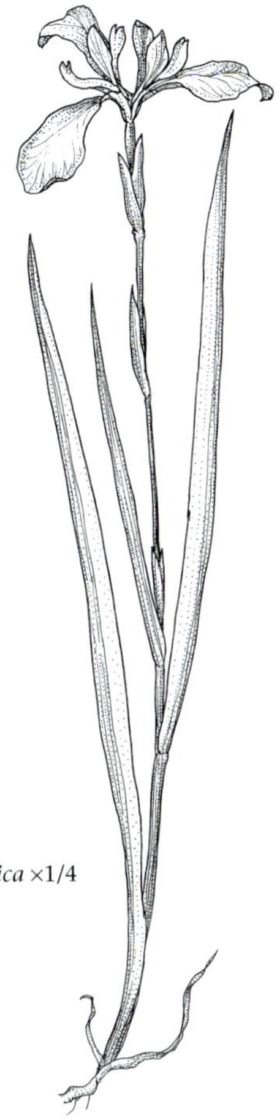

Iris virginica ×1/4

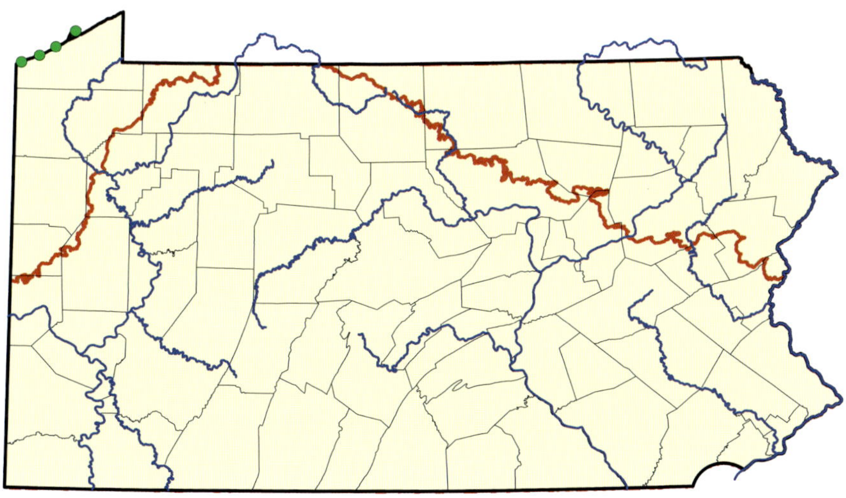

Iris virginica

LIZARD'S-TAIL

SAURURUS L.

Lizard's-tail Family—Saururaceae

Lizard's-tail ***Saururus cernuus* L.**

Native
Rooted, emergent perennial

In July, when its slender curved spikes of small white flowers are present, lizard's-tail forms a striking display. The plants may be from 0.5–1.2 m tall, with the inflorescences extending above the leafy lower stems. Only the base of the plant is normally below the water surface. The leaves are alternate, heart-shaped, and dull, dark green. The flowers are perfect with 6–8 stamens and a single superior ovary. The fruit is a dry nutlet.

Lizard's-tail spreads by rhizomes to form colonies in shallow water. It is more common along the margins of slow-moving streams, but may also grow at the edges of lakes and ponds, where it can form a dense band. Although highly ornamental, lizard's-tail may grow too vigorously for inclusion in a small water garden.

Native Americans used lizard's-tail to treat specific ailments ranging from wounds to rheumatism and spider bites. In addition, an infusion of pounded plants was considered a panacea for general illnesses.

In Pennsylvania lizard's-tail occurs at scattered locations across the state, mostly in major river systems. Its total range extends from New England to Minnesota and south to Florida and Texas.

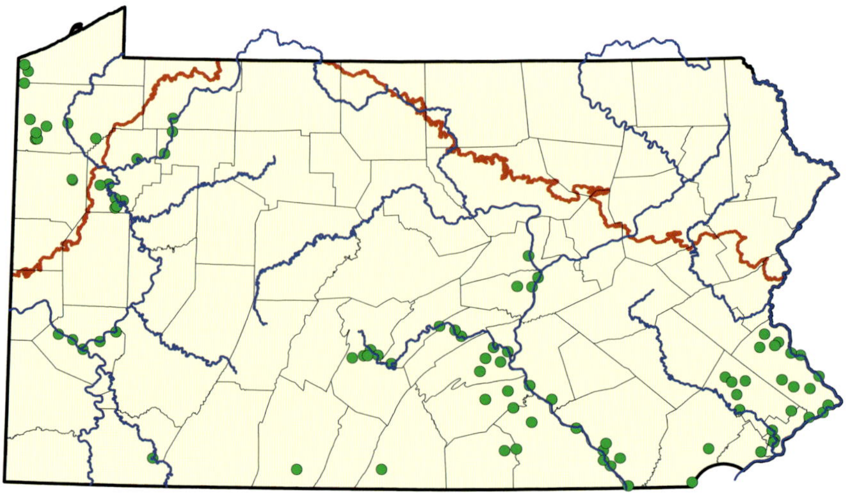

Saururus cernuus

Saururus cernuus—with inflorescence ×1/2

Saururus cernuus

MUD-PLANTAIN

HETERANTHERA RUIZ & PAVON

Pickerel-weed Family—Pontederiaceae

The mud-plantains are submersed or emergent perennials. Leaf blades are broadly heart- or kidney-shaped and floating or emergent. Flowers are produced singly or in a short spike that emerges from a basal spathe; each opens for one day only. The fruit is a capsule.

Key to the Species of Mud-plantain

A. flower spike barely elongating beyond the spathe; hairs on the filaments white .... *Heteranthera reniformis*

A. flower spike elongating well beyond the spathe; hairs on the filaments purple.. .... *Heteranthera multiflora*

Mud-plantain *Heteranthera multiflora* (Griseb.) Horn

PNHP
Native
Rooted, emergent annual

Heteranthera multiflora is an annual; its heart-shaped leaves are floating, submersed, or decumbent on mud flats. The spicate inflorescence contains 3–16 white flowers, each opening for a single day. The filaments of the anthers are densely covered with purple hairs.

The range of *H. multiflora* includes the coastal areas from New Jersey to Virginia and an interior zone in the Mississippi drainage extending from Iowa to Oklahoma. It is also known in South America. In Pennsylvania this species is a rare plant of the freshwater tidal shores of the Delaware River system.

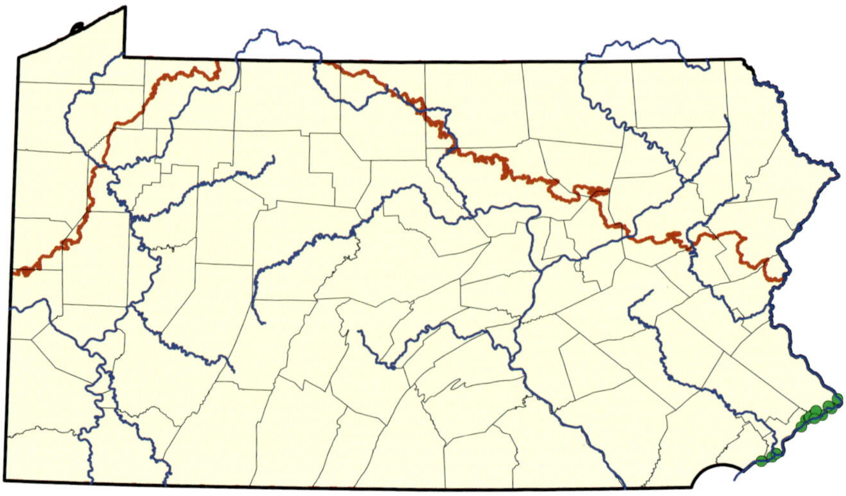

Heteranthera multiflora

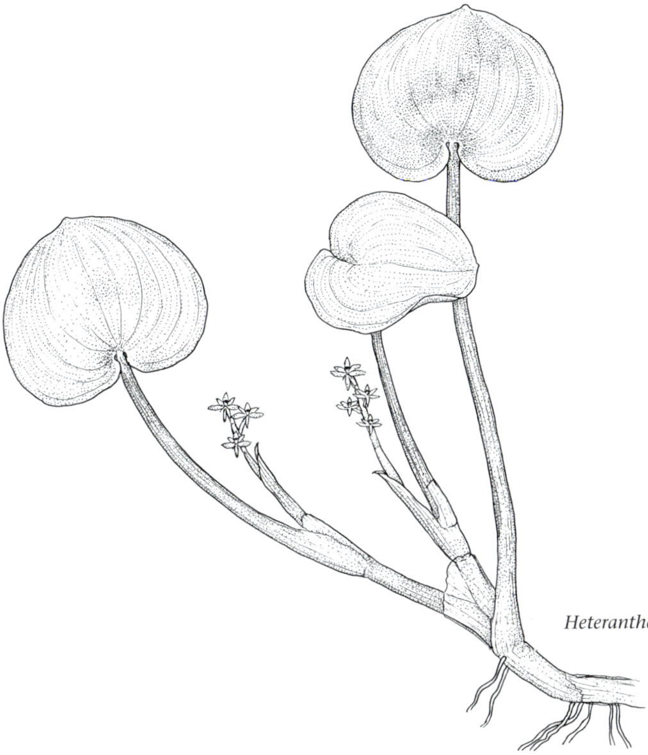

Heteranthera multiflora ×1/2

Heteranthera multiflora

Mud-plantain *Heteranthera reniformis* Ruiz & Pavon

Native

Rooted, emergent annual or short-lived perennial

Heteranthera reniformis, an annual or short-lived perennial, is very similar to *H. multiflora.* The leaves are broader, more kidney-shaped than heart-shaped, and the hairs on the filaments are white. In addition the flower spike remains enclosed in the spathe. *H. reniformis* ranges from New York to Louisiana; it also grows in Mexico and Central America and at scattered locations in South America. In Pennsylvania it occurs primarily in the southeastern region, where it grows in shallow water and on muddy shores.

In addition to its similarity to *H. multiflora,* non-flowering plants of *H. reniformis* could also be confused with sterile plants of wild calla (*Calla palustris*), which has a very similar growth habit. The flowers and fruits are quite different, however.

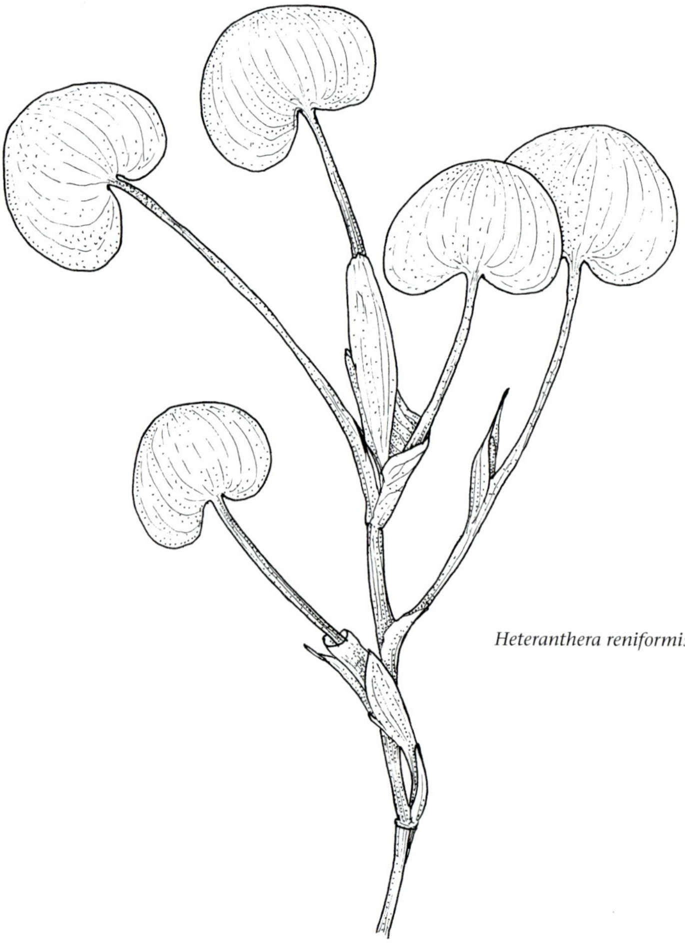

Heteranthera reniformis ×3/4

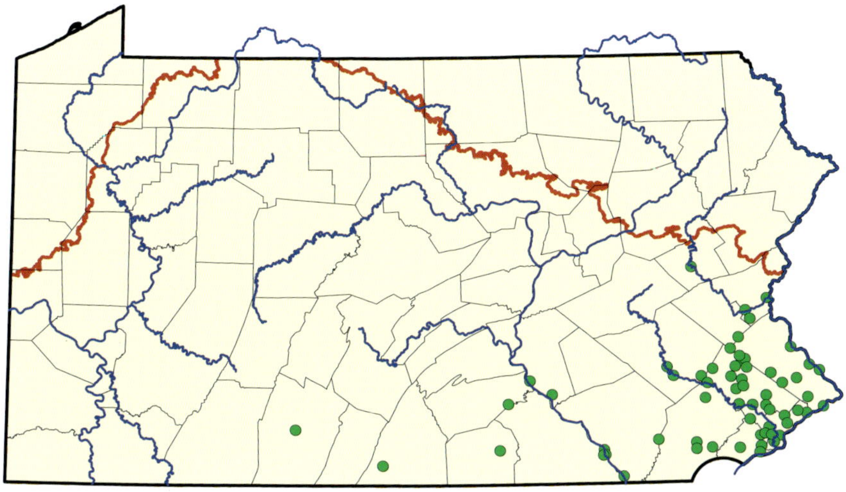

Heteranthera reniformis

Heteranthera reniformis

PICKEREL-WEED

PONTEDERIA L.

Pickerel-weed Family—Pontederiaceae

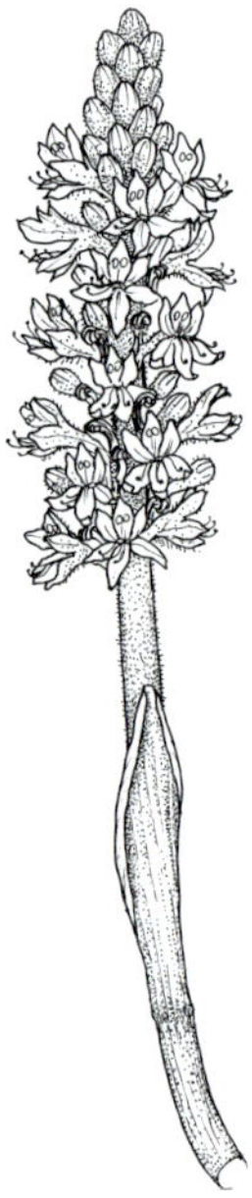

Pontederia cordata—inflorescence in flower ×1/2

Pickerel-weed ***Pontederia cordata* L.**

Native
Rooted, emergent perennial

Pickerel-weed often forms large colonies in shallow water along the edges of lakes, streams, and tidal shores; the numerous spikes of purple flowers are conspicuous throughout most of the summer. This common native plant is an erect emergent aquatic to 1 m in height. The stems arise from creeping rhizomes rooted in sediments beneath the water; both the leaves and the inflorescence extend well above the water surface.

The leaves are broadly heart-shaped, 10–25 cm in length with long petioles; they are indented at the base and taper to a point, with veins radiating from the base and curving upward. The inflorescence is a dense terminal spike of purple flowers 5–15 cm long, subtended by a loosely sheathing spathe. The flowers are slightly bilabiate, with a 3-lobed upper lip; the middle lobe is marked with 2 yellow spots. Flowering occurs from June to October and is followed by clusters of angular achenes with a highly textured surface.

The range of pickerel-weed includes the eastern half of the United States and extends south to Mexico, Central America, and South America. It is common in the glaciated regions of Pennsylvania and along the Delaware and lower Susquehanna Rivers. Ducks and muskrats eat the fruits.

Pontederia cordata in the emergent zone of a lake in Pike County, Pennsylvania

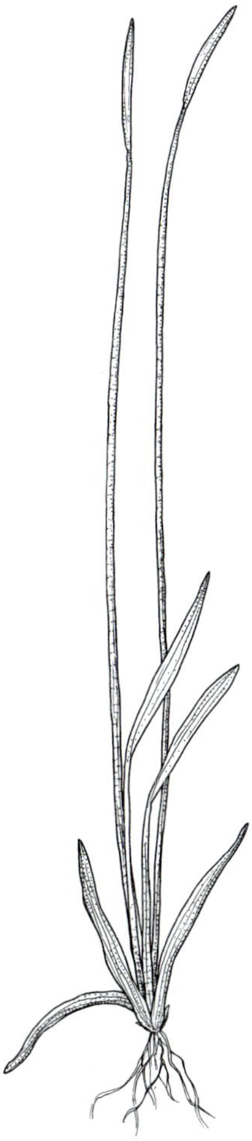

Pontederia cordata—juvenile plant ×1/2

Pontederia cordata—inflorescence in fruit ×1/2

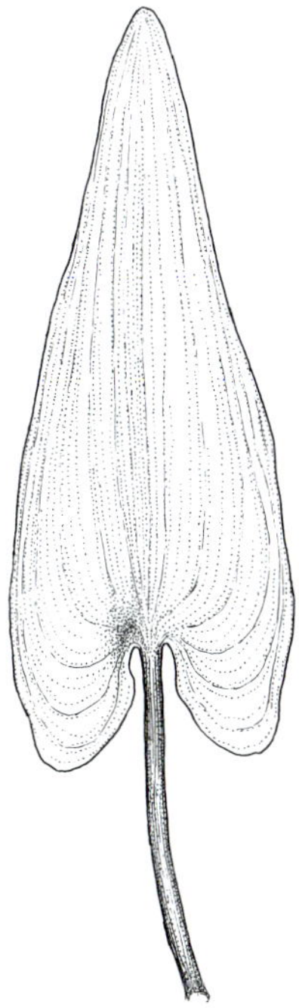

Pontederia cordata—leaf ×1/4

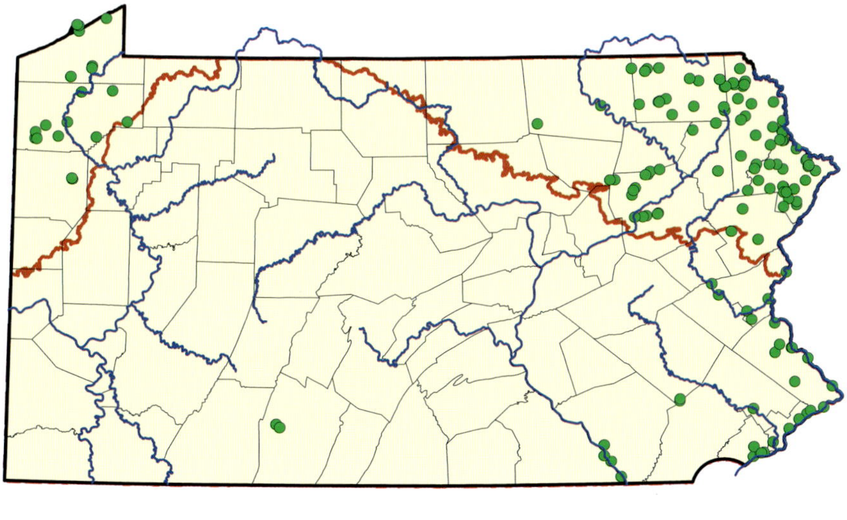

Pontederia cordata

BAYONET RUSH

JUNCUS BIGEL.

Juncaceae—Rush Family

Bayonet rush — *Juncus militaris* Bigel.

PNHP
Native
Rooted, emergent perennial

Bayonet rush is a stiffly erect, emergent plant to 1 m tall; its round stem bears a single stiff pointed leaf about half way from the base to the inflorescence, giving the plant its common name. Round in cross-section, the leaf has hardened partitions at intervals within which can be felt by pulling the leaf between your fingers while pressing gently.

A branched inflorescence is produced at the stem tip. Individual flowers are in small clusters; each flower has 6 stamens and 6 tepals. The fruit is a 3-sided capsule containing numerous seeds.

The stems of bayonet rush grow from a rhizome, sometimes forming dense colonies. In addition to the emergent stems, in deep water the rhizome may also bear long, slender, flexible underwater leaves which also have hardened cross partitions at regular intervals.

Bayonet rush is a plant of shallow water (to 0.7 m) of natural glacial lakes. In deeper water it may remain sterile.

Juncus militaris ×1/8

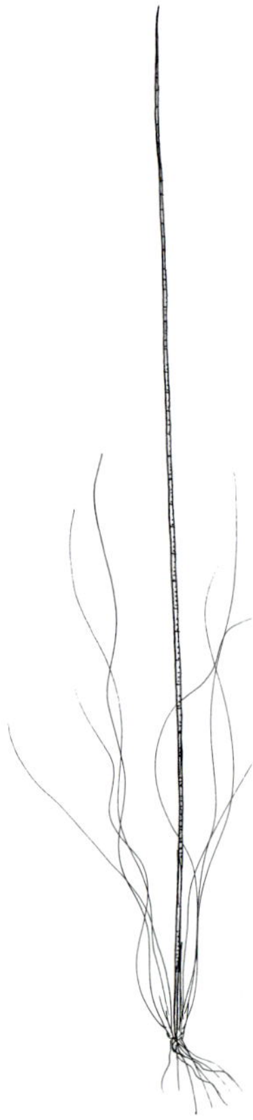

Juncus militaris—
juvenile plant ×1/6

The range of bayonet rush extends from Nova Scotia to Delaware along the coast; it is also known from a few inland locations in New York, southern Ontario, and northern Michigan. In Pennsylvania it is confined to a few natural glacial lakes in the northeastern corner of the state. Plants in shallow water near the shoreline appear to be vulnerable to grazing by deer and/or geese at some sites.

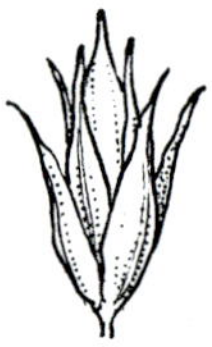

Juncus militaris—
capsule surrounded by tepals ×5

All the 29 species of *Juncus* that grow in Pennsylvania are wetland plants and may occur at the edges of lakes or ponds, but most do not consistently grow in standing water. There are no other emergent species of *Juncus* in our range that could be confused with bayonet rush; its size and single stiff leaf midway along the stem are distinctive. Sterile plants might be confused with twig-rush (*Cladium mariscoides*); however, leaves of twig-rush are channeled at the base and flattened in the middle, unlike the consistently round leaves of bayonet rush. In addition, the hardened cross partitions in leaves of bayonet are not present in twig-rush; nor does twig-rush produce slender, flexible underwater leaves.

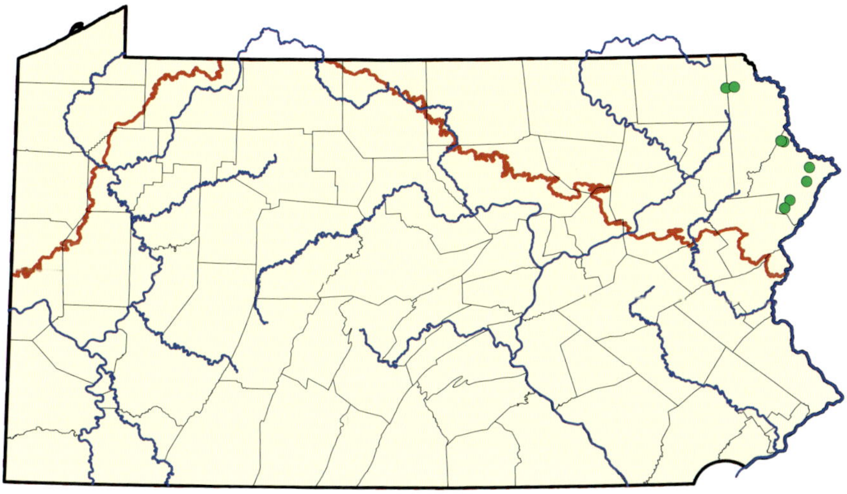

Juncus militaris

Juncus militaris in the emergent zone of a lake in Pike County, Pennsylvania

SMARTWEED
Persicaria (L.) Miller
Smartweed Family—Polygonaceae

The smartweeds are perennial (or annual) herbs with simple alternate leaves; the nodes are often swollen and bear distinctive tubular stipules (ocreae) and inflorescence bracts (ocreolae). Flowers are small, and are borne in terminal spike-like inflorescences. Petals are lacking; the 4–6 sepals are fused in a cup or tube that is persistent in fruit. Stamen number is 5–8. The fruit is an achene.

Waterfowl, marsh and shore birds, and even upland game birds and song birds eat smartweed achenes. Muskrats and raccoons eat the plants and seeds.

Key to the Aquatic Species of Smartweed

A. ocreae of the upper and middle leaves with a ring of bristles along the upper margin
  B. flowers dotted with numerous yellow glands
    C. ocreolae lacking cilia at the margins; leaves 2–4 cm wide .... *Persicaria robustior*
    C. ocreolae with ciliate margins; leaves less than 2.5 cm wide
      D. flowers greenish-white .... *Persicaria punctata*
      D. flowers creamy, green- or red-tinged .... *Persicaria hydropiper*
  B. flowers lacking glands .... *Persicaria hydropiperoides*
A. ocreae entire or shattering but lacking bristles .... *Persicaria amphibia*

Water smartweed — ***Persicaria amphibia*** **(L.) S.F. Gray [synonym: *Polygonum amphibium* L.]**

Native

Rooted, emergent or floating-leaf perennial

Persicaria amphibia

Water smartweed is variable in growth form; along the shoreline it may be erect and 1 m or more in height. In deeper water the stems are sprawling with the leaves floating and only the inflorescence extending above the surface. The rhizomes are stout with roots at the nodes. Leaves are hairy, becoming glabrous, elliptical; the ocreae lack bristles. The pink flowers are borne in dense, erect, terminal spikes. The dark, shiny, flattened achenes are an important food for waterfowl.

Water smartweed is a plant of shallow lakes and lake and stream margins throughout Pennsylvania; it is wide-ranging in the northern hemisphere and is naturalized in South Africa and South America.

Persicaria amphibia ×1/2

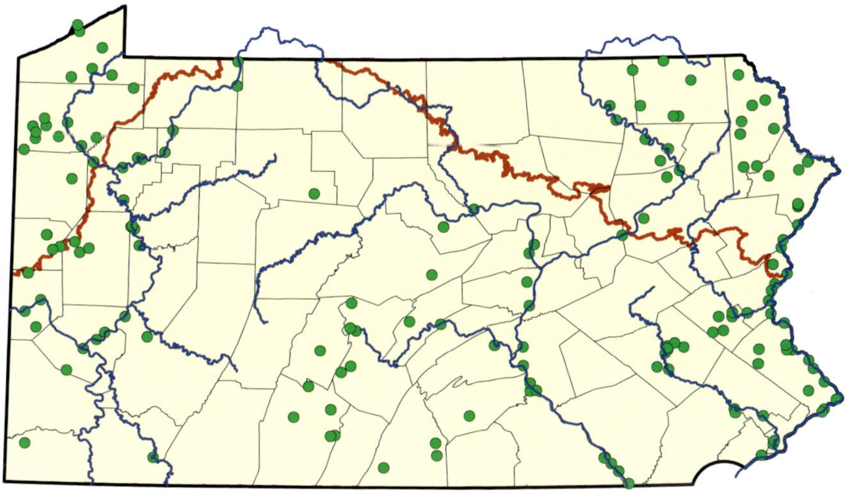

Persicaria amphibia

Smartweed, water-pepper *Persicaria hydropiper* (L.) Spach [synonym: *Polygonum hydropiper* L.]

Introduced
Rooted, emergent annual

Persicaria hydropiper is an erect-ascending, tap-rooted annual that forms large colonies in shallow water of lake, pond, and stream margins throughout Pennsylvania.

Stems and leaves are often reddish or bronze-tinged. The flowering/fruiting spikes are slender and interrupted at the base; the ocreae are both bristle-fringed and pubescent. The creamy white flowers are densely covered with flattened, golden-brown glands.

Persicaria hydropiper is very similar to *Persicaria punctata*. It can be distinguished by the tendency of *P. hydropiper* to bear flowers enclosed in the ocreae, resulting in the inflorescence appearing leafy at the bottom.

Persicaria hydropiper is native to Europe but is widely naturalized in North America. Long used medicinally in Europe, all parts of this plant have an acrid taste which accounts for the common name, smartweed. Native Americans also used the plant medicinally and as a food.

Persicaria hydropiper—
portion of inflorescence ×2

Persicaria hydropiper ×1/2

Persicaria hydropiper—flower showing glandular-punctate dots ×3

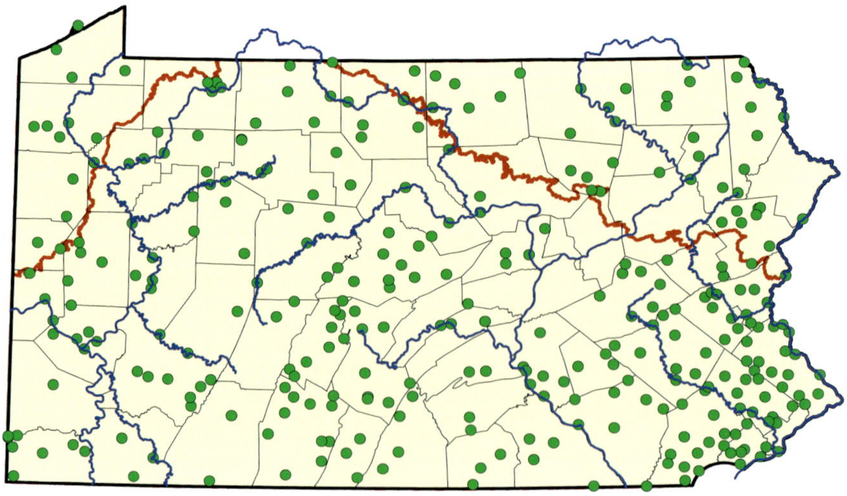

Persicaria hydropiper

Mild water-pepper, water smartweed

Persicaria hydropiperoides (Michx.) Small [synonym: *Polygonum hydropiperoides* Michx.]

Native
Rooted, emergent perennial

Mild water-pepper is a rhizomatous perennial of shallow water at the margins of lakes, ponds, and streams. Leaves are lanceolate; the ocreae bear a ring of bristles on the upper margin. The flowers are white or more often pale pink, in loose terminal or axillary spikes. The achenes are brownish or black, and 3-sided; as with other smartweeds, they are an important food source for waterfowl.

Mild water-pepper is widespread in eastern North America from Newfoundland to Florida and west to the Dakotas and Texas. It also occurs on the Pacific coast. In Pennsylvania it is a plant of lake and stream margins throughout, except for areas of highest elevation along the Allegheny Front.

Persicaria hydropiperoides ×1/2

Persicaria hydropiperoides

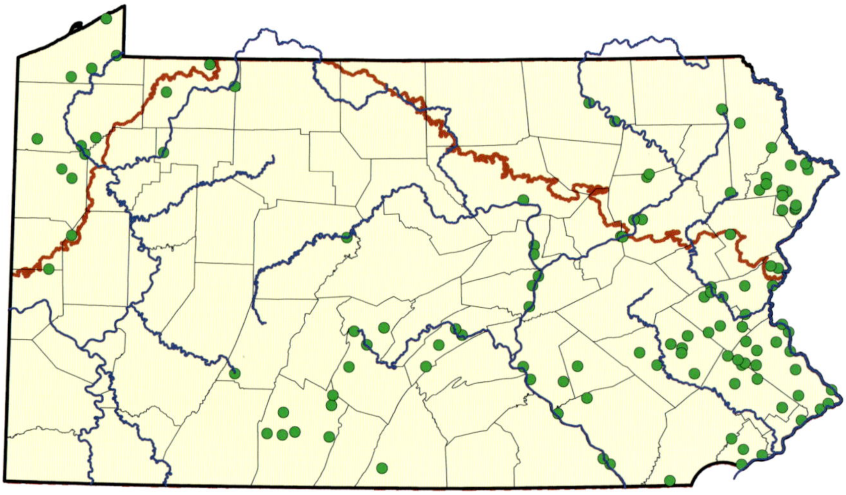

Persicaria hydropiperoides

Dotted smartweed *Persicaria punctata* (Elliott) Small [synonym: *Polygonum punctatum* Elliott]

Native
Rooted, emergent annual or perennial

Dotted smartweed is an emergent plant of lake and stream margins. The stems are erect, up to 1 m tall, and glandular punctate, as are the lanceolate leaves.

Flowers are greenish-white and covered with flat yellow glands. The inflorescences are loose terminal or axillary spikes in which the lower flowers are well separated. The shiny brownish-black achenes are 3-angled or, rarely, flat. As in other species the fruits are an important food for waterfowl.

Dotted smartweed occurs throughout the eastern and northern United States and southern Canada. It is common throughout Pennsylvania in shallow water and on muddy shores.

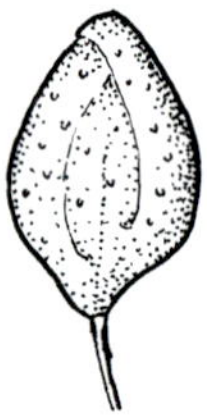

Persicaria punctata—flower showing glandular-punctate dots ×5

Persicaria punctata ×1/2

Persicaria punctata

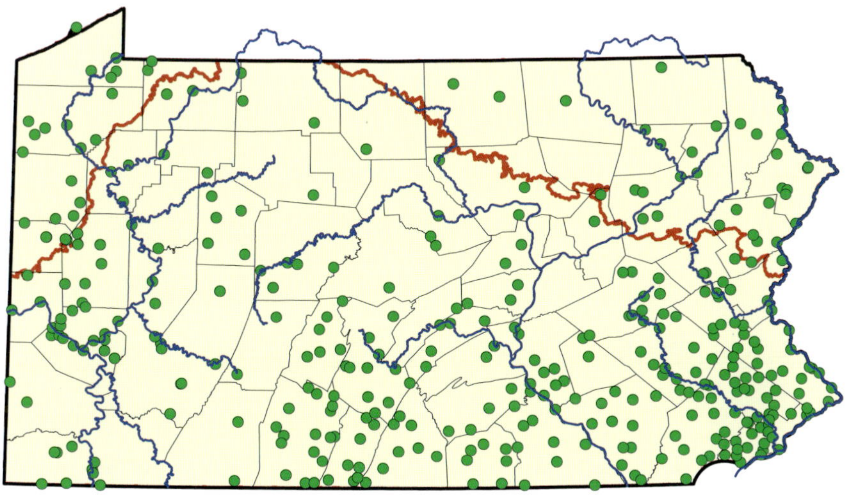

Persicaria punctata

Large dotted smartweed — *Persicaria robustior* (Small) E.P. Bicknell [synonym: *Polygonum robustius* (Small) Fernald]

Native

Rooted, emergent perennial

Persicaria robustior ×1/2

Large dotted smartweed is a robust perennial up to 2 m tall that grows in shallow water and along wet shores. Roots often grow from the lower stem nodes of plants in standing water. The leaves are lanceolate and 2–4.5 cm wide; both leaves and stems are glandular-punctate. The ocreae have a ciliate margin. Flowers are greenish-white and are covered with yellow glands; the achenes (fruits) are dark brown or brownish-black and 3-sided.

The main range of large dotted smartweed is the northeastern states and adjacent Canada, but it also occurs in the West Indies and South America. In Pennsylvania it is known primarily from the northwestern corner of the state with a few scattered locations in the east. The achenes of large dotted smartweed are eaten by waterfowl.

Persicaria robustior—ocreolae in the inflorescence ×3

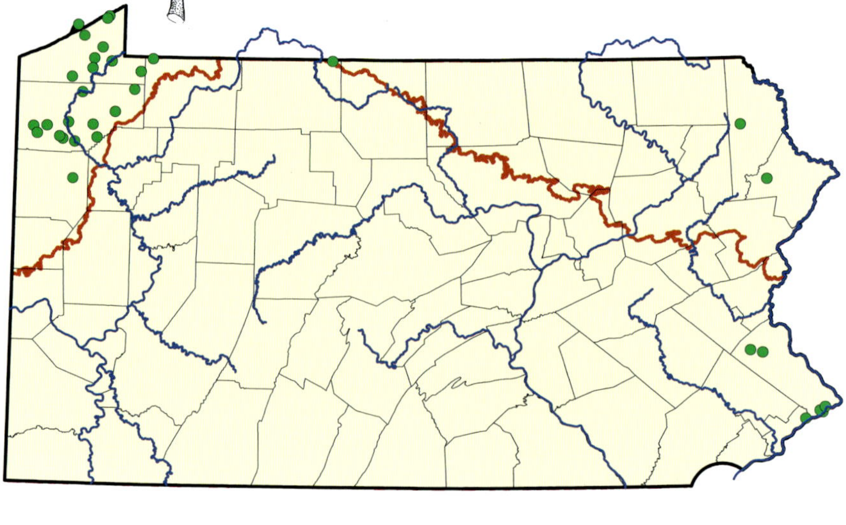

Persicaria robustior

SPIKE-RUSH
ELEOCHARIS R.BR.
Sedge Family—Cyperaceae

The spike-rushes are emergent plants of lake margins. They consist of stems with a single terminal spikelet. Leaves are reduced to bladeless sheaths at the base of the stem. In some species the stems are densely clustered; in others they arise singly along a rhizome. The flowers are perfect; each is subtended by a scale. In addition the spikelet has 1 or more empty scales at the base. The fruit is an achene surrounded by bristles. A distinctive feature of the spike-rushes is the tubercle, or top part of the achene; it usually differs from the main body of the achene in color and texture and sometimes is separated by a distinct groove or constriction.

Fruits, stems, and tubers of the spike-rushes are food for waterfowl, muskrat, and other wildlife.

Key to the Aquatic Species of Spike-rush

A. spikelet about the same diameter as the stem or only slightly thicker; scales persistent
- B. stems 4-sided .... *Eleocharis quadrangulata*
- B. stems triangular........ *Eleocharis robbinsii*

A. spikelets distinctly thicker than the stem; scales deciduous
- C. tubercle continuous with the achene except that it is a different color or texture........ *Eleocharis parvula*
- C. tubercle separated from the achene by a narrow constriction as well as by color and texture
  - D. plants with obvious hard, cord-like rhizomes........ *Eleocharis palustris*
  - D. plants with slender or short rhizomes or rhizomes lacking or inconspicuous
    - E. styles 2-branched; achenes flattened
      - F. bristles present........ *Eleocharis obtusa* var. *obtusa*
      - F. bristles absent........ *Eleocharis obtusa* var. *peasii*
    - E. styles 3-branched; achene 3-angled .... *Eleocharis acicularis*

Needle spike-rush — *Eleocharis acicularis* (L.) Roem. & Schult.

Native

Rooted, emergent or submergent perennial

Eleocharis acicularis ×1/2

Needle spike-rush is a small, grass-like plant that often forms a fine turf along moist shorelines and in water to 1–2 m deep. In deep water it grows as a submergent plant and does not produce flowers or fruits. Under these conditions the sterile stems may be as long as 25–30 cm. At the water's edge where the plants are partially or completely emergent, the sterile stems rarely exceed 5–8 cm and flowering is common.

The slender, flexible stems of needle spike-rush arise in small clusters from a thread-like rhizome. As in all the spike-rushes, leaves are reduced to an inconspicuous bladeless sheath at the base of the stem. Flowering occurs at the stem tip.

Although it is not mentioned in previous published descriptions of the species, at several locations in Pennsylvania we have found non-flowering submergent plants of *Eleocharis acicularis* that have small tubers growing along the rhizomes. We have never seen tubers on emergent plants.

In Pennsylvania needle spike-rush is common and abundant except in the north central region. Its total range extends from boreal areas in the north to Florida and Mexico.

Needle spike-rush sometimes grows intermixed with sterile submergent populations of horned bladderwort (*Utricularia cornuta*). Examination with a hand lens is necessary to distinguish the two; the linear leaves of the bladderwort have minute bladders along their edges. In deeper water, needle spike-rush could also be confused with sterile plants of *Schoenoplectus subterminalis*. Close examination of the stem or leaf bases is necessary to distinguish the two species.

Eleocharis acicularis—submergent plant with tubers ×1/2

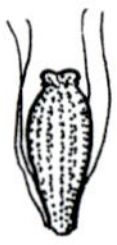

Eleocharis acicularis—achene ×15

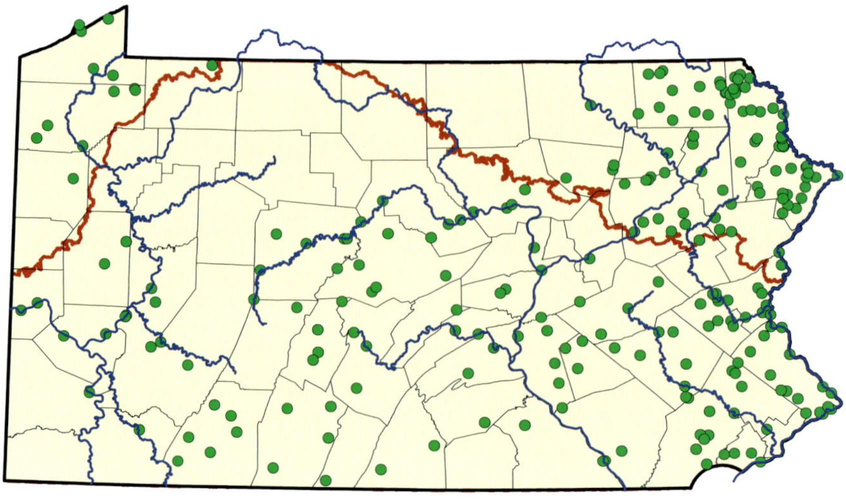

Eleocharis acicularis

Blunt spike-rush — *Eleocharis obtusa* (Wild.) Schult. var. *obtusa*

Native
Rooted, emergent perennial

Blunt spike-rush is a common plant of lake and pond margins, marshes, wet meadows, and ditches. This native, rooted, emergent perennial has densely tufted stems 3–50 cm tall. Each stem bears a single terminal spike that is ovoid to cylindrical and contains up to 100 or more flowers. Achenes are about 1 mm long with a broad tubercle that is as wide as the achene and one-third to one-half as high.

Blunt spike-rush is a wide-ranging plant that occurs from Newfoundland to British Columbia and south to Florida, Texas, California, and Hawaii. It is common throughout Pennsylvania.

Eleocharis obtusa var. *obtusa*—achene ×10

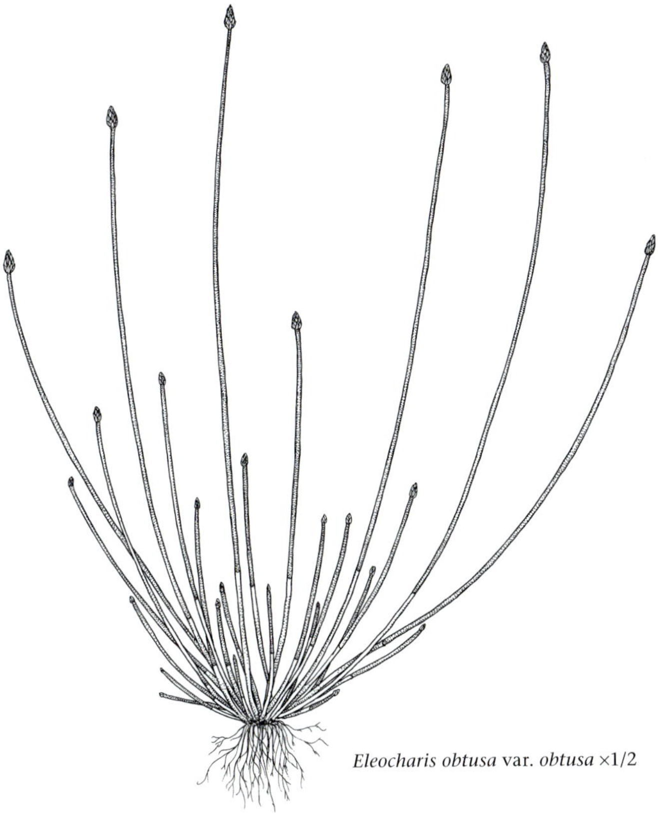

Eleocharis obtusa var. *obtusa* ×1/2

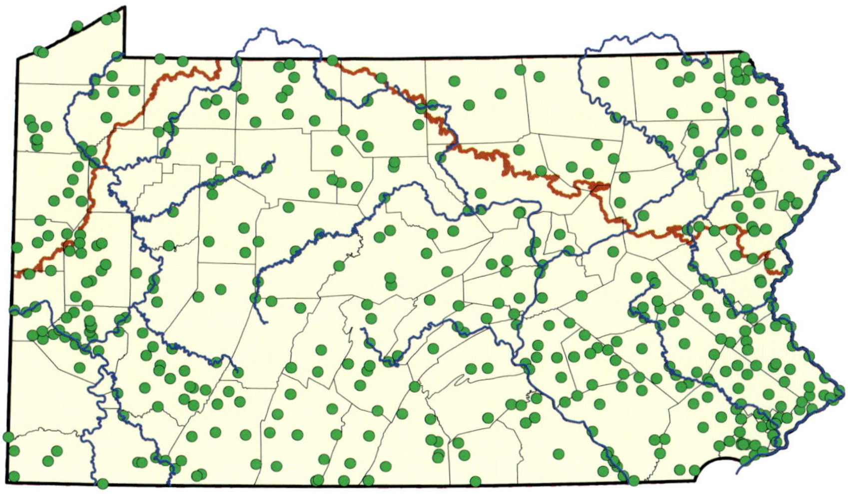

Eleocharis obtusa var. *obtusa*

Wright's spike-rush — *Eleocharis obtusa* (Wild.) Schult. var. *peasei* Svenson

PNHP
Native
Rooted, emergent perennial

Similar to blunt spike-rush, Wright's spike-rush is distinguished by the absence of bristles surrounding the achenes. It is a rare plant of freshwater tidal marshes.

Eleocharis obtusa var. *peasei*—
achene ×15, note lack of bristles

Eleocharis obtusa var. *peasei*—
inflorescence ×3

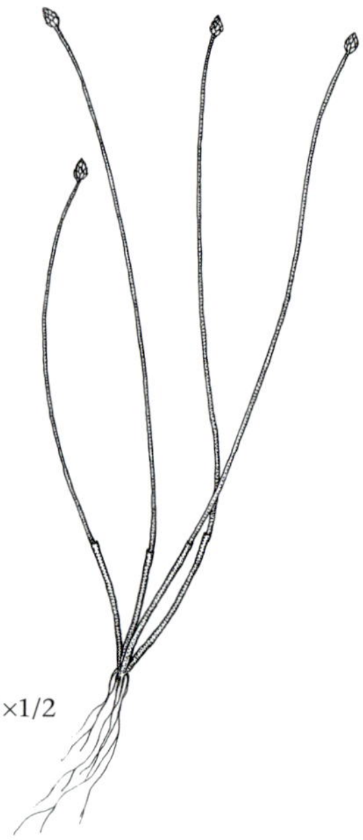

Eleocharis obtusa var. *peasei* ×1/2

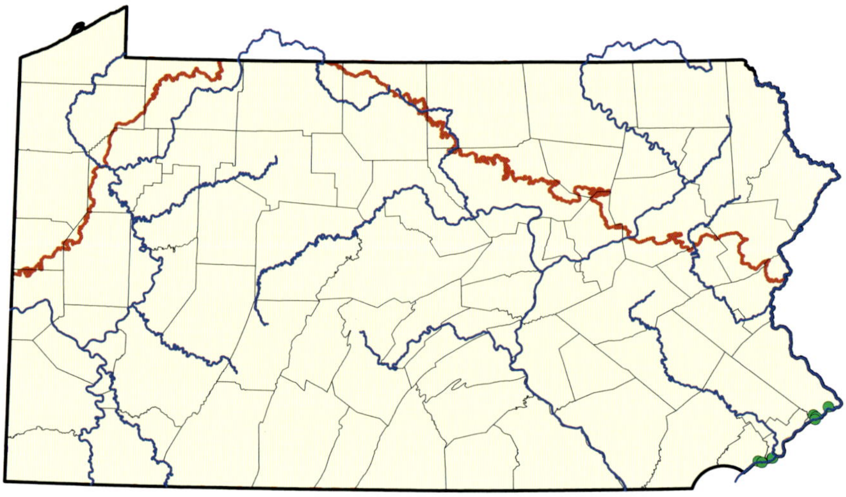

Eleocharis obtusa
var. *peasei*

Creeping spike-rush — *Eleocharis palustris* (L.) Roem. & Schult.

Native
Rooted, emergent perennial

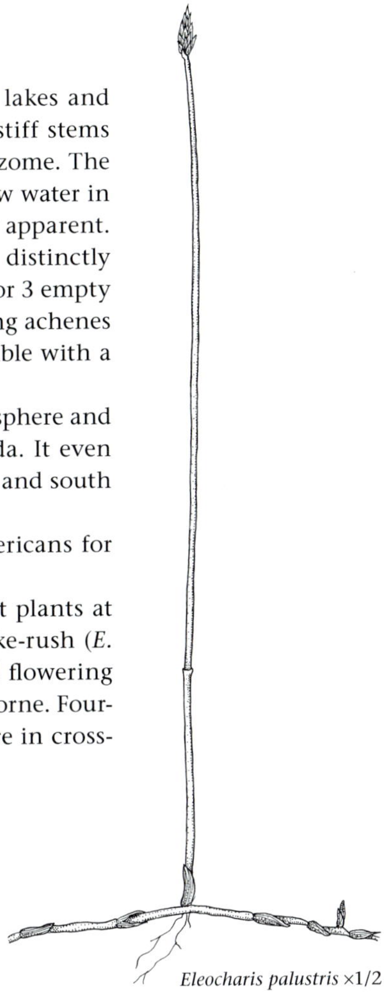

Eleocharis palustris ×1/2

Creeping spike-rush forms large colonies at the margins of lakes and streams. It is a very common, emergent plant with round, stiff stems up to 1 m tall that arise singly from a cord-like perennial rhizome. The leafless, unbranched, erect stems are often lined up in shallow water in such a way that the course of the buried rhizome is readily apparent. The inflorescence is a single terminal spike 0.5–2.5 cm long, distinctly thicker than the stem, with spirally overlapping scales and 2 or 3 empty scales surrounding the base of the spikelet. The 1.1–2 mm-long achenes are lens-shaped, and surrounded by tiny barbed bristles (visible with a 10X hand lens).

Creeping spike-rush occurs throughout the northern hemisphere and is common and widespread in the United States and Canada. It even grows in New Zealand. In Pennsylvania it occurs both north and south of the glacial boundary.

Fibers from creeping spike-rush were used by Native Americans for weaving mats, rugs, and bedding.

Several other species of spike-rush also occur as emergent plants at the margins of lakes and ponds in our region. Robbins' spike-rush (*E. robbinsii*) has stems that are triangular in cross-section and flowering spikes that are no thicker than the stems on which they are borne. Four-angled spike-rush (*E. quadrangulata*) has stems that are square in cross-section and a spikelet barely thicker than the stem.

Eleocharis erythropoda, a wetland plant of river and stream banks and wet meadows, is very similar to creeping spike-rush. The main difference between the two species is that the tips of the scales of *E. erythropoda* are broadly rounded or obtuse as opposed to acute in *E. palustris*, and the spikelet is encircled at the base by a single empty scale in *E. erythropoda* in contrast to several empty scales present at the base of *E. palustris* spikelets. *Eleocharis erythropoda* is not treated separately here because it does not ordinarily grow in standing water.

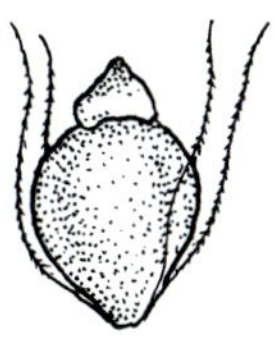

Eleocharis palustris—
achene with bristles ×10

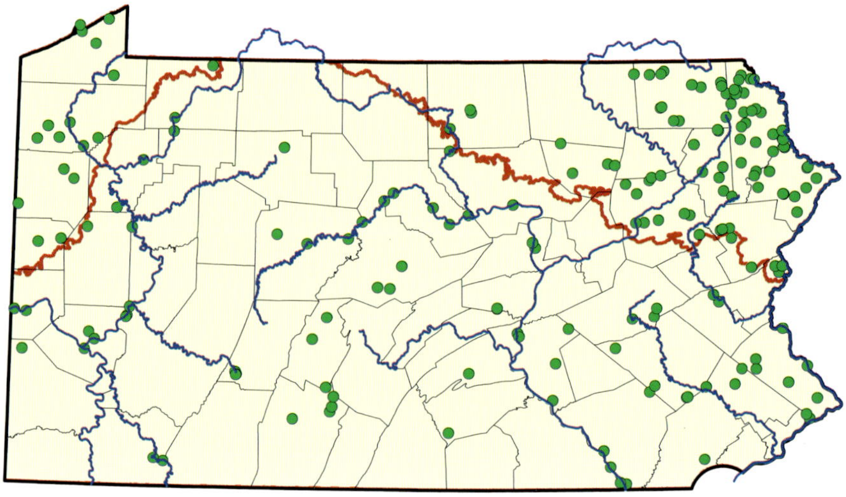

Eleocharis palustris

Dwarf spike-rush

Eleocharis parvula (Roem. & Schult.) Link. ex Bluff, Nees, & Schauer

PNHP
Native
Rooted, emergent or submergent perennial

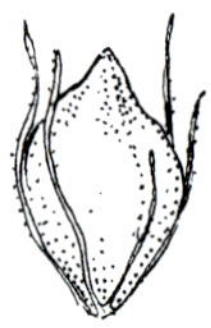

Eleocharis parvula—achene with bristles ×20

Dwarf spike-rush is a recent arrival in Pennsylvania; it was first discovered in 1984 on the tidal shore of the Neshaminy Creek in Bucks County. The species appears to be undergoing a natural range extension.

Its 2–6 cm-tall stems arise from slender, tuber-bearing rhizomes. Achenes are 3-angled, about 1 mm long, and have a short, thickened tubercle which differs from the body of the achene in texture, but is not separated by a constriction. Dwarf spike-rush closely resembles needle spike-rush (*Eleocharis acicularis*). Mature achenes are necessary to distinguish the two species with certainty.

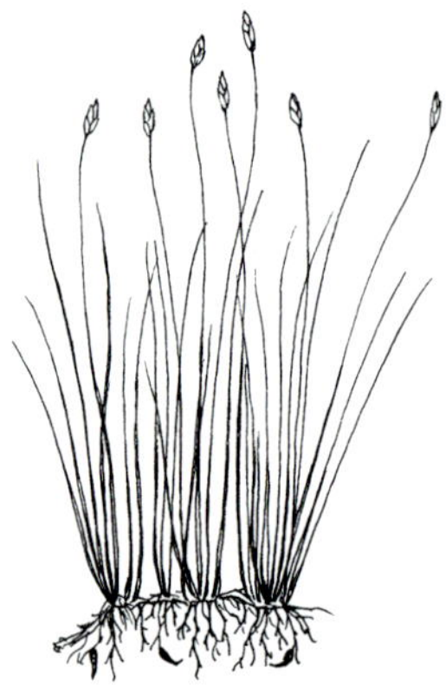

Eleocharis parvula ×3/4, note tubers on roots

In North America dwarf spike-rush is a plant of coastal areas from Newfoundland to Louisiana and along the Pacific coast. Its range also includes the Mississippi River valley, Mexico, Central America, and Eurasia. In Pennsylvania it is a plant of freshwater intertidal shores and mudflats along the Delaware River and its tributaries.

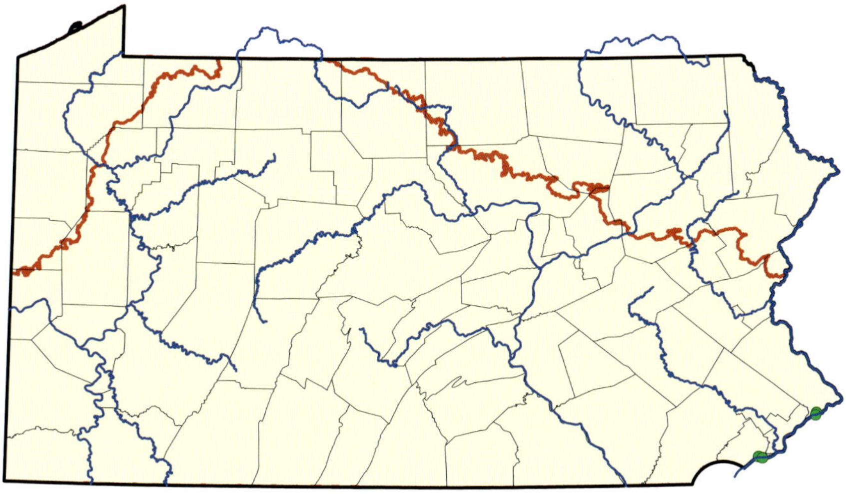

Eleocharis parvula

Eleocharis quadrangulata ×1/2

Four-angled spike-rush ***Eleocharis quadrangulata* (Michx.) Rom & Schult.**

PNHP
Native
Rooted, emergent perennial

Four-angled spike-rush is the only sedge in our flora with stems that are square in cross-section. It is a rare plant of lake margins, swamps, and ponds, where it forms colonies in water 0.1–0.5 m in depth. Four-angled spike-rush is a robust plant; the stems arise singly from a rhizome and grow to a height of 1 m. Spikelets are about as wide as the stem and 2–7 cm long. The achenes are 1.8–3 mm long and up to 2 mm wide with a tubercle that is much narrower.

Four-angled spike-rush occurs in several lakes and ponds in western Pennsylvania, where it may be locally abundant. Populations in the southeastern part of the state are probably of planted origin. Four-angle spike-rush is occasionally cultivated in ornamental ponds and water gardens.

The full range of four-angled spike-rush extends from Massachusetts to Wisconsin and south to Florida and Texas. It is also present along the Pacific coast and in Mexico.

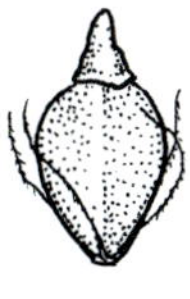

Eleocharis quadrangulata — achene with bristles ×5

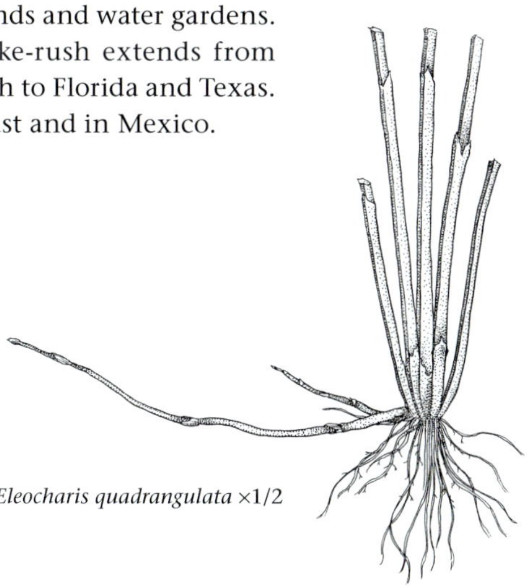

Eleocharis quadrangulata ×1/2

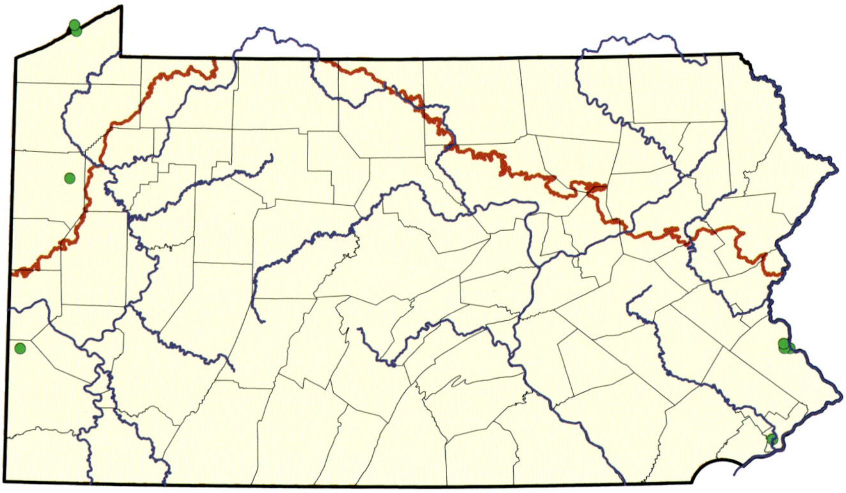

Eleocharis quadrangulata

Robbins' spike-rush — *Eleocharis robbinsii* Oakes

PNHP
Native
Rooted, emergent perennial

Robbins' spike-rush is a rare plant of shallow water at the edges of natural glacial lakes. It has erect, flowering stems that are sharply triangular in cross-section and can reach 0.6–1 m in height. In addition, this species has slender, flexuous underwater stems; both types of stems arise from a buried rhizome. As in all the spike-rushes, leaves are represented only by bladeless sheaths. Flowers are borne in a single terminal spike that is no thicker than the stem itself. The fruits, tiny achenes, are biconvex.

The range of this species extends along the coast from Nova Scotia and New Brunswick to Florida and inland to the Great Lakes region. It is very rare in Pennsylvania, occurring in only a handful of natural glacial lakes in the northeastern corner of the state.

Other species of spike-rush that also form colonial patches in shallow water at the edges of lakes include creeping spike-rush (*E. palustris*) and four-angled spike-rush (*E. quadrangulata*). The former has stems that are round in cross-section, and the latter has square stems, unlike the 3-edged stems of Robbins' spike-rush. Neither of the other two species produces flexible underwater stems.

Water bulrush (*Schoenoplectus subterminalis*) has a growth habit similar to Robbins' spike-rush in that it has erect, emergent flowering stems and slender, flexible underwater leaves. However, water bulrush has round stems and an inflorescence which projects from the side of the stem below the tip.

Eleocharis robbinsii—flowering stem ×1/2

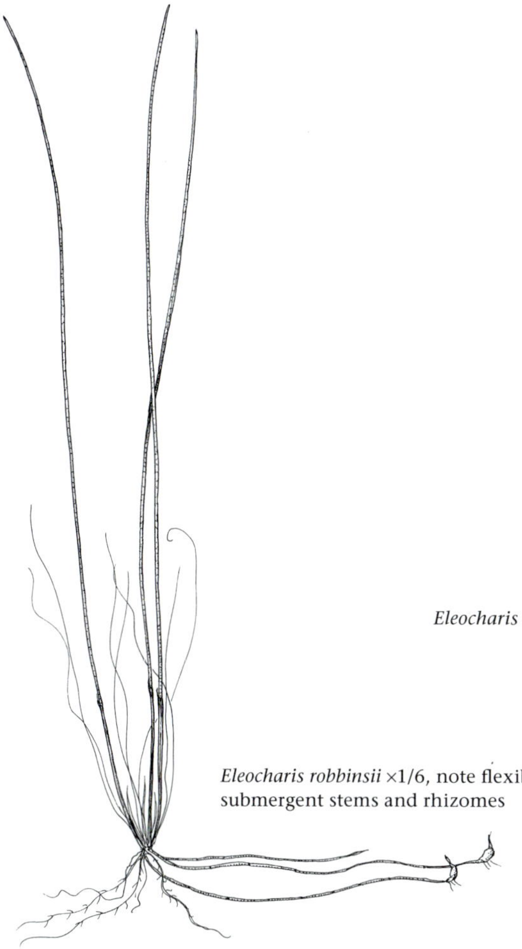

Eleocharis robbinsii ×1/6, note flexible submergent stems and rhizomes

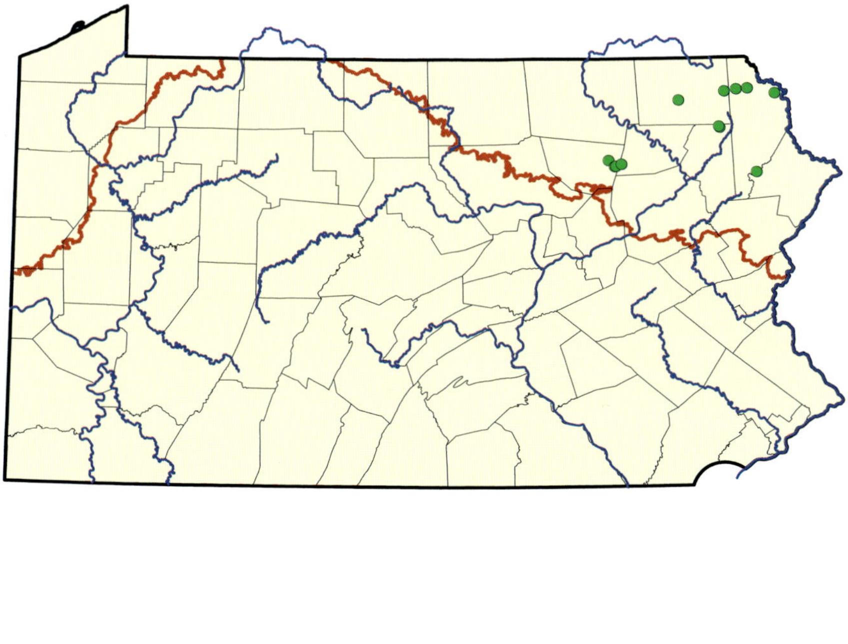

Eleocharis robbinsii

Eleocharis robbinsii with flowering inflorescence (below) and fruiting inflorescence (above)

Eleocharis robbinsii growing with fragrant water-lily in the emergent zone of a natural glacial lake in Wayne County, Pennsylvania

THREE-WAY SEDGE
DULICHIUM PERS.
Sedge Family—Cyperaceae

Three-way sedge ***Dulichium arundinaceum* (L.) Britt.**

Native
Rooted, emergent perennial

Three-way sedge is a very common emergent aquatic species in Pennsylvania. This native perennial has erect, unbranched stems bearing three vertical rows of narrow, spreading leaves which are best viewed by looking straight down on the plants from above. The bright green stems, which grow from a rhizome, are 0.5–1 m tall. Spikelets of perfect flowers are formed in the upper leaf axils in July and August. The fruit is a small achene.

Although other sedges or even some grasses may superficially resemble the growth habit of three-way sedge, upon close examination none will be found to have the same distinctive leaf arrangement coupled with inflorescences produced in the axils of the upper leaves.

Three-way sedge forms colonies at the edges of lakes, ponds, and slow-moving streams and rivers in water to 0.6 m deep over a soft substrate. It also sometimes grows in ditches and swampy fields.

This species is common and abundant throughout Pennsylvania. Its total range extends from Newfoundland to Minnesota and south to Florida and Texas. It is also found in the Pacific Northwest from British Columbia to California and Montana.

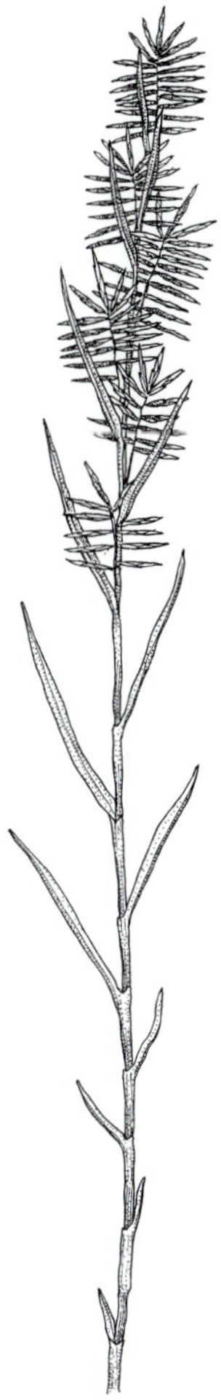

Dulichium arundinaceum ×1/4

Dulichium arundinaceum—spikelet ×2

Dulichium arundinaceum—flower ×3

Dulichium arundinaceum

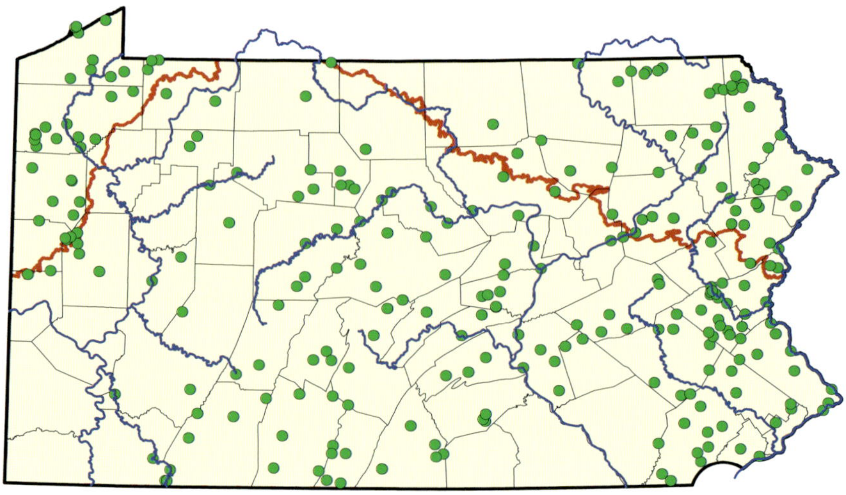

Dulichium arundinaceum

TWIG-RUSH

CLADIUM R.BR.

Sedge Family—Cyperaceae

Twig-rush *Cladium mariscoides* (Muhl.) Torrey

PNHP
Native
Rooted, emergent perennial

Twig-rush is not a true rush, but rather a member of the sedge family. It is a rooted, emergent perennial with round, upright stems up to 1 m tall. The stems arise from a flaccid, whitish rhizome that grows buried in sandy or peaty sediments. Branches are produced in the upper third of the stem and terminate in the inflorescence, which consists of clusters of spikelets enclosed by chestnut-brown scales. The narrow leaves of twig-rush are channeled at the base, flat toward the middle, and rounded at the tip. The fruit is an achene.

Twig-rush is a rare plant that forms colonies in shallow water of lake margins; it also grows in bogs, where its rhizomes may be a major structural component of a floating mat of vegetation. It is most often found in areas where the pH is somewhat alkaline. Although it is of limited occurrence in Pennsylvania, it can be locally abundant.

The range of twig-rush extends from Newfoundland to Saskatchewan and south to Tennessee, western Florida, and Texas. It also occurs in the West Indies. In Pennsylvania it is found primarily in the glaciated northeastern corner of the state; however, a few scattered locations have been recorded across the state, including along the Lake Erie shoreline.

Non-flowering plants of twig-rush might be mistaken for some of the true rushes (*Juncus* spp.), such as bayonet rush. However, the leaves of bayonet rush are round in cross-section throughout and contain internal cross partitions at intervals.

Cladium mariscoides—achene ×5

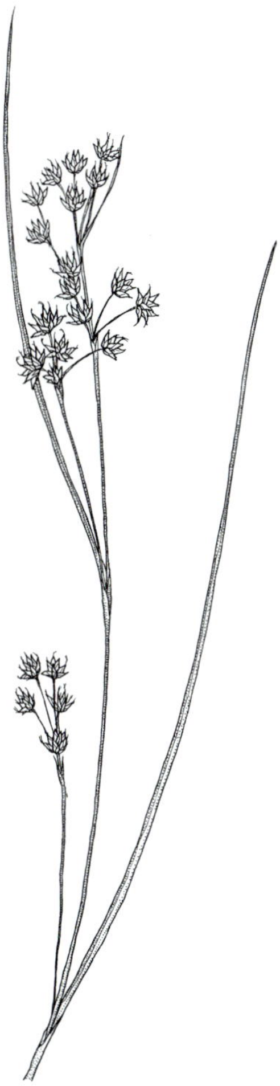

Cladium mariscoides ×1/2

Cladium mariscoides in the emergent zone of a lake in Wayne County, Pennsylvania

Cladium mariscoides—inflorescence

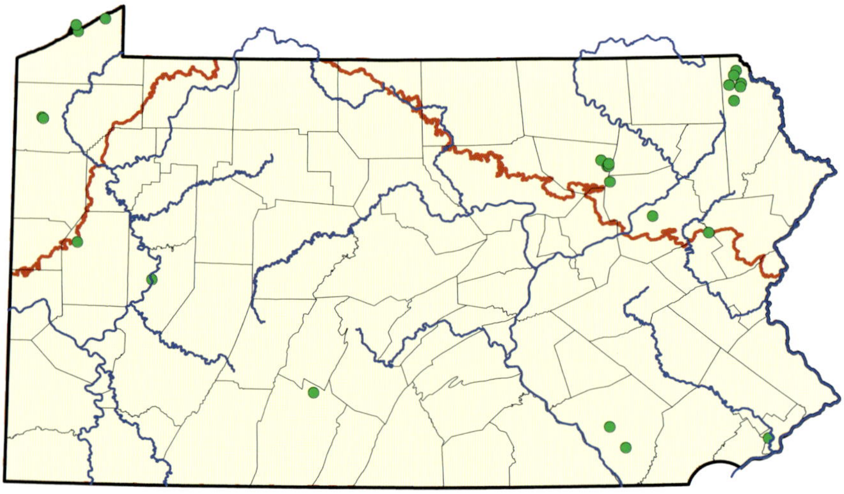

Cladium mariscoides

WATER-HEMP RAGWEED

AMARANTHUS L.

Amaranthus Family—Amaranthaceae

Water-hemp ragweed, saltmarsh water-hemp — *Amaranthus cannabinus* (L.) Sauer

PNHP
Native
Rooted, emergent annual

Water-hemp ragweed is a plant of the freshwater tidal marshes of the Delaware Estuary in Pennsylvania. It is an annual that can reach 3 m in height; the ascending branches bear simple alternate leaves that are narrow and pointed at both ends.

Male and female flowers are produced on separate plants, making the species dioecious; the individual flowers are inconspicuous and wind-pollinated. The fruit is a nutlet. Water-hemp ragweed typically grows near the upper edge of the tidal marsh, where its roots are inundated at high tide but the upper stem and flowers remain above the water level. Frequent associates are wild-rice (*Zizania aquatica*) and swamp beggar-ticks (*Bidens bidentoides*).

The range of water-hemp ragweed extends along the east coast of the United States from Maine to Florida. The fruits of amaranths have been used as food by many Native American groups and wildlife.

Amaranthus cannabinus with female flowers on the tidal shore of the Delaware River

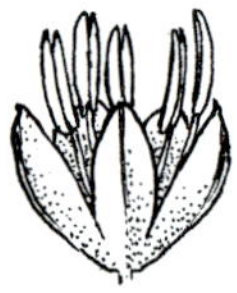

Amaranthus cannabinus—
male flower ×5

Amaranthus cannabinus—
female flower ×10

Amaranthus cannabinus ×3/4

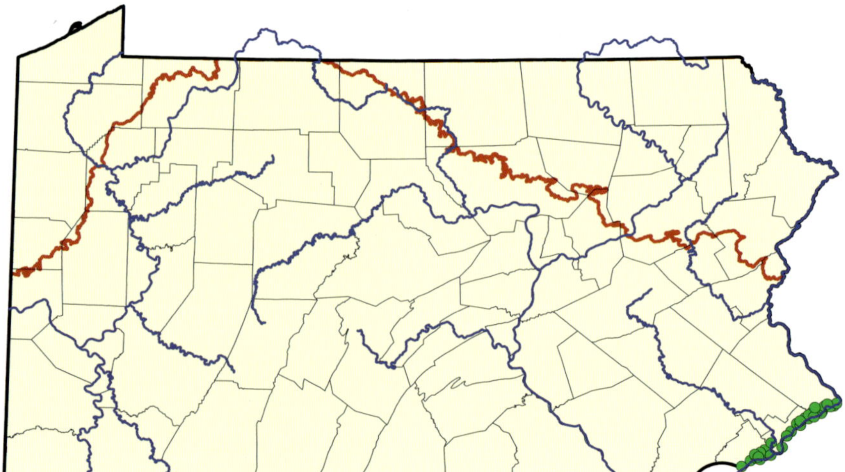

Amaranthus cannabinus

WATER-HOREHOUND

Lycopus L.

Mint Family—Lamiaceae

Water-horehound *Lycopus rubellus* Moench.

PNHP
Native
Rooted, emergent perennial

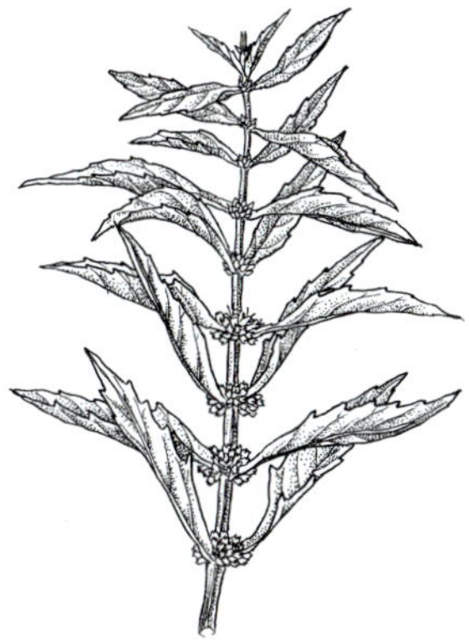

Lycopus rubellus—flowering stem ×1/2

Lycopus rubellus is a plant of the freshwater tidal marshes along the Delaware River in southeastern Pennsylvania and scattered locations elsewhere across the state. Along the Delaware, it often grows in crevices of tree roots or stumps at the high tide line where its roots and lower stem are inundated at high tide, but the upper stem and flowers remain above the water surface.

Water-horehound has opposite simple, toothed leaves and crowded whorls of tiny white flowers at the nodes. There are five species of *Lycopus* in our flora; all are wetland plants. Only *Lycopus rubellus* is treated here because of its association with freshwater tidal marsh habitats.

Lycopus rubellus is distinguished from the very similar *Lycopus americanus* by the presence of sharp-pointed calyx lobes that are 1–2 mm long but lack a rigid, needle-like (subulate) tip and a corolla tube that is twice as long as the calyx. In addition, the stolons of *Lycopus rubellus* frequently end in tubers, whereas *Lycopus americanus* does not produce tubers. Both species have leaves that are glabrous or only slightly pubescent.

Lycopus rubellus—flower ×5

The range of *Lycopus rubellus* extends from Maine to Florida and Texas on the Coastal Plain and inland to Michigan, Illinois, Missouri, and Texas. Water-horehound species were used by Native Americans in combination with other plants to treat colds in children.

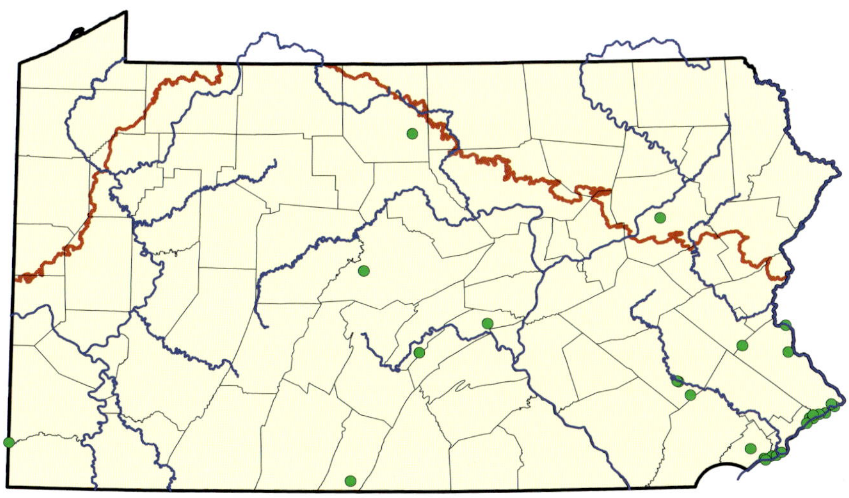

Lycopus rubellus

WATER-PARSNIP

Sium L.

Parsley Family—Apiaceae

Water-parsnip — *Sium suave* Walter

Native

Rooted, emergent perennial

Sium suave ×1/4

Water-parsnip is a stout, erect plant to 2 m tall. Leaves are pinnately compound with finely toothed, linear-lanceolate leaflets, but submersed basal leaves may be more finely divided. The size and form of the cauline leaves are progressively reduced upward. The tiny white flowers are borne in terminal and axillary compound umbels. Flowering occurs from July to September.

Water-parsnip is a plant of pond and stream margins, bogs, and swamps throughout Pennsylvania. Its total range extends from Newfoundland to Alaska and Siberia and south to Florida and California.

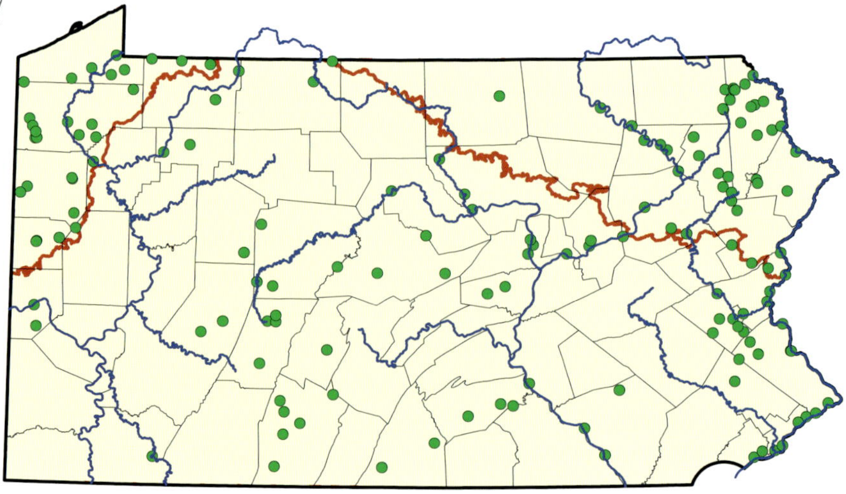

Sium suave

WATER-PLANTAIN
ALISMA L.
Water-plantain Family—Alismataceae

The water-plantains are rooted, emergent or submergent aquatics. They have a basal rosette of petiolate leaves with elliptical blades. The numerous small, white flowers are produced in a panicle that is nearly twice as tall as the leaves. The flowers are perfect and contain 3 sepals, 3 white petals, 6–9 stamens, and 15–20 pistils. The fruit is a cluster of achenes. Water-plantains are typically plants of shallow water; however, mature plants occasionally are found on exposed mud, and juvenile plants may occur completely submersed in deeper water. Water-plantain has been used medicinally by Native Americans to treat heart problems, kidney pain, and tuberculosis.

Key to Species of Water-plantain

A. flowers 4–5 mm wide; fruiting heads 3–4 mm .... *Alisma subcordatum*
A. flowers 7–13 mm wide; fruiting heads 4–7 mm .... *Alisma triviale*

Southern water-plantain — *Alisma subcordatum* Raf.

Native
Rooted, emergent perennial

Ranging in height from 10 cm to 1 m or more, southern water-plantain consists of a basal cluster of long-petioled leaves surrounding an erect flowering stem. The inflorescence has numerous whorled branches bearing tiny white flowers 4–5 mm wide. Flowering occurs late July to late August.

Southern water-plantain is the common species in Pennsylvania; it is found in lake and pond margins, muddy shores, marshes, and ditches throughout. Its total range extends from New Brunswick west to North Dakota and south to Georgia and Oklahoma.

Alisma subcordatum

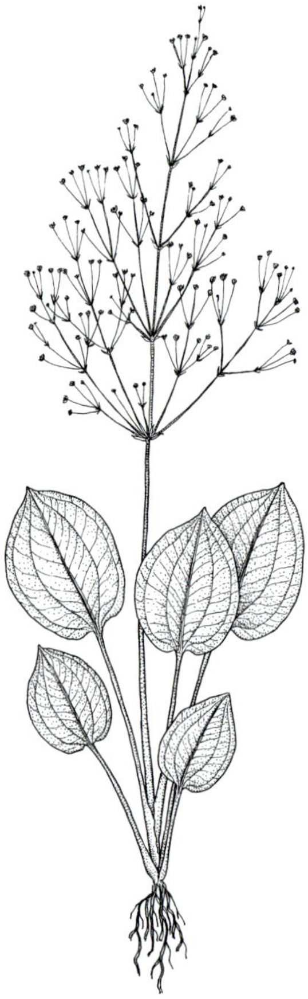

Alisma subcordatum ×1/4

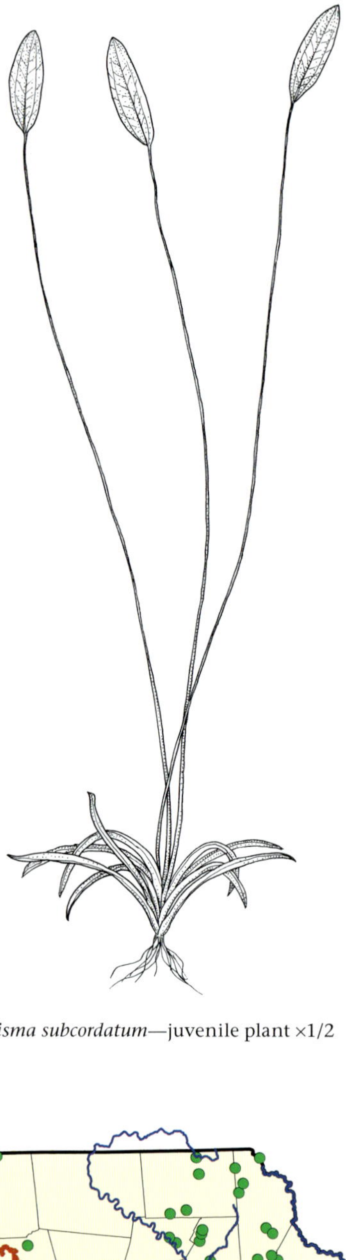

Alisma subcordatum—juvenile plant ×1/2

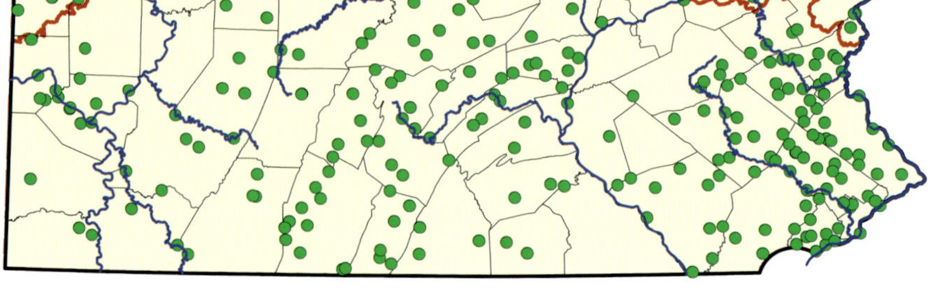

Alisma subcordatum

Northern water-plantain *Alisma triviale* Pursh

PNHP
Native
Rooted, emergent perennial

Northern water-plantain is very similar to the much more common southern water-plantain; however, the flowers are larger (7–13 mm).

The range of this species extends from southern Quebec and Ontario south to New York, northwestern Pennsylvania, and Ohio in the east. In the west it occurs from Saskatchewan and Alberta to California, Arizona, and New Mexico. In Pennsylvania it is a very rare species of lake margins and stream edges; most recorded sites are in the western third of the state.

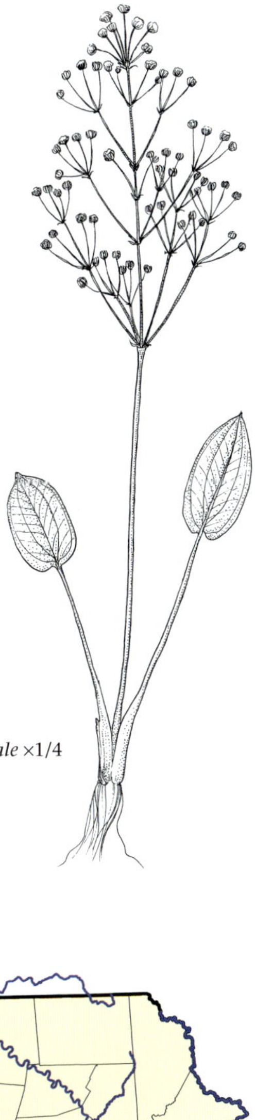

Alisma triviale ×1/4

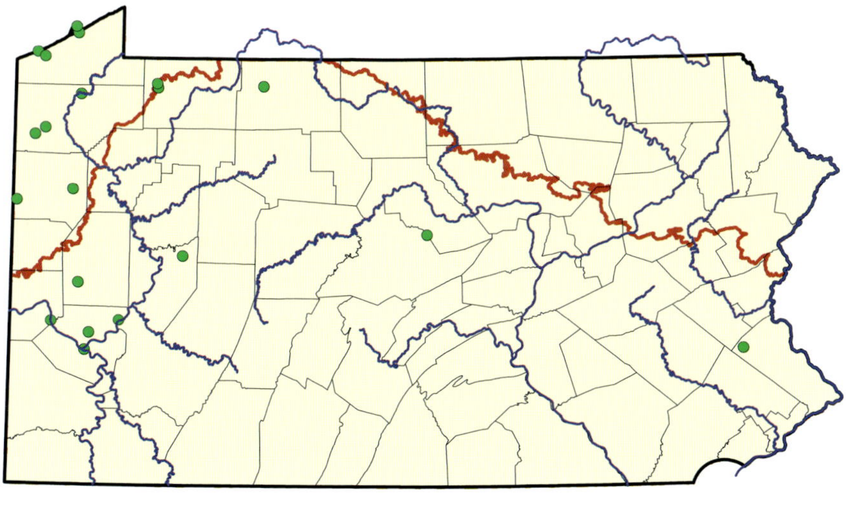

Alisma triviale

WATER-PRIMROSE

Ludwigia L.

Evening-primrose Family—Onagraceae

The water-primroses are showy herbaceous perennials with opposite, or alternate, simple leaves. The flowers are axillary, 4-merous, and have an inferior ovary. The fruit is an elongate capsule with 4 persistent sepals at the tip.

In some situations water-primrose forms modified roots, composed of aerenchyma tissue, which grow upward or outward from the horizontal stems to the water surface. These spongy, white structures allow for movement of oxygen and carbon dioxide from the atmosphere into stems and roots below the water surface (Ellmore 1981).

Both species of water-primrose may be more widespread than the maps indicate due to recent spread and under-collecting. Both have shown invasive tendencies in several parts of the state.

Key to the Species of Water-primrose

A. flowering stems creeping or floating; petals 1–1.5 cm long.....*Ludwigia peploides*
A. flowering stems erect; petals 1.5–2.5 cm long.......................*Ludwigia hexapetala*

Water-primrose — *Ludwigia hexapetala* (Hook. & Arn.) Zardini, H.Gu & P.H.Raven

Introduced
Rooted, emergent perennial

Water-primrose is a rhizomatous perennial with long floating stems bearing erect branches on which the showy yellow flowers are produced. Its fruits are slender, 2–4 cm-long capsules.

Native from the southeastern United States south to Argentina, water-primrose appears to be introduced in Pennsylvania, where it grows, often aggressively, in shallow water of streams, ponds, and canals.

Ludwigia hexapetala ×1/6

Ludwigia hexapetala

Ludwigia hexapetala growing vigorously in a shallow pond

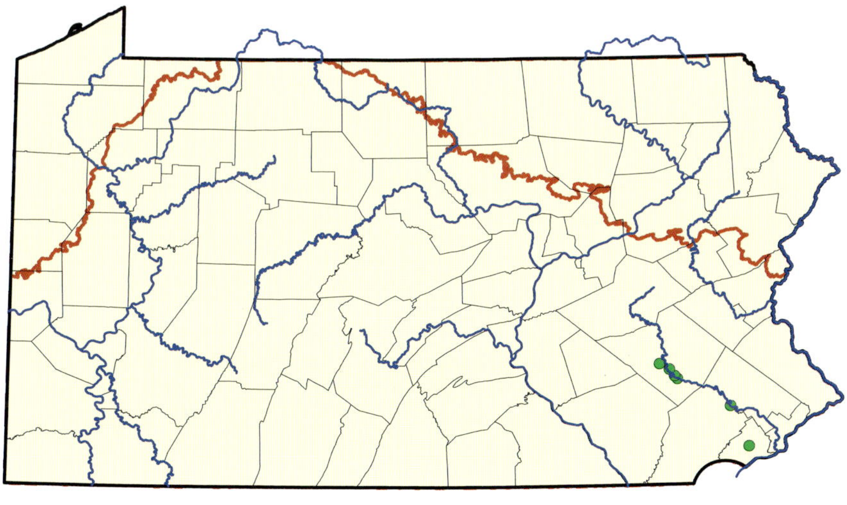

Ludwigia hexapetala

Primrose-willow *Ludwigia peploides* (Kunth) P.H.Raven ssp. *glabrescens* (Kuntze) P.H.Raven

Introduced
Rooted, floating-stem perennial

Primrose-willow is neither a primrose nor a willow; it is, however, a striking plant because of the showy yellow flowers on long stalks that are produced from July through September. The flowers are followed by slender, elongate fruits that are 2–4 cm in length.

The stems of primrose-willow float at or below the surface, often extending several meters in length and forming roots at the nodes.

Native from South Carolina to Kansas and south to Texas and Louisiana, primrose-willow is introduced in Pennsylvania. It grows in shallow ponds and canals and appears to have invasive tendencies.

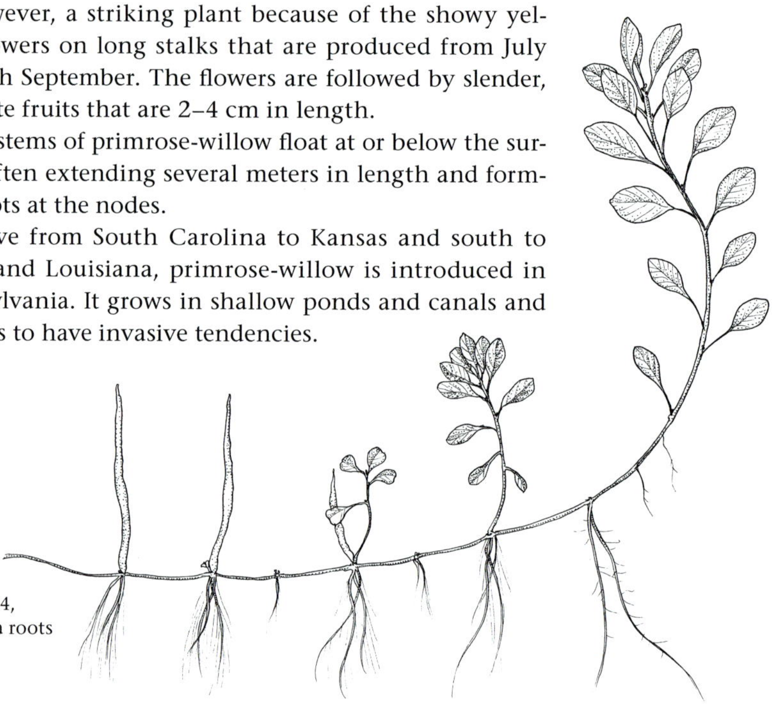

Ludwigia peploides ×1/4, note erect arenchyma roots

Ludwigia peploides—flower ×1/2

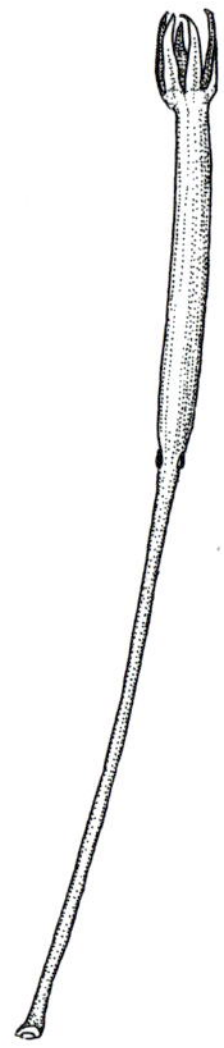

Ludwigia peploides—fruit ×1/2

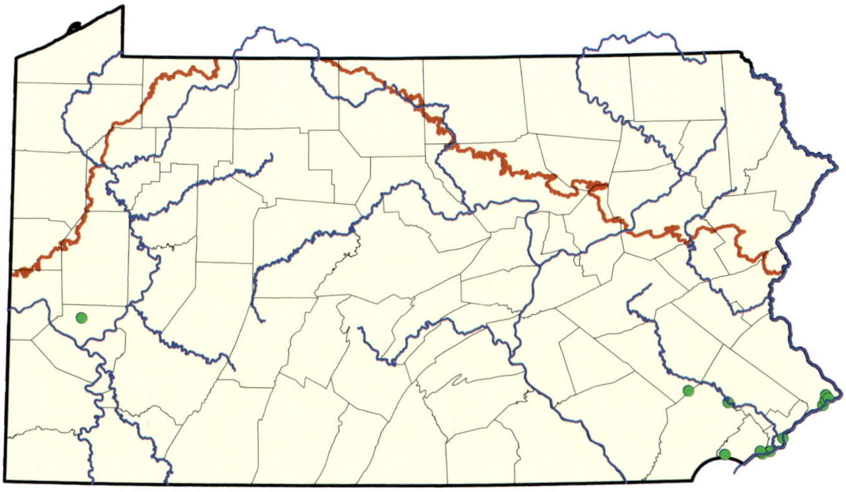

Ludwigia peploides

Ludwigia peploides

WATER-PURSLANE
Ludwigia L.
Evening-primrose Family—Onagraceae

Water-purslane ***Ludwigia palustris* (L.) Elliott**

Native

Rooted, emergent or partially floating perennial

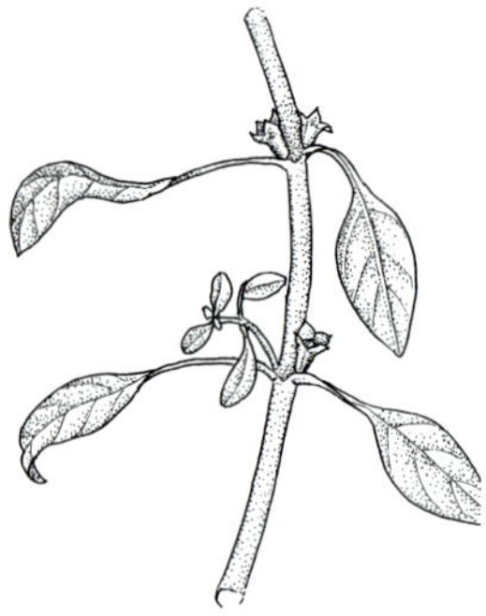

Ludwigia palustris—
stem with flowers
at the nodes ×1/2

Water-purslane's prostrate or creeping stems bear a pair of leaves at each node. The inconspicuous flowers are borne singly in the leaf axils; they do not have petals. The inferior ovary matures to form a 4-sided fruit topped with 4 small, triangular sepals.

Water-purslane is a very common plant of shallow water, shores, marshes, and wet soil throughout Pennsylvania. It has a wide range in the northern hemisphere.

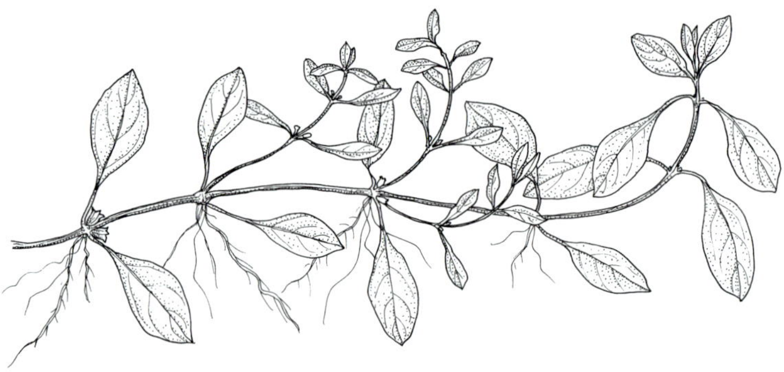

Ludwigia palustris ×1/2

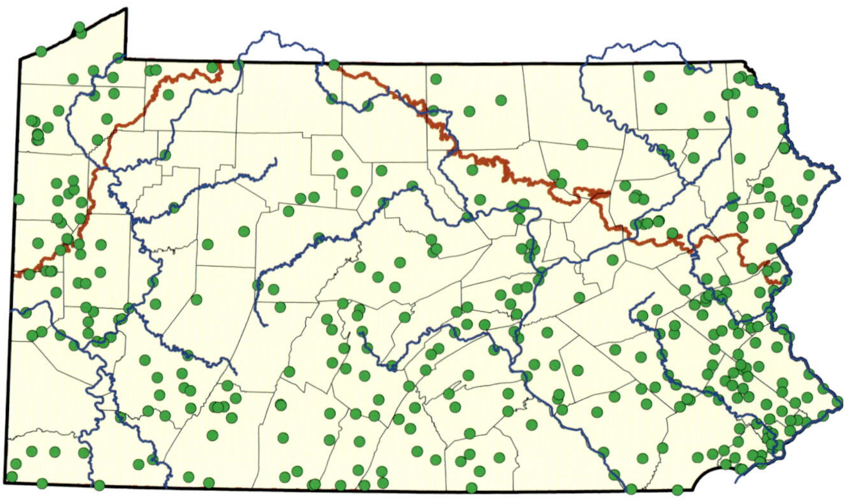

Ludwigia palustris

WATER-WILLOW
DECODON J.F. GMEL.
Loosestrife Family—Lythraceae
JUSTICIA L.
Acanthus Family—Acanthaceae

Despite having identical common names, *Justicia americana* and *Decodon verticillatus* are very different plants and not closely related at all.

While both are rooted emergents, *Decodon* has a woody base with long arching stems that often form roots when they touch the water; in addition, it has opposite or whorled leaves and axillary whorls of magenta flowers. *Justicia,* on the other hand has erect stems and a rhizomatous growth habit.

Key to the Species of Water-willow

A. stems semi-woody at the base, arching and sometimes rooting at the tips; flowers magenta .... *Decodon verticillatus*
A. stems completely herbaceous; erect; flowers white .... *Justicia americana*

Water-willow — *Decodon verticillatus* (L.) Elliott

Native
Rooted, emergent perennial

Water-willow is a robust perennial with arching stems 2–2.5 m long that arise from a semi-woody base. Submersed portions of stems develop inflated spongy aerenchyma tissue which facilitates movement of oxygen and carbon dioxide into stems and roots. The arching stems also produce adventitious roots at the tips when they contact water. The stems bear simple, entire leaves in whorls of three at the nodes. The purple flowers, which are borne in the leaf axils, are of three forms varying in the length of the style and stamens. They are insect-pollinated.

Water-willow grows in shallow water at the edges of lakes and ponds, frequently at the margin of a swamp or bog. Its long arching stems contribute to extension of the land surface, similar to mangroves at the margin of the sea.

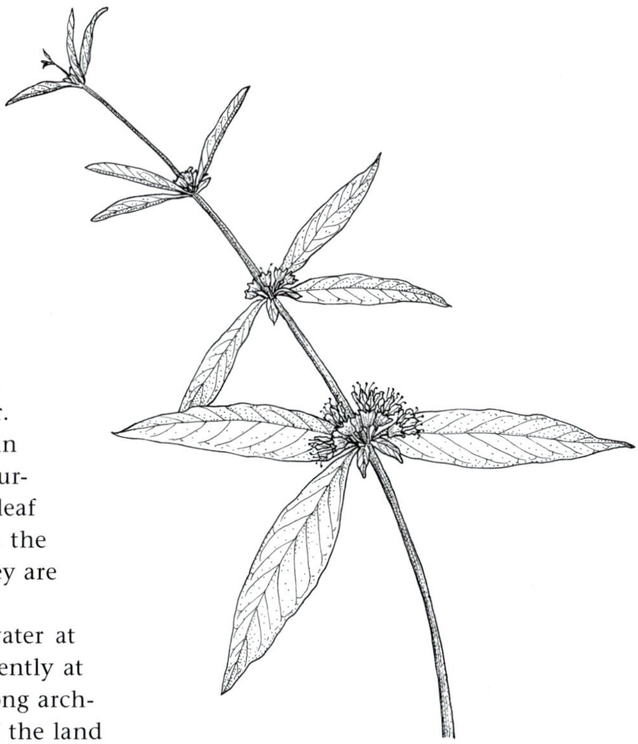

Decodon verticillatus—stem with flowers at the nodes ×1/2

Decodon verticillatus—flowers

In Pennsylvania water-willow is found mainly in the glaciated northeast and northwestern regions. Its total range extends from Maine to Florida and Louisiana and north in the Mississippi Valley to Indiana and Missouri.

It would be hard to confuse this water-willow with anything else. Purple loosestrife also has purple flowers and forms semi-woody hummocks in shallow water, but the stems are upright rather than arching and the leaves are opposite, not whorled.

Stems of *Decodon verticillatus* arching out from the lake margin at a lake in Wayne County, Pennsylvania

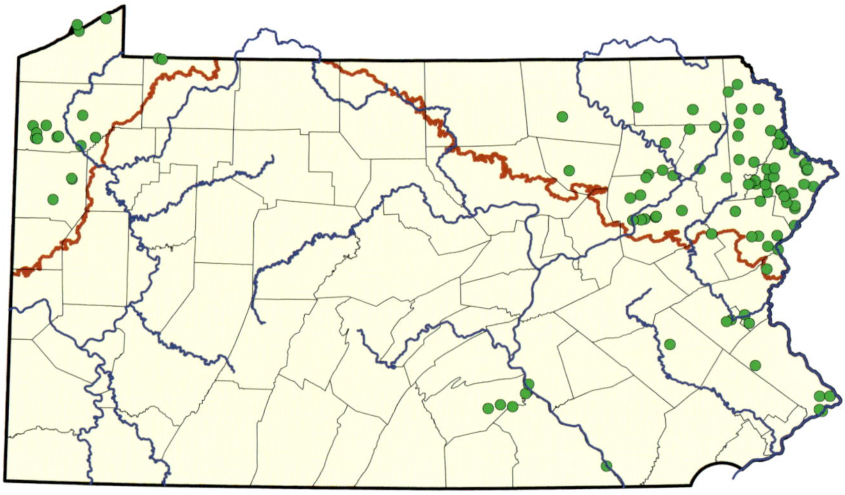

Decodon verticillatus

Water-willow

Justicia americana (L.) Vahl.

Native
Rooted, emergent perennial

Conspicuous colonies of this water-willow are present in shallow water along the edges of rivers throughout Pennsylvania, especially in the Susquehanna and Allegheny watersheds. In recent years this species has also become increasingly common in the Delaware and Schuylkill Rivers in the eastern part of the state. Water-willow is also occasionally found along the marshy shores of lakes or ponds.

A rooted emergent with unbranched stems 0.5–1 m tall, water-willow spreads by rhizomes to form large colonies. The leaves are opposite, linear to lanceolate, and tapered at the base. Flowers are borne in dense, short axillary spikes on stalks that are longer than the subtending leaves. The 5 calyx lobes are narrowly lanceolate. The 2-lipped corolla is 8–13 mm long, pale violet to nearly white with purple markings on the lower lip; the upper lip is concave. Each flower has 2 stamens and a single superior ovary. Flowering occurs from late June to early September; the blooms are insect-pollinated. The fruit is a capsule that opens explosively, propelling the seeds some distance from the parent plant.

Water-willow is wide-ranging in eastern North America, where it occurs from western Quebec to Wisconsin and south to Georgia and Texas.

Justicia americana ×1

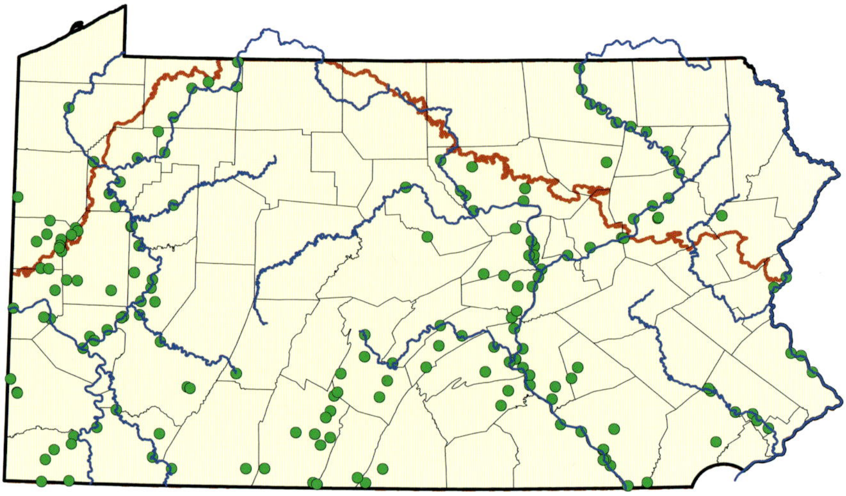

Justicia americana

Justicia americana—flowers

WATERCRESS
NASTURTIUM R.BR.
Mustard Family—Brassicaceae

Watercress — *Nasturtium officinale* R.Br.

Introduced
Rooted, emergent perennial

Watercress is a dark green, leafy plant of clear, cool, shallow water of springs, seeps, and quietly flowing streams throughout Pennsylvania. Its hollow herbaceous stems are erect; roots often form at the lower nodes. The leaves are pinnately compound and alternately arranged on the stems. Tiny white flowers with 4 petals are borne in a terminal raceme; the fruits are narrowly elongate and 1–1.5 cm long.

Watercress is native to Europe but introduced here through cultivation and now widely naturalized. It frequently grows in cold water of springheads and seeps. The shoots are valued for their pungent taste and for use as a garnish.

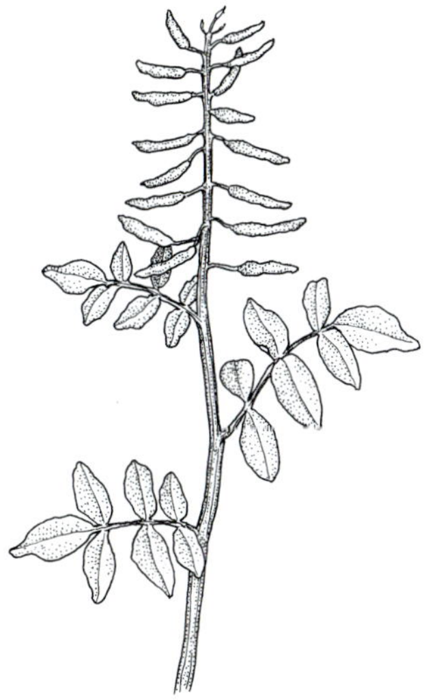

Nasturtium officinale—stem with fruits ×1/2

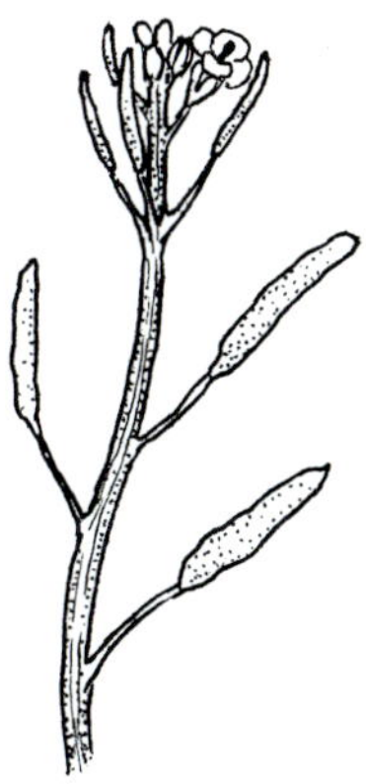

Nasturtium officinale—inflorescence ×1

Nasturtium officinale—fruit ×2

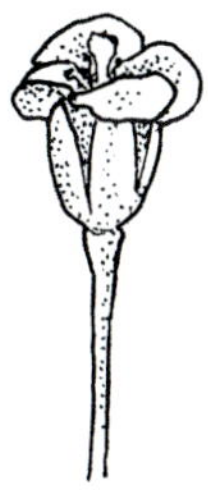

Nasturtium officinale—flower ×3

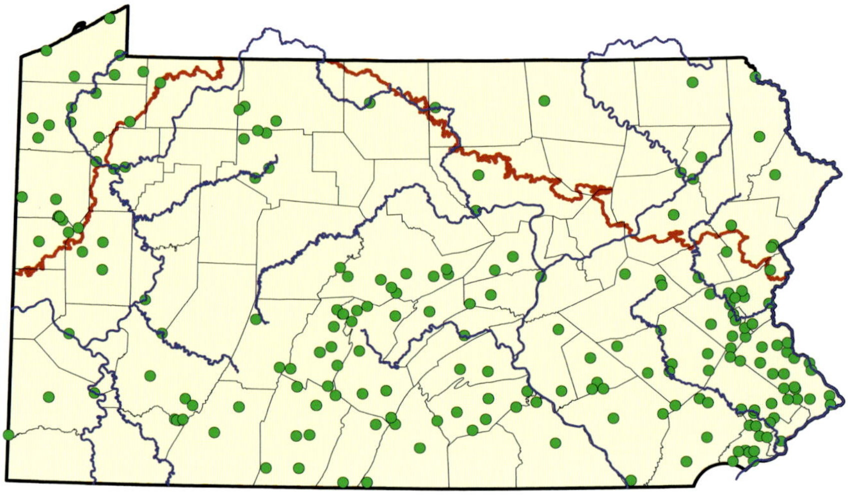

Nasturtium officinale

Chapter 4. Floating-leaf Plants

This group of species is characterized by leaves that float on the water surface, although some members may also have underwater leaves. Flowering typically occurs at or just above the surface of the water.

FLOATING-HEART
NYMPHOIDES HILL.
Buckbean Family—Menyanthaceae

The floating-hearts are rooted aquatics with a long stem that bears leaves and flowers at the water surface. Although the floating leaves look superficially like those of water-lily, the plants are not closely related at all. The flowers have 5 fused petals and 5 stamens attached to the corolla tube. The fruit is a capsule.

Key to the Species of Floating-heart

A. flowering stem with a single leaf at the base of the inflorescence; flowers white .... *Nymphoides cordata*

A. flowering stem with 2 leaves at the base of the inflorescence; flowers yellow..... .... *Nymphoides peltata*

Floating-heart ***Nymphoides cordata*** **(Elliott) Fernald**

PNHP
Native
Rooted, floating-leaf perennial

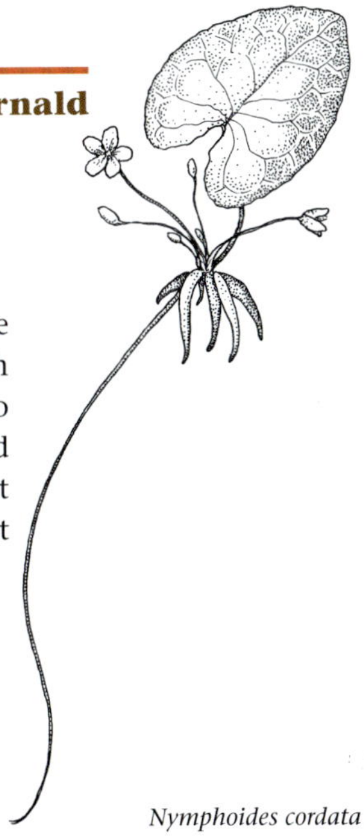

Nymphoides cordata ×1/2

Floating-heart is named for its heart-shaped floating leaves that are about 5–8 cm long and look like miniature water-lily leaves. Although some leaves arise directly from the basal rhizome, floating-heart also produces erect, buoyant stems from which a leaf, several flowers, and a cluster of stubby banana-shaped roots arise. The flowers are about 12 mm wide and consist of a 5-lobed corolla surrounding an ovary that has 2 stigma lobes. The fruit is a capsule.

The range of floating-heart extends from Newfoundland to Quebec and Ontario and south to New York, Connecticut, and Pennsylvania; it also occurs locally on the coastal plain to Florida and Louisiana. In Pennsylvania it is a rare native plant that is limited to natural glacial lakes, where it grows in water to 1.6–2 m deep, often occurring with water-lilies and other floating-leaf plants.

Juvenile leaves of water-lilies and spatterdock might be confused with floating-heart, although examination of the stem structure should resolve any doubt about the correct identification. When the plants are flowering there is no confusing floating-heart with anything else.

Native Americans employed floating-heart as a treatment for trembling, shortness of breath, and cough.

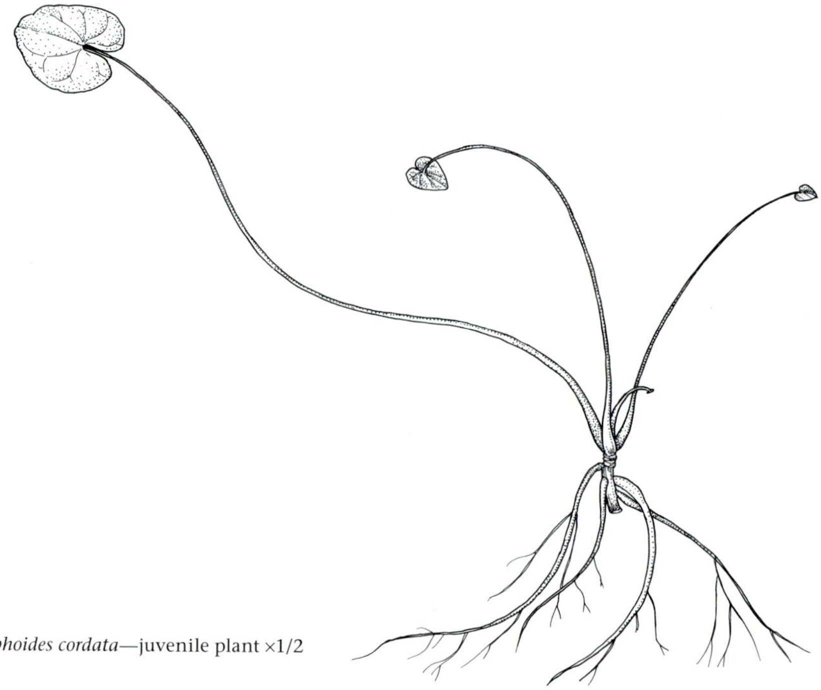

Nymphoides cordata—juvenile plant ×1/2

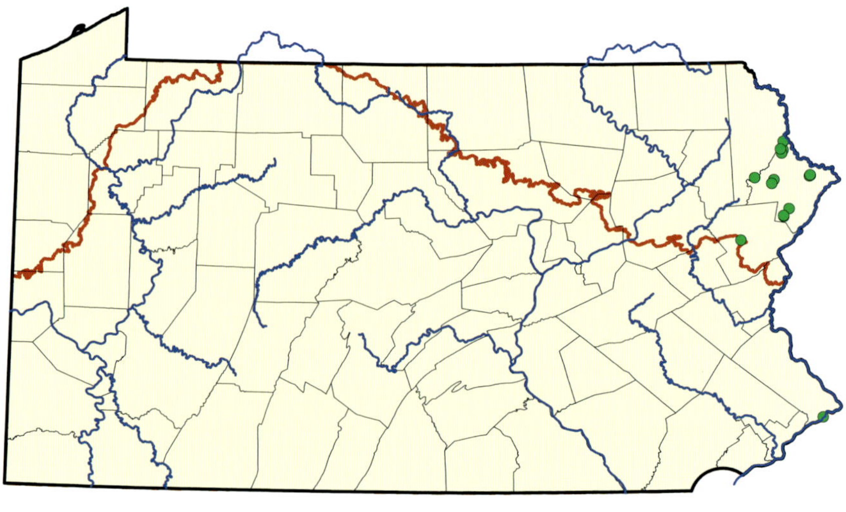

Nymphoides cordata

Nymphoides cordata

Yellow floating-heart — *Nymphoides peltata* (J.G. Gmel.) Kuntze

Introduced
Rooted, floating-leaf perennial

Yellow floating-heart has floating, heart-shaped leaves that are green with reddish blotches and a wavy margin. Flowering occurs on stout upright stems bearing 2 unequal, nearly round floating leaves below the inflorescence. The flowers are 2–2.5 cm wide and bright yellow, with corolla lobes that are fringed near the base. Flowering occurs from June to October.

The fruits split open when ripe, releasing disk-shaped seeds that float on the surface of the water. Stiff marginal hairs on the seeds cause them to adhere to the external surfaces of waterfowl, which disperse the seeds as they move from site to site (Cook 1990).

Yellow floating-heart is native to Europe and Asia Minor; it was introduced for use in ornamental ponds and water gardens. The earliest recorded occurrence in Pennsylvania is from a pond in Berks County in 1905 (Stuckey 1974). It has since become naturalized at additional sites, primarily in the southeastern corner of the state.

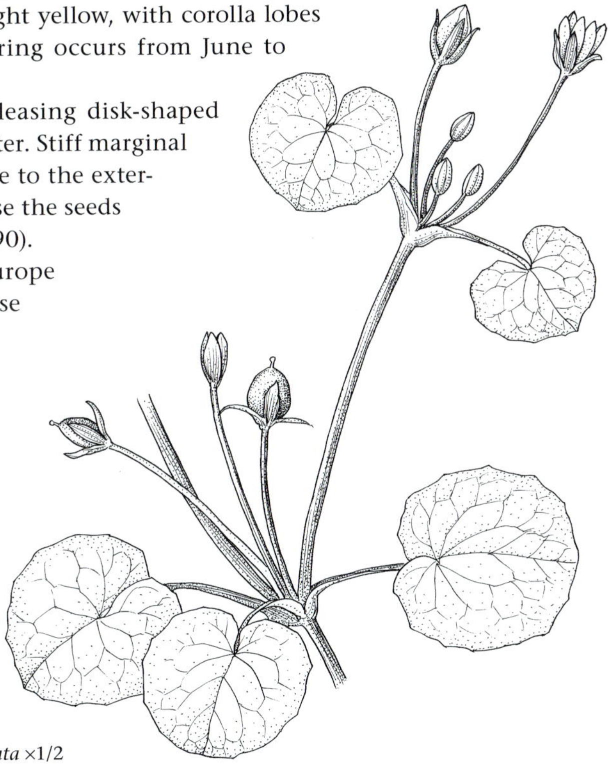

Nymphoides peltata ×1/2

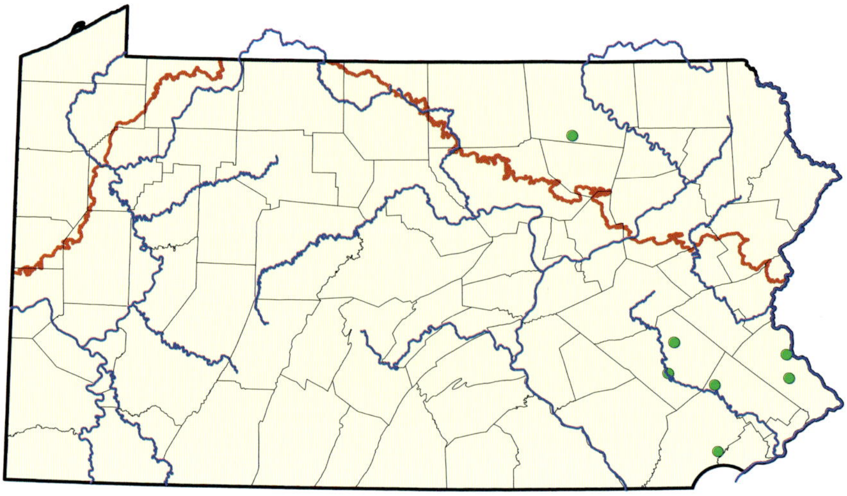

Nymphoides peltata

Nymphoides peltata

AMERICAN LOTUS

NELUMBO ADANS.

Lotus-lily Family—Nelumbonaceae

American lotus — *Nelumbo lutea* Willdenow

PHNP
Native
Rooted, floating-leaf or emergent perennial

American lotus is an emergent aquatic perennial with large leaves that are usually elevated several inches above the water surface, but may occasionally float. The leaf blade, which can be up to 0.6 m in diameter, is circular and is attached to the supporting petiole at its center. Lotus flowers are also borne above the water surface; each pale yellow flower can be 10–20 cm wide.

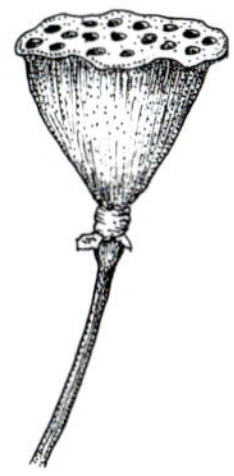

Nelumbo lutea—fruit ×1/4

American lotus is a plant of quiet water; it is believed to be native to floodplains of major rivers in the east central United States. Native Americans gathered lotus seeds and tubers for food and may have transplanted them to more eastern areas to assure a regular supply. American lotus is also grown as an ornamental; several known occurrences in Pennsylvania today are at sites of early amusement parks where it was likely planted. The sacred lotus (*Nelumbo nucifera*) is also grown as an ornamental, and has become naturalized at at least one site in the state. It has pink, rather than yellow, flowers.

Dried lotus fruiting heads are used in flower arrangements.

Recently, characteristics of the leaf surface of American lotus, which is extremely hydrophobic, have been used a model for the design of paint and other industrial products that shed dirt and water, an example of biomimicry.

No other aquatic plant has such large leaves, or a fruit structure that could be confused with lotus.

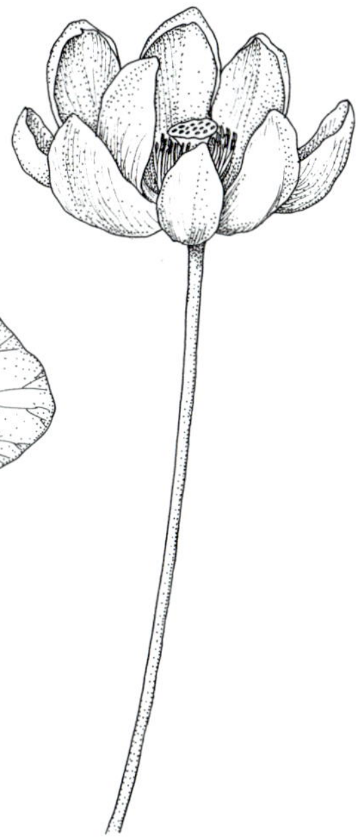

Nelumbo lutea—flower ×1/4

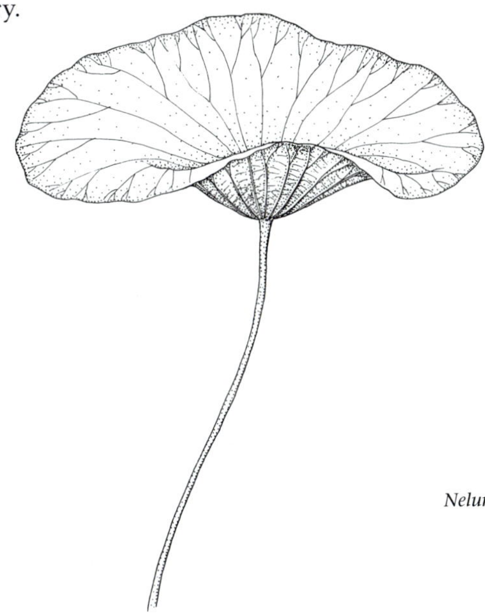

Nelumbo lutea—leaf ×1/4

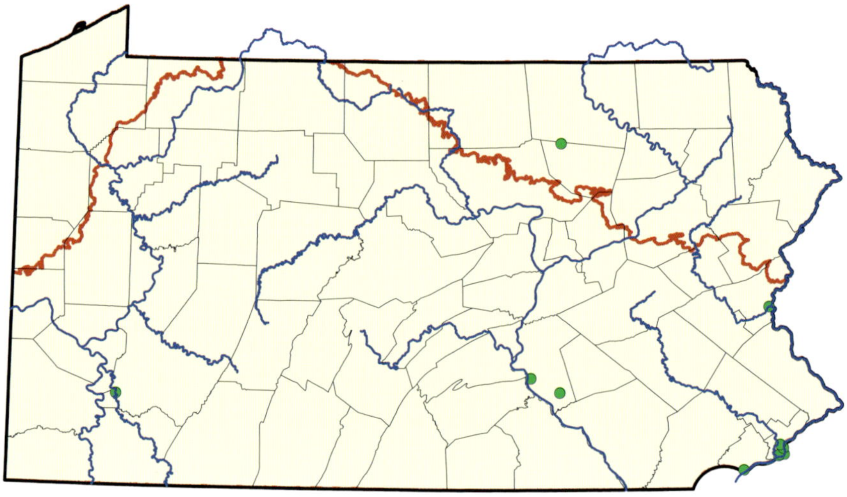

Nelumbo lutea

Nelumbo lutea growing at a site in Dauphin County, Pennsylvania

WATER-CHESTNUT

TRAPA L.

Loosestrife Family—Lythraceae

Water-chestnut, caltrop *Trapa natans* L.

Introduced
Rooted, floating-leaf annual

Water-chestnut is a distinctive plant with a floating rosette of triangular leaves; each leaf has a swollen portion in the middle of the petiole that adds to the buoyancy of the plant.

An elongate, often branching, stem bearing finely divided underwater leaves trails down through the water. The base of the stem may be rooted in the substrate.

Water-chestnut is an annual in our climate; farther south it is perennial. Flowers are borne in the axils of the floating leaves, they are perfect, 4-merous, and have an inferior ovary. The distinctive hard, black, spiny fruits are frequently found littering the shoreline of infested bodies of water. They are 3–4 cm wide and bear 4 stout spines that could inflict a nasty puncture wound, hence the common name, caltrop.

Water-chestnut is native to Eurasia and has become naturalized throughout the northeastern and mid-Atlantic regions of the United States. In some areas it has become sufficiently abundant to interfere with navigation; it is designated as a noxious weed in several states. In Pennsylvania water-chestnut is found in scattered lakes, ponds, and river backwaters in the eastern half of the state. Populations grow quickly, as each seed can give rise to 10–15 rosettes; in addition individual rosettes can break loose and be carried downstream to initiate new infestations.

Despite its name, *Trapa* is not the water-chestnut of culinary uses. That plant is a species of spike-rush (*Eleocharis dulcis*). European water-chestnut is sometimes grown in ornamental ponds or water gardens; however, this practice is not recommended due to the risk that it might escape and establish additional naturalized populations.

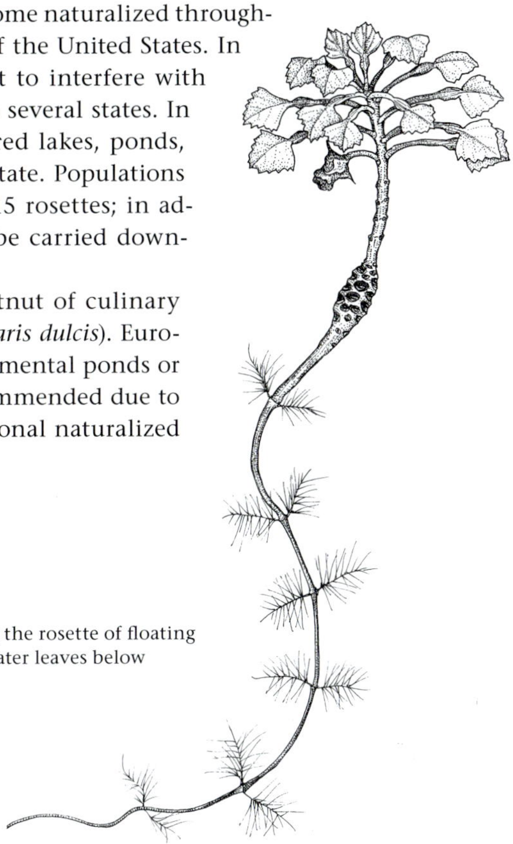

Trapa natans ×1/4—stem showing the rosette of floating leaves and finely divided underwater leaves below

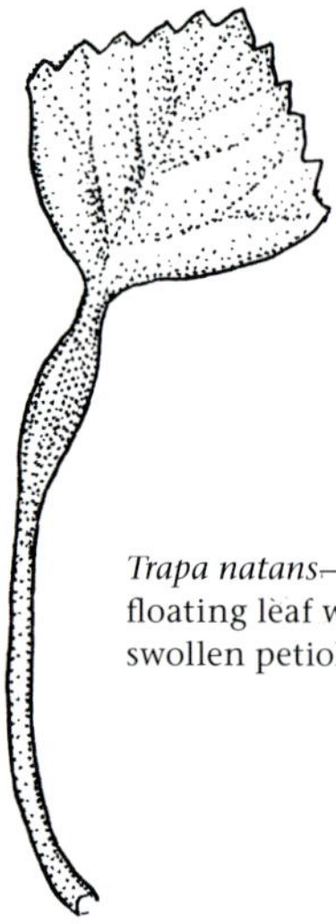

Trapa natans—
floating leaf with
swollen petiole ×1

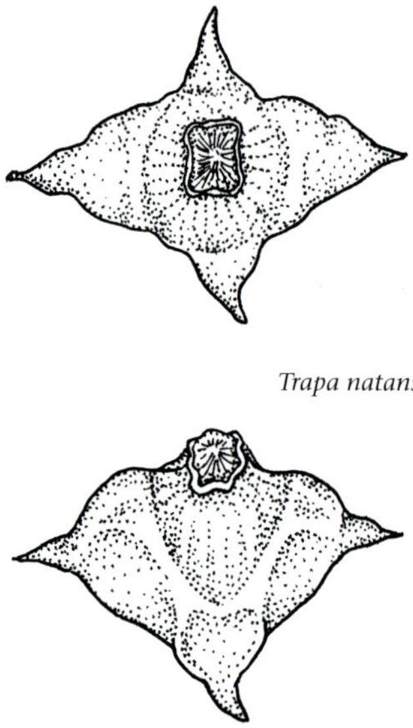

Trapa natans—fruit ×1

Trapa natans

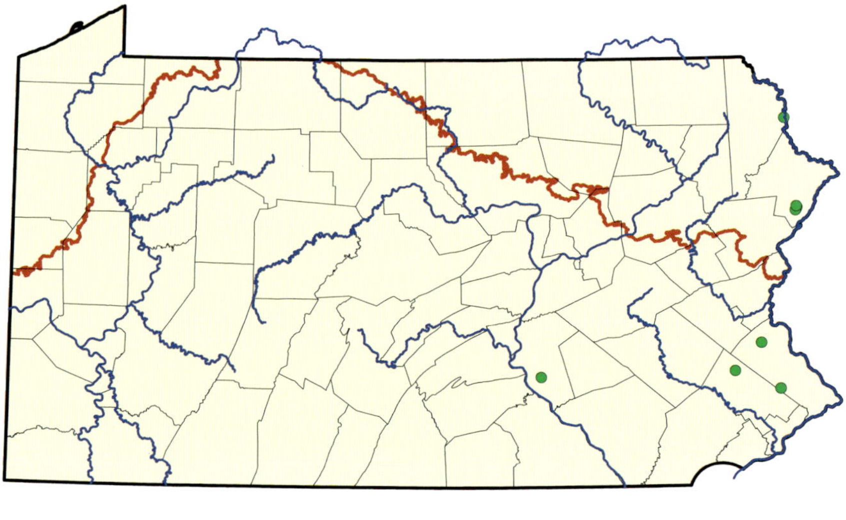

Trapa natans

EUROPEAN WATER-CLOVER

Marsilea L.

Water-clover Family—Marsileaceae

European water-clover, pepperwort — *Marsilea quadrifolia* L.

Introduced

Rooted, floating-leaf or emergent perennial

Water-clover, an aquatic fern, is named for its four-lobed fronds (leaves) that resemble the leaves of clover. Usually found in shallow water, the plants have rhizomes rooted in mud from which fronds and sporocarps arise. The fronds are borne on stalks 5–15 cm long that often extend above the water surface. The sporocarps (spore-producing structures) also arise from the rhizomes, but are on much shorter stalks than the leaves and frequently remain below the water surface.

European water-clover is a rare naturalized species in Pennsylvania (Givens and Ferren 1972). However, the increasing popularity of water gardening may result in additional introductions. Its native range includes southeastern Europe and Asia.

Marsilea quadrifolia naturalized at a site in Chester County, Pennsylvania

Marsilea quadrifolia—leaf ×3/4

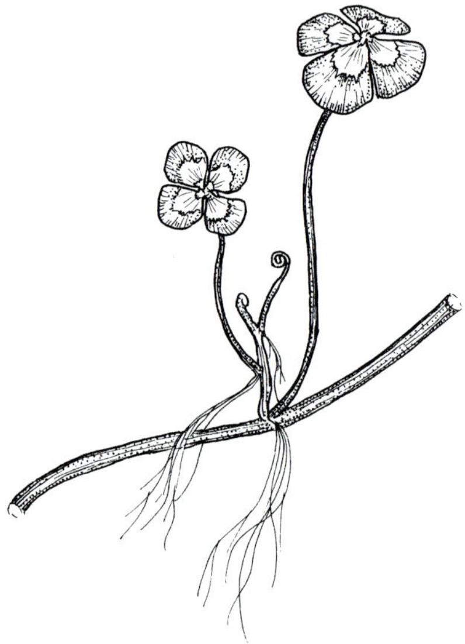

Marsilea quadrifolia ×3/4

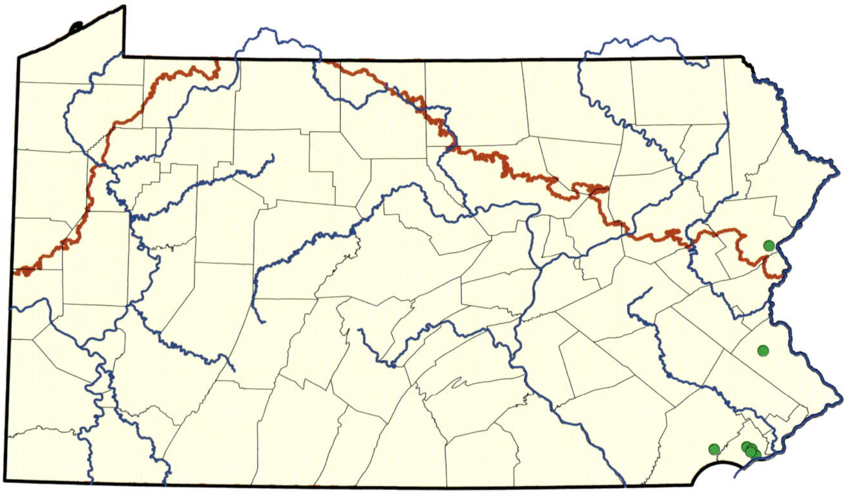

Marsilea quadrifolia

WATER-LILY

NUPHAR SM. AND *NYMPHAEA* L.

Water-lily Family—Nymphaeaceae

Water-lilies are perhaps the quintessential aquatic plants. Characterized by large floating leaves (lily-pads) and showy flowers produced at the water surface, they are familiar to all. The common name water-lily is applied to two different genera of aquatic plants, *Nymphaea* and *Nuphar*. Both are common in lakes, ponds, and streams in Pennsylvania.

Key to the Species of Water-lily

A. leaf longer than wide, corners of the sinus lobes rounded
  B. leaf blades less than 10 cm long, flowers 1.2–2 cm.......... *Nuphar microphylla*
  B. leaf blades and flowers larger
    C. leaf petioles round or oval in cross-section........................ *Nuphar advena*
    C. leaf petioles flattened on one side and somewhat winged.......................... *Nuphar variegata*
A. leaf circular in outline, sinus lobes with sharp corners..........*Nymphaea odorata*

Yellow water-lily, spatterdock, pond-lily, cow-lily — ***Nuphar*** **Aiton**

The spatterdocks are conspicuous members of the floating-leaf aquatic community. The leaves are oval with a deep sinus extending to the point of attachment of the petiole. Their bright yellow, 1.2–4 cm-wide flowers extend 2–5 cm above the water surface. They consist of a stout central pistil topped by a stigmatic disk and surrounded by numerous stamens and 6 yellow or greenish sepals. Leaf petioles and flower stalks arise directly from a stout rhizome that is rooted in lake or river bottom sediments.

Native Americans had many uses for *Nuphar*: the rhizomes and seeds were used as food; preparations of roots were used to treat a variety of ailments, including pain, swelling, cuts, inflammation, fever, rheumatism, and gynecological problems. The rhizomes, which are also a favored food of beaver, are often seen floating or cast up along lakeshores where these aquatic mammals are active. Various waterfowl eat the seeds.

There has not been complete agreement among botanists about how many species of *Nuphar* should be recognized. In earlier treatments of the Pennsylvania flora, *N. variegata* and *N. advena* were lumped as *N. lutea* (Rhoads and Block 2000). However, recent studies support the recognition of three species of spatterdock in Pennsylvania as described here (Padget 1997; Flora of North America Editorial Committee 2000).

Yellow water-lily, spatterdock — *Nuphar advena* (Aiton) Aiton f.

Native

Rooted, floating-leaf or emergent perennial

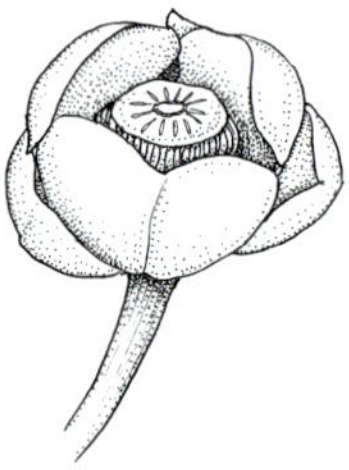

Nuphar advena—
flower ×1/2

Unlike the more northern *Nuphar variegata*, the leaves of *N. advena* are mostly emergent, extending above the water surface. However, in freshwater tidal marshes along the Delaware River, where this species is frequently the dominant plant, the leaf blades float at high tide. The petioles of *Nuphar advena* are more or less round or oval in cross-section rather than flattened on one side, as is the case with *N. variegata*. The globular yellow flowers have a yellow or greenish stigmatic disk.

Nuphar advena is a plant of coastal regions from New England to Florida and Texas and inland to central New York, western Pennsylvania, Ohio, Wisconsin, and Nebraska. It grows in lake margins, wetlands, slow-moving portions of rivers and streams, and freshwater intertidal marshes.

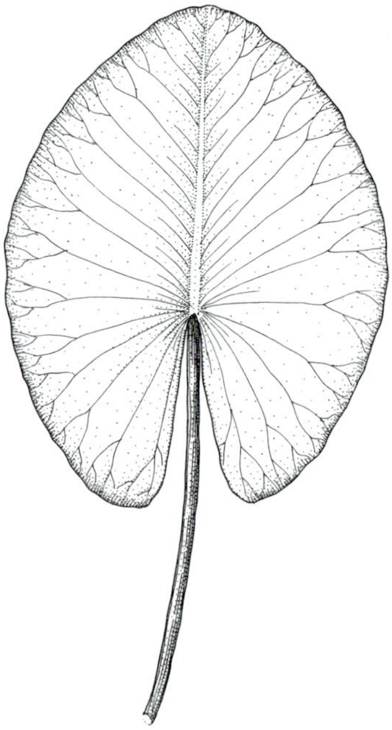

Nuphar advena—leaf ×1/4

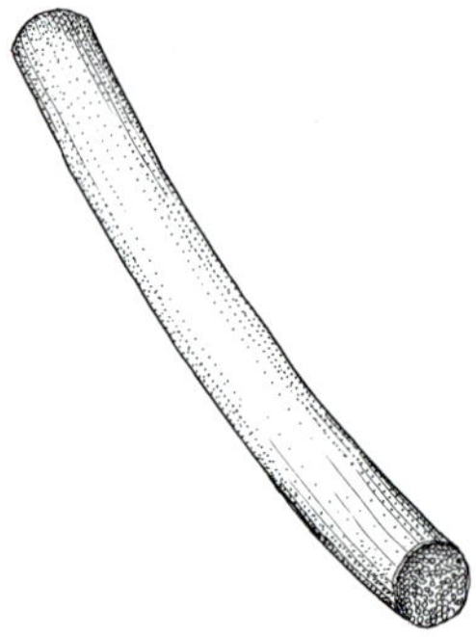

Nuphar advena—
petiole ×1/2

Nuphar advena

Nuphar advena—
immature fruit ×1/2

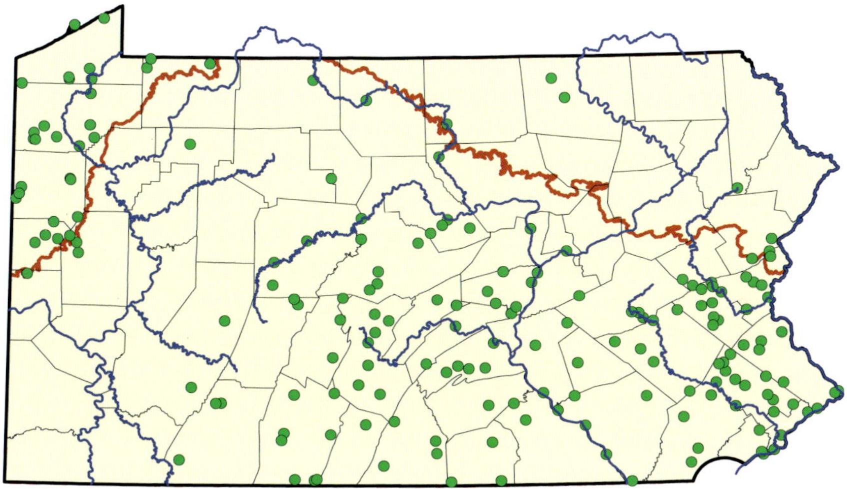

Nuphar advena

Small yellow water-lily — *Nuphar microphylla* (Persoon) Fernald

PNHP
Native
Rooted, floating-leaf perennial

Small yellow water-lily is a very rare plant in Pennsylvania, known from only a few scattered sites. Although similar in form to the larger *Nuphar variegata* and *N. advena*, it is distinctly more delicate. Its 5–10 cm-long leaves are floating or occasionally submergent. Flowers are 1.2–2 cm in diameter with 5 yellow sepals and a reddish stigmatic disk. The horizontal rhizome from which the flowers and leaves arise is 1.2–2 cm thick.

Small yellow water-lily is native from Newfoundland to Wisconsin and south to northern New Jersey and Pennsylvania. It grows in water to 1 m deep along the edges of lakes, ponds, and slow-moving streams. Some eastern Pennsylvania populations may have originated from plantings.

Nuphar microphylla ×1/4

Flowers are necessary for accurate identification, as the leaves of juvenile plants of the larger species could be confused with those of the small-leaved spatterdock. A hybrid with *Nuphar variegata* (*Nuphar* x *rubrodisca*) is intermediate in size and has a deep red stigmatic disk.

Nuphar microphylla

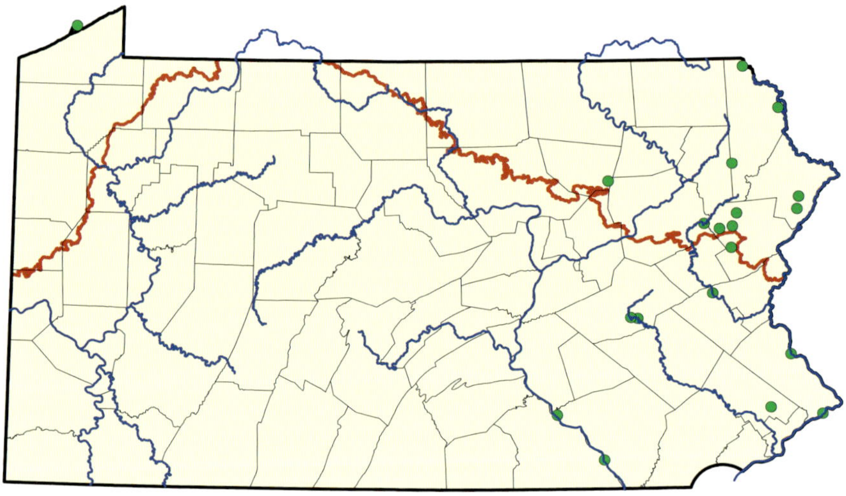

Nuphar microphylla

Yellow water-lily, spatterdock — *Nuphar variegata* Durand

Native

Rooted, floating-leaf perennial

Nuphar variegata is a common plant of Pennsylvania's lakes, ponds, and slow-moving river backwaters especially in the northeastern part of the state. The leaves are 15–30 cm long with a basal sinus extending about one-third the length of the blade; they usually float, but may occasionally be elevated above the water surface. The petioles are distinctly flattened on the upper side. Early in the season, thin, irregularly shaped, almost lettuce-like submersed leaves are produced at the rhizome. These leaves may appear pinkish in color when viewed from above the water surface. The rhizome is up to 10 cm in diameter and is marked with prominent scars where leaf and flower stalks were attached.

Nuphar variegata flowers are 2.5–4 cm in diameter with 6 yellow (or green) sepals and a yellow or greenish stigmatic disk.

Spatterdock is usually found in water up to 2 m deep and is often very abundant in eutrophic lakes. It often grows in areas with soft bottom sediments. *Nuphar variegata* has a more northern range than *Nuphar advena*; it extends from Newfoundland to Alaska and south to Oregon, Nebraska, Iowa, Illinois, Indiana, Ohio, and Maryland.

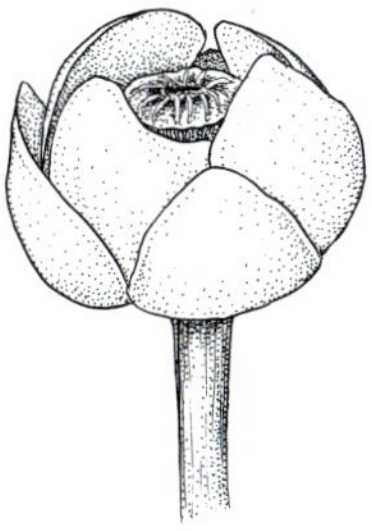

Nuphar variegata—flower ×1/2

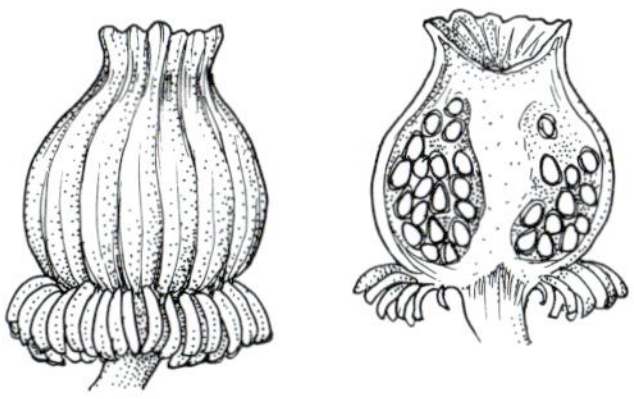

Nuphar variegata—fruit ×1/2

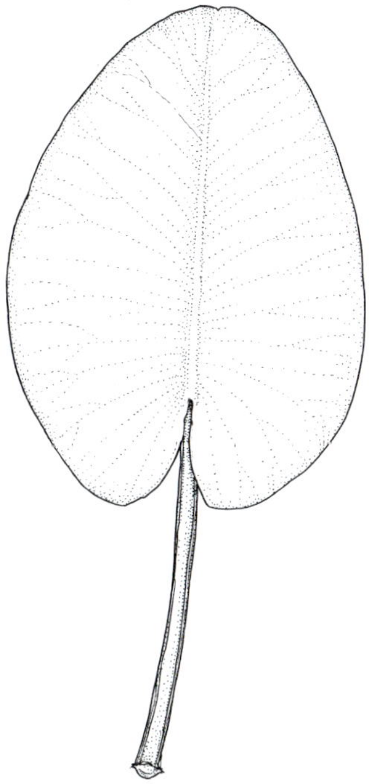

Nuphar variegata—leaf ×1/4

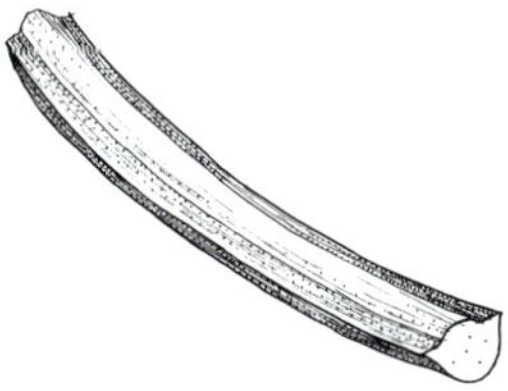

Nuphar variegata—petiole ×1/2

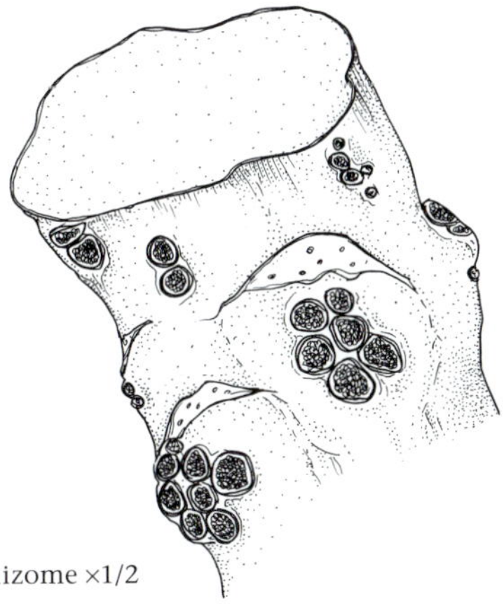

Nuphar variegata—rhizome ×1/2

Nuphar variegata

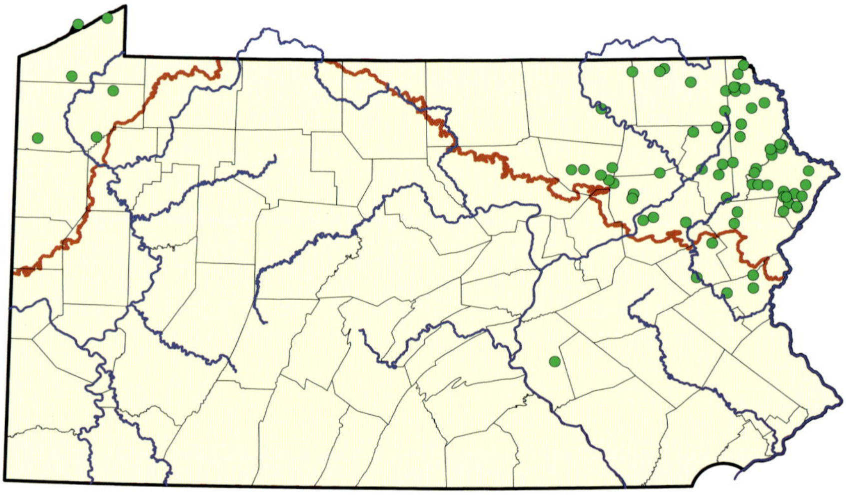

Nuphar variegata

Fragrant water-lily — *Nymphaea odorata* L.

Native

Rooted, floating-leaf perennial

Nymphaea odorata—fruit with seeds ×1/2

Water-lilies are one of the best known aquatic plants. The circular floating leaves (lily pads) have a v-shaped sinus terminating at the point of attachment of the leaf stalk. Leaf diameter can vary from 8–30 cm depending on the growing conditions. The flowers, with their numerous white petals and yellow center, also float on the water surface. Flowers open in the morning and close by early afternoon on successive days. The first day is characterized by a receptive stigma; not until the second day is pollen released (Wiersema 1988). Pollinators are primarily Hymenoptera (bees) and Diptera (flies).

Both flowers and leaves arise directly from a horizontal rhizome that is rooted in the lake or pond bottom; they are attached by long, flexible stalks that allow for movement caused by waves. The native water-lily is white-flowered or sometimes has a slight pink tinge; deep pink to red forms found in some lakes are the result of introduction of cultivated forms.

Nymphaea odorata—flower ×1/2

Fragrant water-lily grows naturally from Newfoundland and Labrador to the Great Lakes states, south along the coast to Florida and Texas, and inland to Missouri and Kansas. In Pennsylvania, it is most common in the glaciated regions, but water-lilies also occur at scattered locations south of the glacial boundary. Found most often in quiet water, fragrant water-lily tolerates a wide range of nutrient levels and pH. Leaves and flowers tend to be smaller in oligotrophic (low nutrient, low pH) lakes.

Native American used preparations of water-lily leaves or roots to treat mouth sores, coughs, and swellings. The seeds and rhizomes were eaten. Ducks eat the seeds, and beaver, muskrat, and porcupine feed on the leaves, rhizomes, and seeds. Farther north moose depend on water-lilies for a major portion of their diet.

Nymphaea odorata—rhizome with developing leaves ×2/3

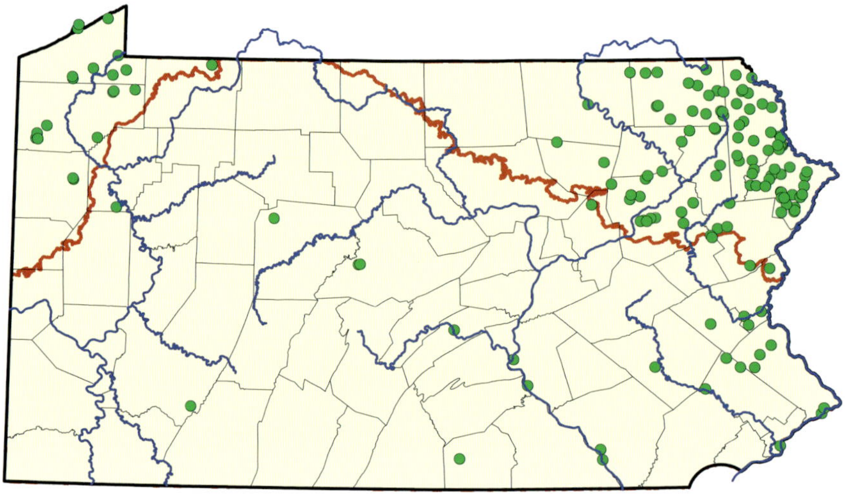

Nymphaea odorata

Nymphaea odorata

WATERSHIELD
BRASENIA SCHREB.
Watershield Family—Cabombaceae

Watershield *Brasenia schreberi* J.F. Gmelin

Native
Rooted, floating-leaf perennial

The oval floating leaves of watershield and the gelatinous sheath that covers the underwater parts of the plant make this common aquatic easy to recognize. Although most abundant in the glaciated region of northeastern Pennsylvania, it occurs in lakes and ponds throughout the state. Watershield is often very abundant in water 1–2 m deep, where it frequently grows with fragrant white water-lily and spatterdock, but the oval-shaped leaves of watershield are distinctive.

Watershield is a rooted floating-leaf perennial with buoyant stems that arise from a slender rhizome rooted in the bottom sediments. The leaves are about 6–12 cm long, oval in shape, and have a centrally attached stalk. The flowers extend 2–3 cm above the water surface; they are small, purple, and perfect. Individual flowers remain open for 2 days; on the first day the stigma is receptive; the anthers do not mature until the second day when the filaments elongate and pollen is released. The flowers are primarily wind-pollinated although some insects do visit to feed on pollen (Osborn and Schneider 1988).

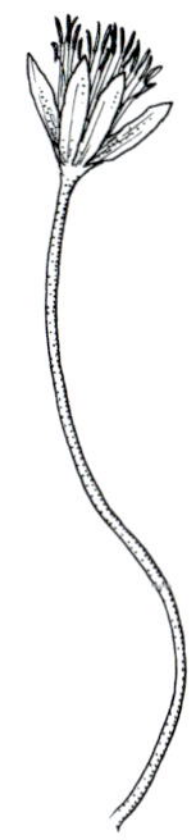
Brasenia schreberi—flower ×1/2

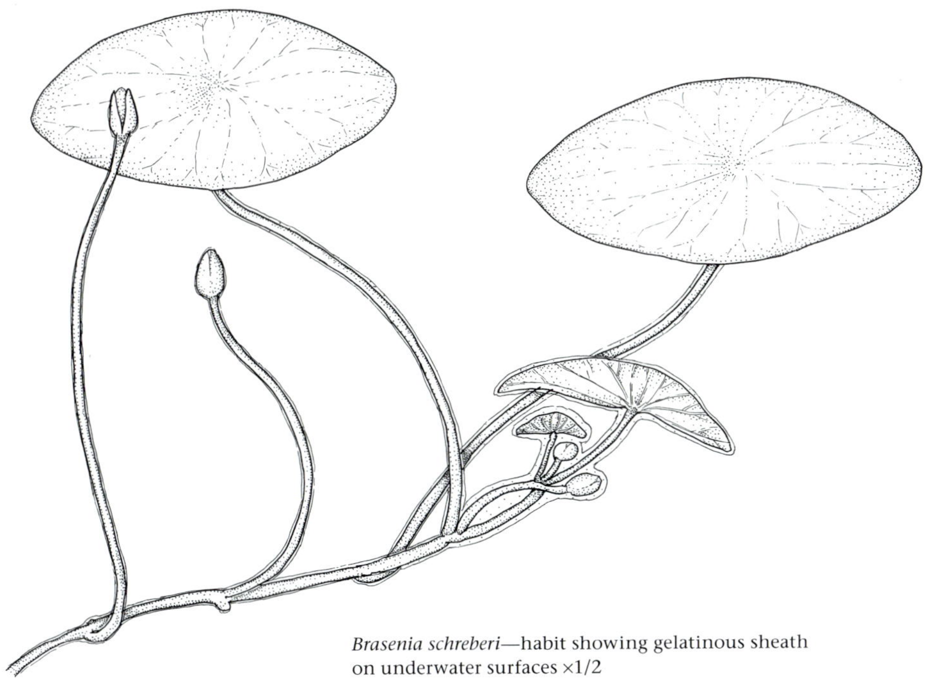
Brasenia schreberi—habit showing gelatinous sheath on underwater surfaces ×1/2

The mucilaginous sheath that is present on underwater surfaces of watershield is thought to prevent the attachment of algae and other aquatic organisms that adhere to most submersed plant surfaces.

Watershield has a worldwide distribution in temperate regions, except Europe, and the high elevation tropics. In the United States it is widespread east of the Mississippi River and also present in the Pacific Northwest.

Native Americans used watershield to treat pulmonary complaints and dysentery. The seeds are eaten by waterfowl.

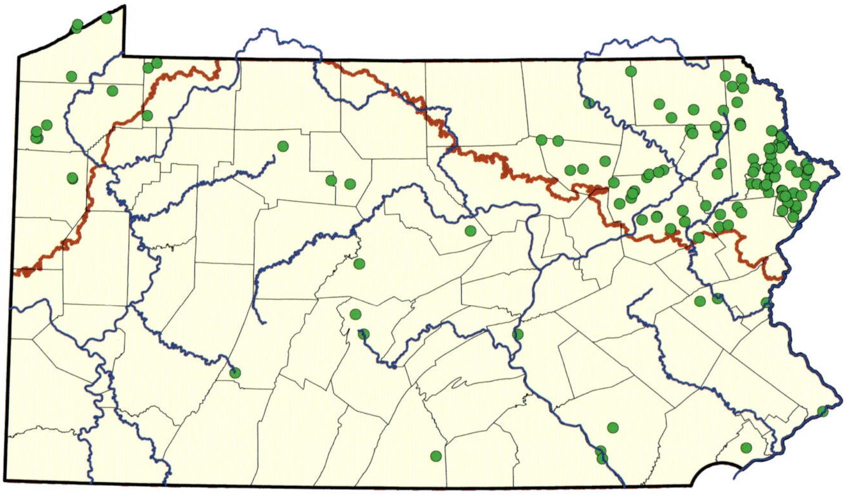

Brasenia schreberi

Brasenia schreberi

Chapter 5. Thalloid and/or Tiny Floating Plants

Plants in this group have a reduced structure lacking well-defined stems and leaves, or have tiny floating stems with minute leaves. They are small plants that float at or just below the water surface, or in a few cases grow attached to submerged rocks or other solid surfaces. Included in this group are flowering plants (the duckweeds and riverweed), and non-flowering plants including aquatic liverworts and a fern.

AQUATIC LIVERWORTS

The aquatic liverworts are small floating plants that may also exist stranded on exposed mud. They are described as thalloid because they lack true roots, stems, or leaves; they also lack specialized vascular tissues. Aquatic liverworts are generally flat and often strap-shaped with dichotomous branching. Rhizoids may be present.

Liverworts do not produce flowers or fruits. They reproduce by spores, but the sporangia are not often seen. Fragmentation is a more common means of reproduction.

Aquatic liverworts are usually found near lake margins, and may be floating among emergent plants.

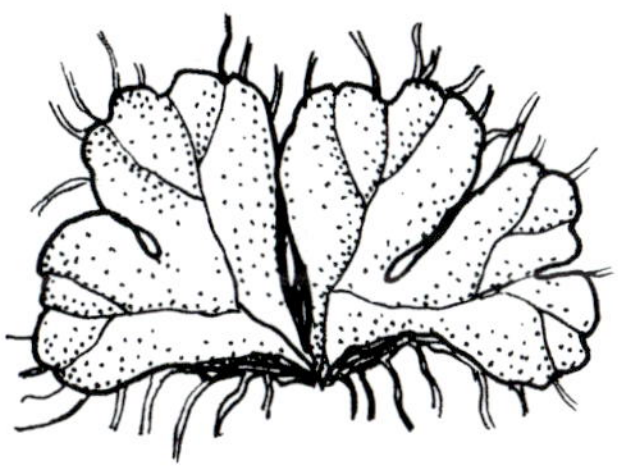

Ricciocarpus natans ×2

We have included descriptions of two floating aquatic liverworts, purple-fringed riccia (*Ricciocarpus natans*) and slender riccia (*Riccia fluitans*). Both are widespread in Pennsylvania (Manville 2001); however, others may also be encountered. Many additional species of liverwort grow on wet rocks or tree bases along the edges of lakes and streams. Additional references should be sought for more specific information on this group of plants.

The plant body of purple-fringed riccia is 2-lobed, often forming a floating fan-shaped thallus about 1 cm wide. Conspicuous dark rhizoids extend downward from the lower surface.

Slender riccia has a strap-like thallus that branches repeatedly in a dichotomous pattern forming a radiating mass 4–8 cm in diameter that floats just below the water surface. Rhizoids are not present.

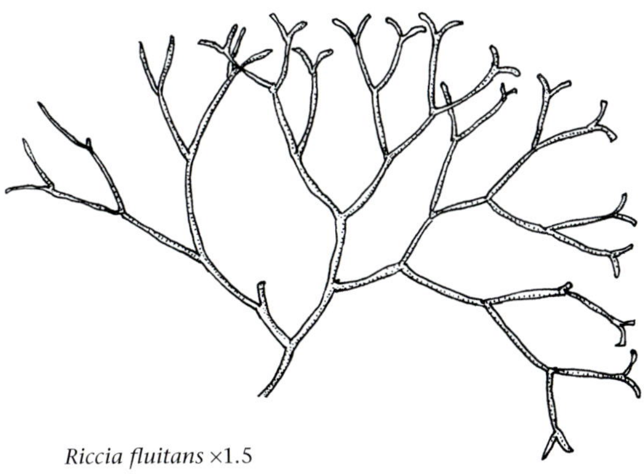

Riccia fluitans ×1.5

CHAROPHYTE ALGAE
MUSK-GRASS (*CHARA* SPP.) AND STONEWORT (*NITELLA* SPP.)

The Charophyte algae (musk-grasses and stoneworts) could be confused with higher plants, due to their upright, branching structure. Close observation will reveal that the "stems" are of a homogeneous consistency and lack vascular tissues. "Stems" are anchored to the substrate and support "branches" arising in whorls along the main axis. The primary stems give rise to additional whorls and a dense tufted appearance in some species. The plants range from light yellow-green to very dark green that can appear almost black.

The Charophyte algae have been grouped with the green algae in the past but are currently considered a sister group to the land plants (McCourt et al. 2004) (Figure 1.1). These plants do not produce flowers or fruits; rather, they reproduce by eggs and sperm borne in special structures located at nodes. The egg- and sperm-producing structures of both the stoneworts and the musk-grasses are pear-shaped, often with a spiral pattern on the surface.

Distinguishing characteristics of the musk-grasses include their stiffness and rough texture, which is due to encrustations of lime, and their garlic-like odor. In addition, the surface of musk-grass "stems" has a longitudinally striated appearance when viewed under 10X magnification (Dillard 1999). Musk-grasses are commonly found in calcium-rich (soft) waters.

Stoneworts, on the other hand, do not have a distinctive odor, they are smooth to the touch, and their cells lack striations. Stoneworts are more often found in soft water (low pH) lakes.

Musk-grasses and stoneworts often form a continuous ground cover at depths of up to 3 m in lakes with clear water. Stands of these plants, which are 0.15–1 m in height, provide cover for small fishes; in addition, they are eaten by waterfowl.

Chara sp.—reproductive structure ×10

Chara sp. ×1

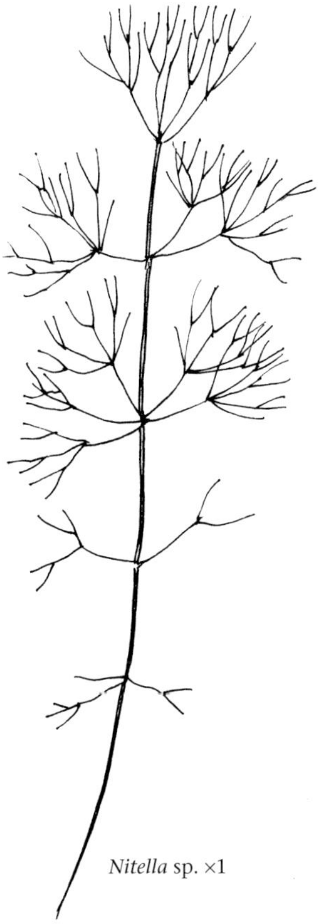

Nitella sp. ×1

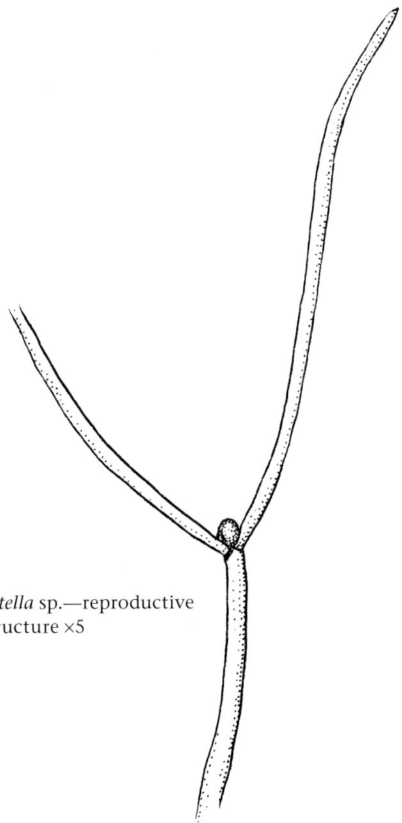

Nitella sp.—reproductive structure ×5

DUCKWEED, WATERMEAL, WATER-FLAXSEED, AND BOGMAT

Lemna L., *Wolffia*, *Wolffiella*, *Spirodela*, and *Landoltia*

Arum Family—Araceae

The duckweeds and their relatives include the smallest flowering plants in the world. They are tiny free-floating plants with a highly simplified growth form that lacks distinct leaves or stems. The plant body consists of more-or-less flattened segments that bear a single unbranched root, a cluster of roots, or no root at all.

Tiny flowers are located in a pouch on the surface of the segment; however, flowering is a very rare occurrence. Most reproduction is asexual, occurring by budding and fragmentation of the segments. Under favorable growing conditions, reproduction can be very rapid, enabling the plants to cover the surface of a small pond within a few weeks.

Key to the Duckweeds and Relatives

A. each plant segment bearing one or more roots
- B. each segment bearing a single root .... *Lemna*
- B. Each segment with a cluster of 2–9 roots
  - C. each segment bearing 5–12 roots; 5–11 conspicuous nerves present .... *Spirodela polyrhiza*
  - C. each segment bearing 2–5 roots; 3–7 obscure nerves present .... *Landoltia punctata*

A. plants lacking roots
- D. segments flat or ribbon-shaped .... *Wolffiella gladiata*
- D. segments globular or boat-shaped .... *Wolffia*

Figure 5.1. Mixed population of water flaxseed (*Spirodela polyrhiza*), duckweed (*Lemna minor*), and watermeal (*Wolffia* sp.) in a pond in Montgomery County, Pennsylvania.

Eastern water-flaxseed

***Landoltia punctata* (G.Meyer) Les & Crawford [synonym: *Spirodela punctata* (G.Meyer) C.H.Thomps.]**

Introduced
Free-floating perennial

Eastern water-flaxseed is very similar to water-flaxseed (*Spirodela polyrhiza*), but smaller; the segments are 2–5.5 mm long, with 2–7 nerves and up to seven roots.

Eastern water-flaxseed is a widespread plant of warm-temperate to subtropical regions, native to South America, Asia, and Africa. In North America it occurs on the Atlantic and Gulf Coastal Plains from New Jersey and Pennsylvania to Florida and Texas and at scattered locations along the Pacific Coast. It is known from only a few scattered locations in Pennsylvania.

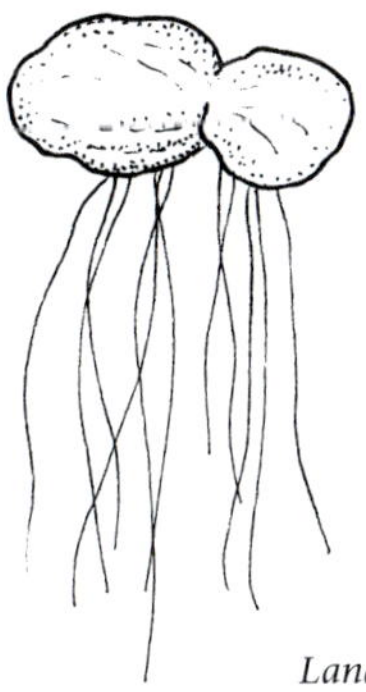

Landoltia punctata ×5

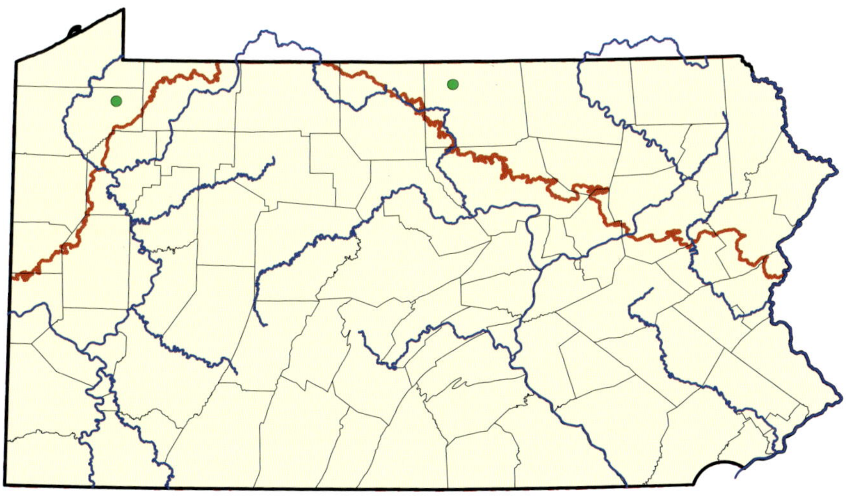

Landoltia punctata

Duckweed *Lemna* L.

In most species of *Lemna* each segment has a single root extending from the lower surface. Segments usually occur in clusters of 2–5. Faint veins are visible on the upper surface.

Duckweeds are considered perennial because either the plants form vegetative over-wintering structures called turions, or the segments themselves sink to the bottom during the winter and resume active growth in the spring. The duckweeds are plants of quiet water of lakes, ponds, and streams; on larger lakes these tiny plants are often found floating among the emergent aquatic vegetation at the lake edge. Duckweeds often grow in mixed populations with species of bogmat, water-meal, and water-flaxseed and are frequently associated with eutrophic conditions.

Duckweeds are consumed by fish, ducks, and other marsh and shore birds and are inadvertently transported from site to site when they adhere to birds' feet or feathers. Some species are abundant enough that they have been harvested for livestock food.

In addition to the five species described below, little water duckweed (*Lemna obscura*) is known to have grown in the southeastern corner of the state in the past, but is believed to be extirpated. It is similar to common duckweed but smaller; in addition, the lower surface the segment is often reddish. It was last collected in Pennypack Park in Philadelphia in 1927.

Key to the Species of *Lemna*

A. segments remaining attached to the parent by a stalk .... *Lemna trisulca*
A. segments without a stalk, forming small colonies but eventually separating completely from the parent plant
  B. segments asymmetrical, somewhat curved
    C. root sheath winged at the base .... *Lemna perpusilla*
    C. root sheath not winged .... *Lemna valdiviana*
  B. segments rotund to elliptical, symmetrical
    D. segments with several equal pimples along the midline on the upper surface .... *Lemna turionifera*
    D. segments with 1 larger pimple (and often several smaller ones), or none
      E. pimple prominent; segments often red underneath....... *Lemna obscura*
      E. pimple not very prominent; segments never red underneath............... .... *Lemna minor*

Common duckweed *Lemna minor* L.

Native
Free-floating perennial

Lemna minor is the most common of the duckweeds; it grows most vigorously in high nutrient (eutrophic) waters and can entirely cover the surface of small ponds by mid-summer. On larger lakes it may concentrate in quiet coves or among emergent vegetation along the shoreline.

Each plant is composed of one or more segments each 1–8 mm long, flat, and rounded to ovate. Each segment has 3–5 faint veins that converge at the point of attachment of the root. Duckweed typically occurs as small floating clusters of 2–5 attached segments.

Duckweed has a worldwide distribution in the north temperate region. In North America it occurs across the United States with the exception of the southern tier of states; to the north duckweed reaches southern Canada. It also occurs in Eurasia. It is found throughout Pennsylvania.

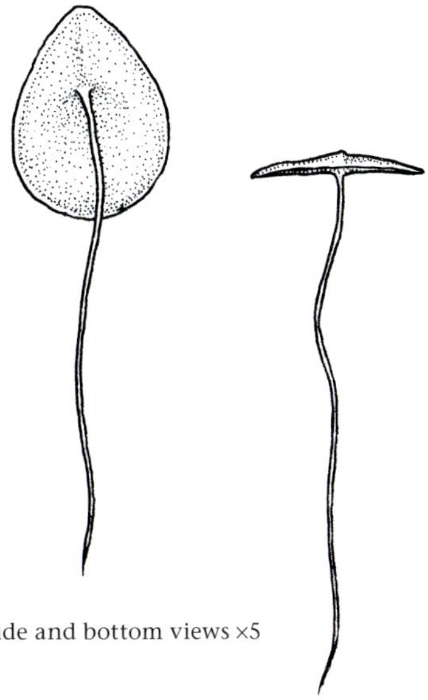

Lemna minor—side and bottom views ×5

Lemna minor ×5

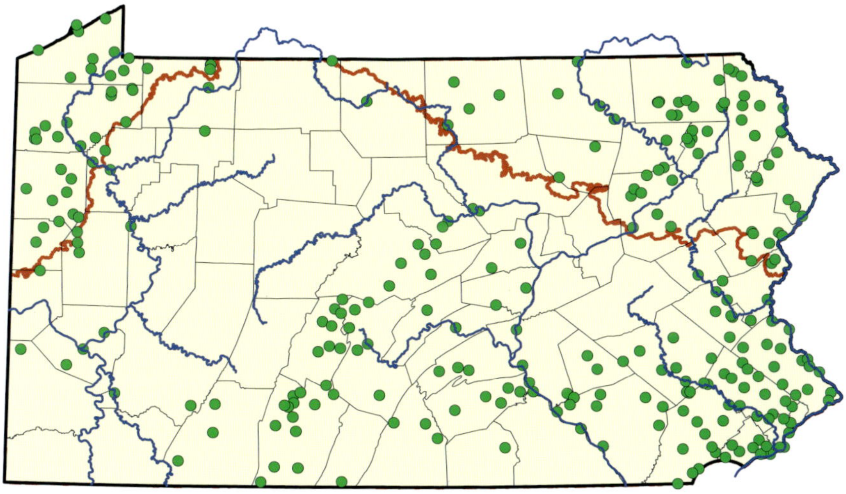

Lemna minor

Minute duckweed — *Lemna perpusilla* Torrey

PNHP
Native
Free-floating perennial

The segments of minute duckweed are approximately 1–4 mm long and asymmetrical or slightly pear-shaped. Each segment has 3 faint veins; a prominent raised papilla is present at the point of convergence of the veins. In addition, smaller papillae occur along the midline. The segments occur in clusters of 2–5; the root is up to 3.5 cm long and has a distinctive winged sheath.

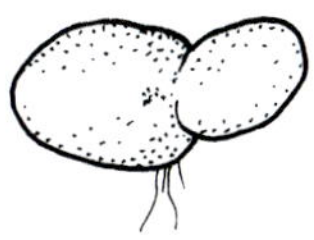

Lemna perpusilla ×5

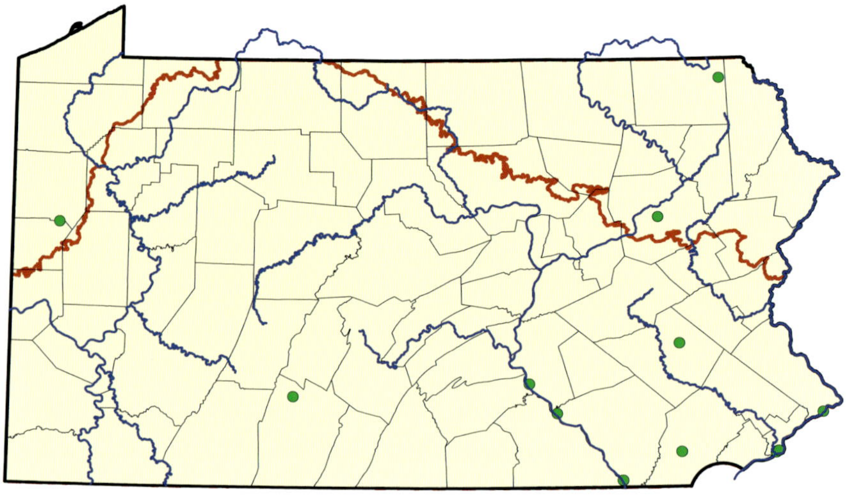

Lemna perpusilla

The range of minute duckweed extends from New York to North Carolina and inland to Nebraska and Oklahoma. In Pennsylvania it is known from a few widely scattered locations.

Star duckweed *Lemna trisulca* L.

Native
Suspended perennial

Star duckweed is distinctive among the duckweeds. It consists of oval or oblong segments that remained attached by a stalk; the result is a tangled colony of interconnected yellowish-green plants that floats just below the water surface. Unlike other species in the genus, star duckweed does not always have roots.

Star duckweed has a worldwide range. In Pennsylvania it is known from scattered sites.

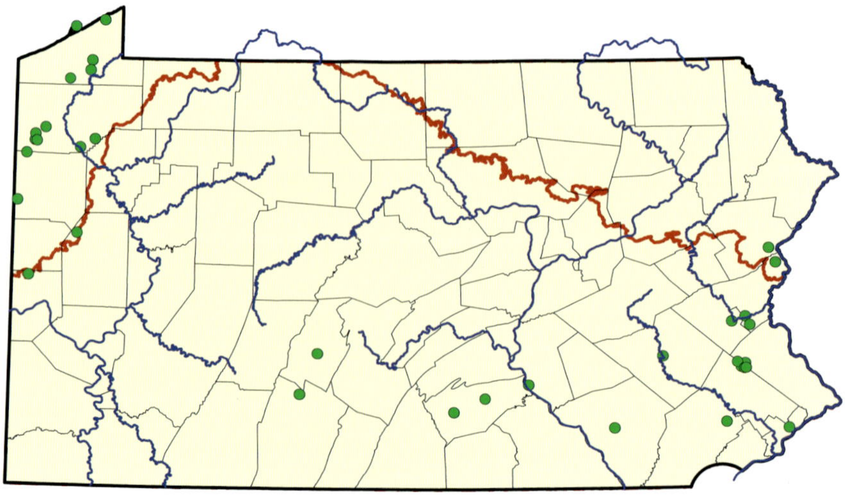

Lemna trisulca

In addition to serving as food for waterfowl and aquatic animals, star duckweed was used by Native Americans to make wet poultices to alleviate swelling.

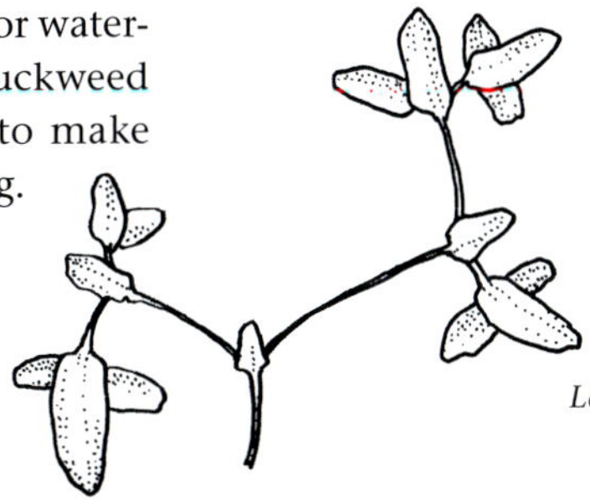

Lemna trisulca ×1

Winter duckweed *Lemna turionifera* Landolt

PNHP
Native
Free-floating perennial

Segments of winter duckweed are 1–4 mm long with several equal papillae (raised pimples) in a row along the midline; the segments occur in floating clusters of 2–5. The lower surface of each bears a single root. The name of this species refers to overwintering turions that are formed; they are similar in appearance to normal segments but lack a root.

Winter duckweed has a worldwide distribution in the northern hemisphere; in North America it occurs from Newfoundland to Alaska and south to New Jersey, California, and Mexico; it also grows in Eurasia. A rare plant in Pennsylvania, winter duckweed is known from a handful of sites in the western part of the state and a few scattered locations elsewhere.

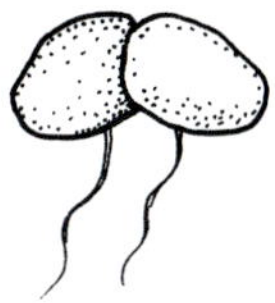

Lemna turionifera ×5

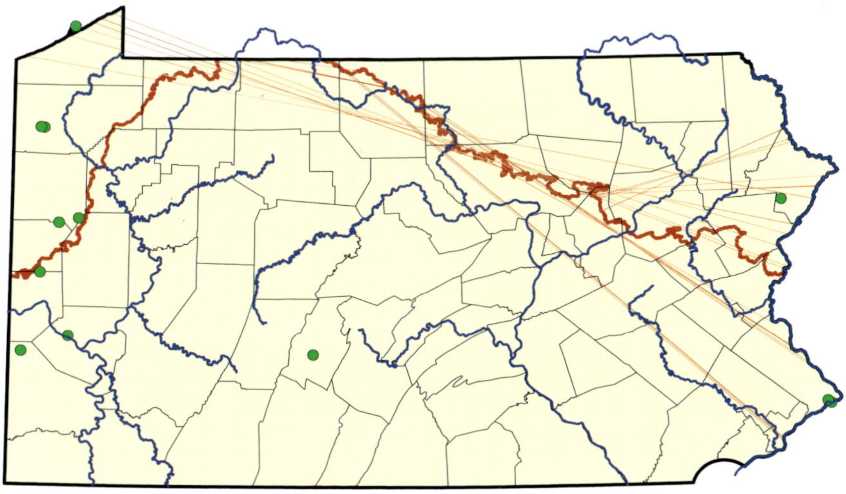

Lemna turionifera

Pale duckweed *Lemna valdiviana* Phil.

PNHP
Native
Free-floating perennial

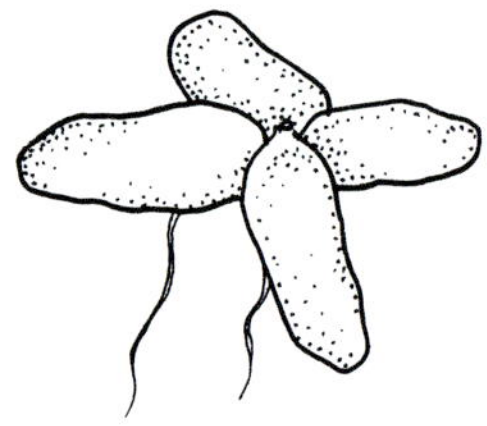

Lemna valdiviana ×5

Pale duckweed has narrowly oblong segments 2.5–5 mm long with 1 faint nerve or nerveless; clusters of 2–10 segments float at the water surface. Each segment has a single root; unlike the otherwise similar *L. perpusilla*, there is no root sheath. Turions are not produced. Pale duckweed occurs floating in shallow water or lodged on submerged debris.

Pale duckweed is widely distributed across the southern United States and south to Central and South America. In Pennsylvania it is a rare plant known historically from only a few scattered locations.

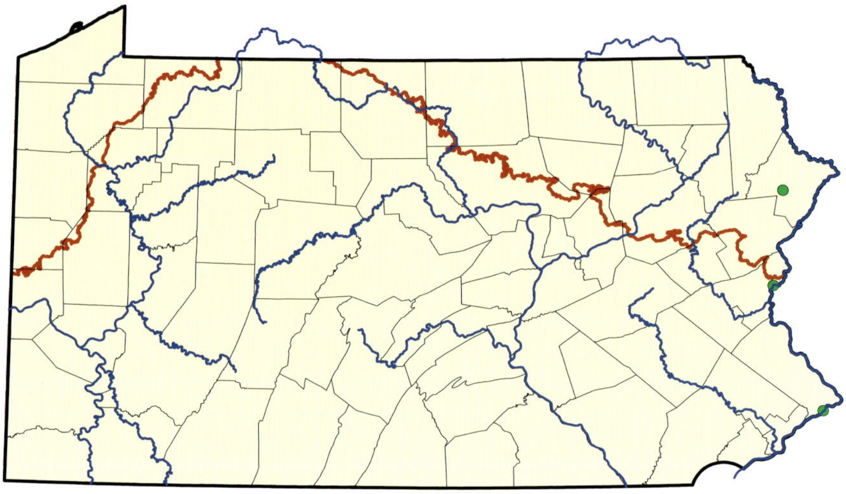

Lemna valdiviana

Water-flaxseed, greater duckweed *Spirodela polyrhiza* Schleiden

Native
Free-floating perennial

Water-flaxseed is the largest of the duckweed group at 2–10 mm long. Each segment has a cluster of simple, unbranched roots dangling from the lower surface. A radiating group of 5–15 purplish veins is visible on the upper surface. The undersurface is also purple. In the autumn, water-flaxseed forms turions, overwintering buds that sink to the bottom until they are stimulated to resume active growth in the spring.

Water-flaxseed is a plant of still water, especially in lakes, ponds, marshes, and slow-moving rivers and streams. It occurs in temperate and tropical regions worldwide; in North America it is found throughout southern Canada and the United States except for the desert southwest. It occurs throughout Pennsylvania, often in mixed populations with other members of the duckweed group.

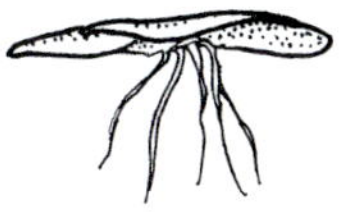

Spirodela polyrhiza—side view ×3

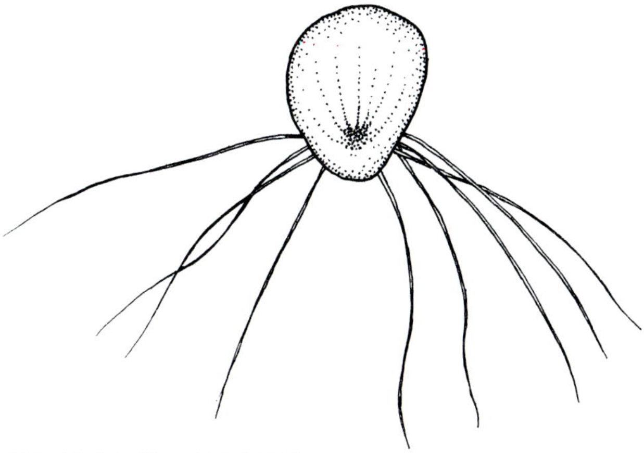

Spirodela polyrhiza—top view ×5

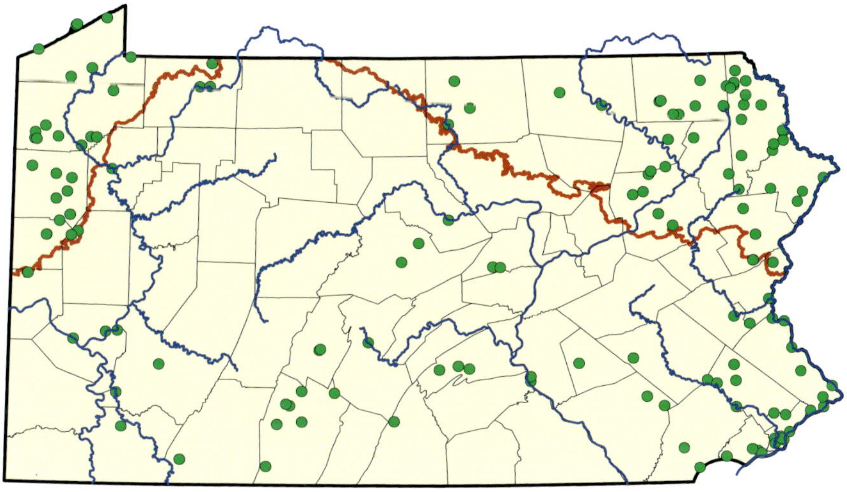

Spirodela polyrhiza

Watermeal ***Wolffia Horkel* ex Schleiden**

The watermeals are tiny, the smallest of the flowering plants; they consist of an oval or spherical segment no more than 1.6 mm long that floats partly below the water surface. No roots are produced. New segments bud off and eventually separate to form a new plant. Flowering is rare; most reproduction is by vegetative budding. Under favorable conditions, growth is so rapid that watermeal can completely cover the surface of a small pond in only a few weeks. Turions are formed late in the season and serve as the overwintering form of the plant.

Like the duckweeds, watermeal is most often found on still water of ponds, lakes, and ditches. It is eaten by fish, ducks, and waterfowl and, because of its small size, it is easily carried on the external surfaces of waterfowl or other animals either in mud or merely adhering to wet fur or feathers. Watermeal is also transported in flowing water.

Three species of watermeal are found in Pennsylvania. Magnification is needed to see the distinguishing features. It is not unusual to find 2 or all 3 species growing together in mixed populations.

Key to the Species of Watermeal

A. segment spherical .... *Wolffia columbiana*
A. segment oval
 B. segment with a raised pimple at one end .... *Wolffia borealis*
 B. segment with a central pimple .... *Wolffia brasiliensis*

Dotted watermeal — *Wolffia borealis* (Engelmann) Landolt

Native
Free-floating perennial

Dotted watermeal has oval segments that are 0.7–1.5 mm long with a raised pimple at one end. It occurs throughout the temperate regions of the world. Dotted watermeal is known from a few widely scattered sites in Pennsylvania but is probably more common than is currently recognized.

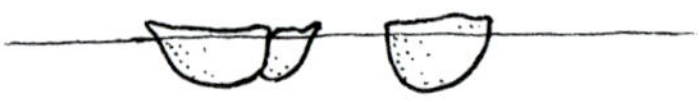

Wolffia borealis—side view ×8

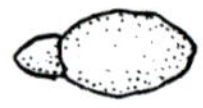

Wolffia borealis—top view ×8

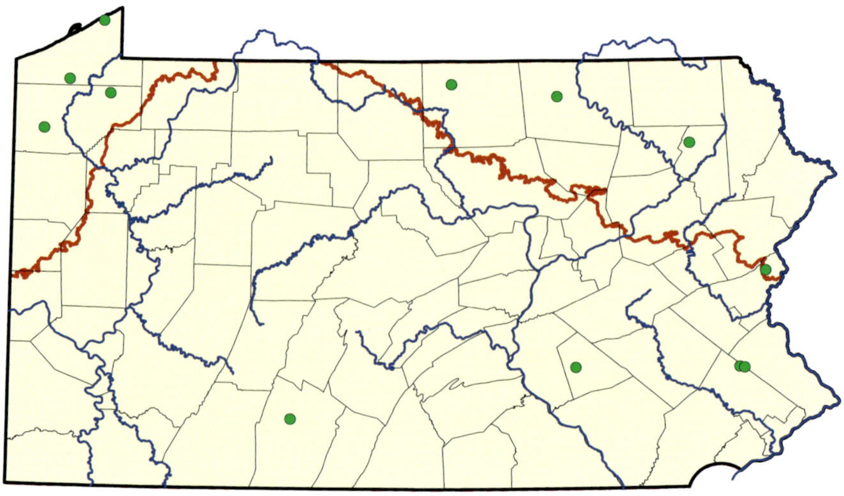

Wolffia borealis

Pointed watermeal *Wolffia brasiliensis* Weddell

Native
Free-floating perennial

Wolffia brasiliensis is oval in shape, 0.5–1.6 mm long and has a raised pimple in the center of each segment. Pointed watermeal occurs throughout the temperate to subtropical regions of the earth. Common on lakes and ponds scattered across Pennsylvania, it is the most frequently encountered of our *Wolffia* species.

Wolffia brasiliensis—top view ×10

Wolffia brasiliensis—side view ×10

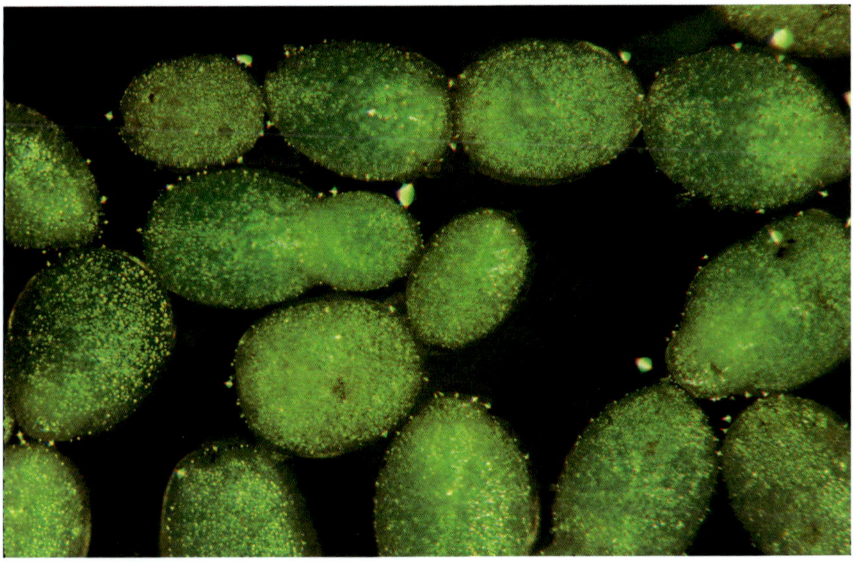

Wolffia brasiliensis

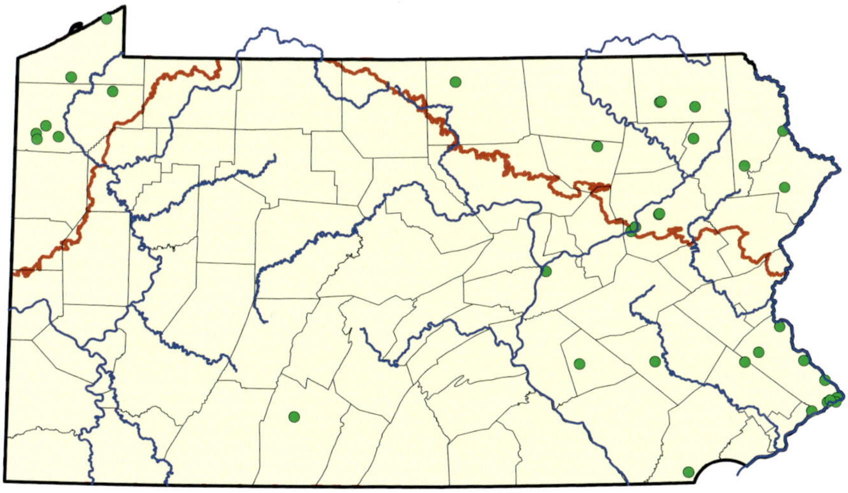

Wolffia brasiliensis

Watermeal *Wolffia columbiana* H. Karsten

Native
Free-floating perennial

This species is spherical and 0.5–1.4 mm in diameter; it floats with only the upper third above water. *Wolffia columbiana* occurs throughout the temperate to subtropical regions of the earth. In Pennsylvania it is scattered throughout the state and usually occurs in mixed populations with one or both of the other *Wolffia* species.

Wolffia columbiana—side view ×5

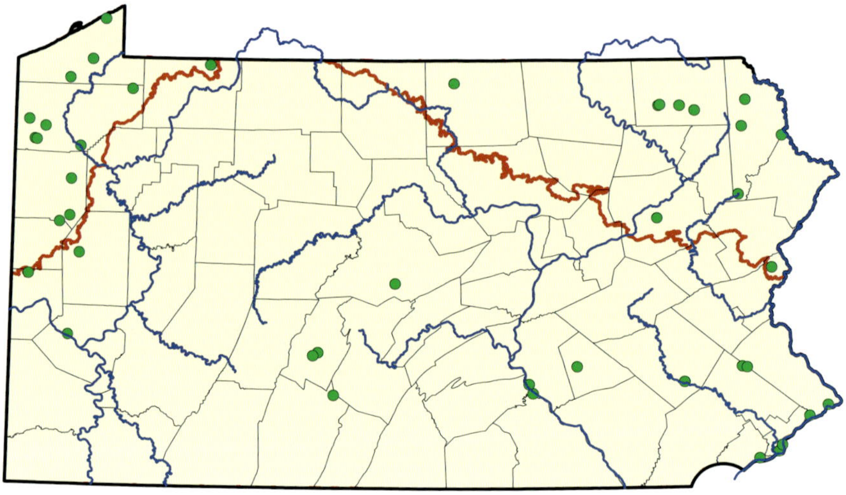

Wolffia columbiana

Bogmat

Wolffiella gladiata (Hegelmaier) Hegelmaier

PNHP
Native
Suspended perennial

Bogmat is a tiny floating plant that lacks well-defined roots, stems, or leaves. The 6–8 mm-diameter colonies consist of tangled, star-shaped clusters of elongate, strap-shaped segments with pointed tips. They form a tangled mass that floats at or just below the water surface. Reproduction is almost entirely by budding; flowering is rare.

A plant of quiet or stagnant waters, bogmat occurs from Pennsylvania south to Florida and Texas along the coast then north in the Mississippi Valley to Illinois and Ohio. In Pennsylvania it is a rare plant known only from the northwestern corner of the state.

Like the duckweeds and watermeals, bogmat is eaten by fish, ducks, and waterfowl.

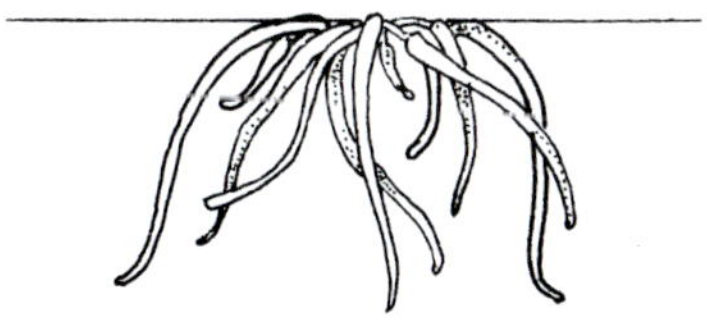

Wolffia gladiata—side view ×2

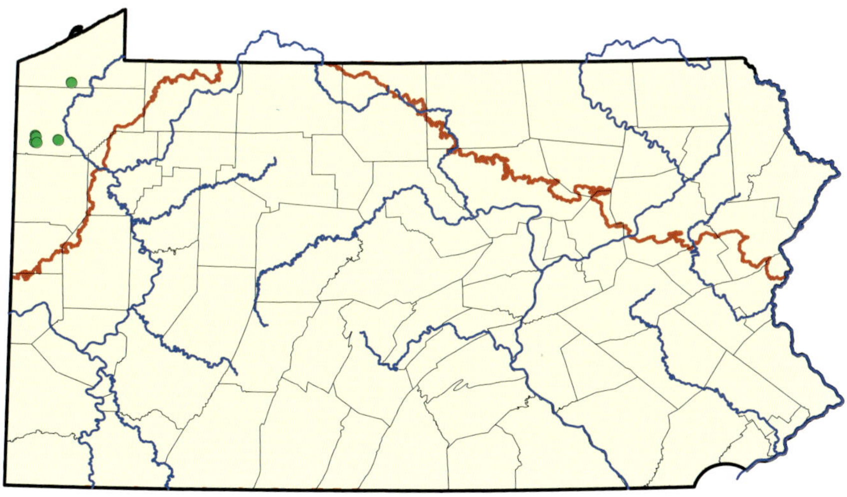

Wolffia gladiata

EASTERN MOSQUITO FERN

Azolla Lam.

Mosquito Fern Family—Azollaceae

Eastern mosquito fern *Azolla caroliniana* Willd.

Introduced, or perhaps undergoing a natural range expansion
Free-floating perennial

Eastern mosquito fern is a tiny, dark green, leafy fern that is free-floating or sometimes grows on exposed mud. Each plant is only 1–2 cm in diameter; they remain green all year, often taking on a reddish tinge in the winter. Spore-bearing sori are produced only rarely.

Found in still water of ponds and canals, eastern mosquito fern has only recently been documented at a few locations in eastern Pennsylvania. Its native range extends throughout eastern and central North America. It is also grown in aquaria and water gardens and may have escaped from cultivation at some sites.

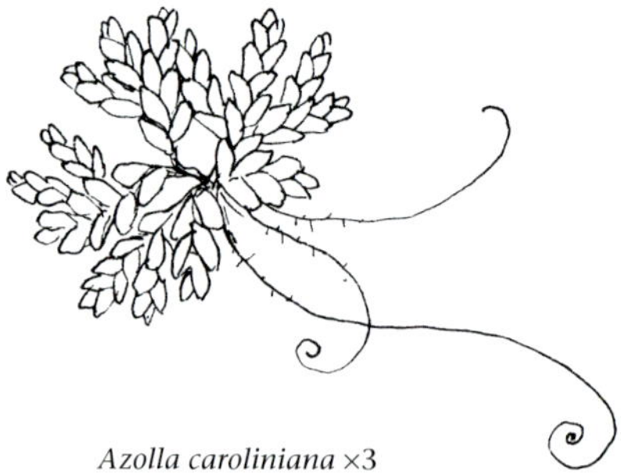

Azolla caroliniana ×3

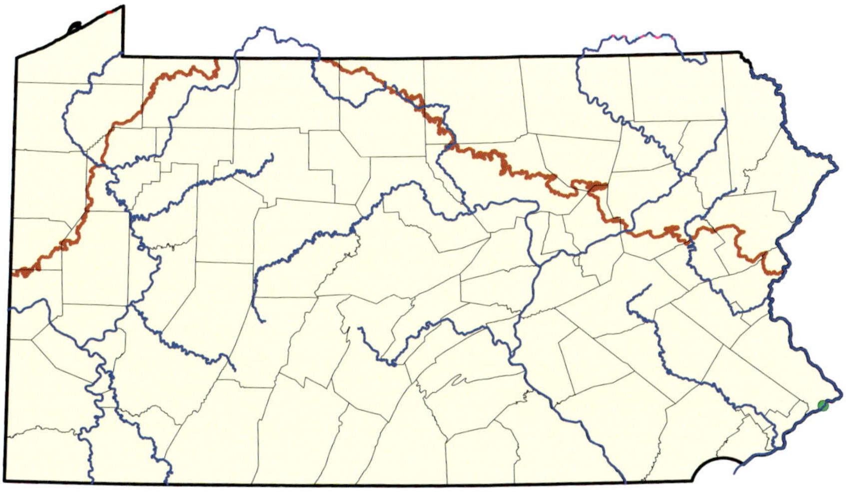

Azolla caroliniana

Azolla caroliniana

RIVERWEED
Podostemum Michx.
Riverweed Family—Podostemaceae

Riverweed ***Podostemum ceratophyllum*** **Michx.**

Native
Attached, submergent perennial

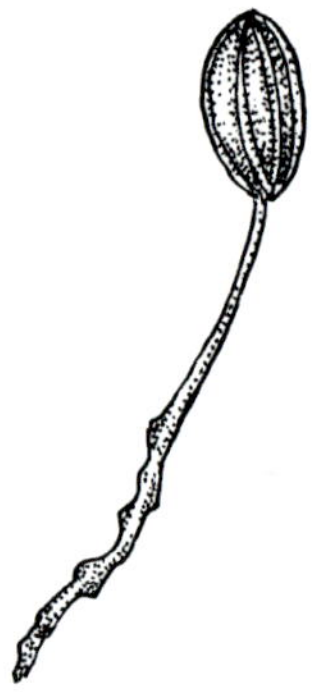

Podostemum ceratophyllum—seed capsule ×4

Riverweed is a very reduced, almost alga-like plant that grows attached to rocks in fast-moving water. It does not have recognizable leaves or roots; its leathery, flattened, branching stems adhere to rocks by means of fleshy disks. Despite its appearance, riverweed is a flowering plant. The flowers are perfect with a superior ovary and 2 stamens; they lack petals and are water-pollinated. The fruit is a small capsule.

Riverweed occurs throughout Pennsylvania, sometimes extending for a considerable distance along a stream or river. Its range extends from eastern Canada to Georgia and west to Oklahoma. It also grows in the West Indies and Central America.

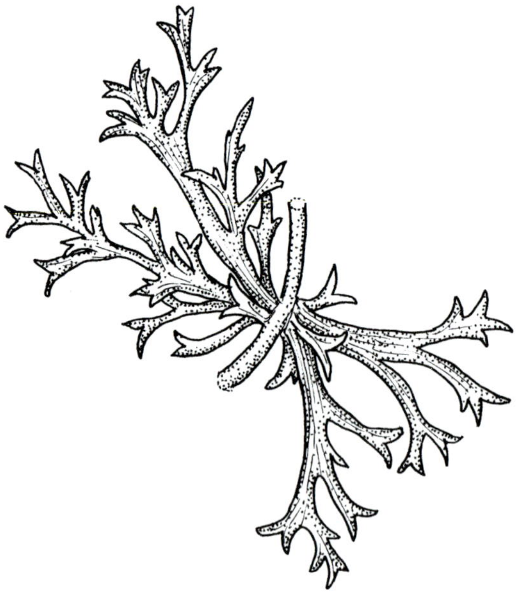

Podostemum ceratophyllum ×2

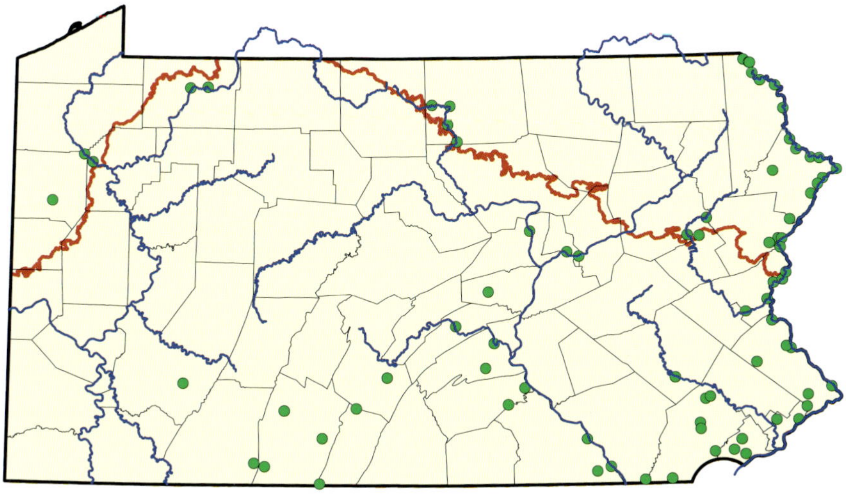

Podostemum ceratophyllum

Podostemum ceratophyllum

Chapter 6. Submergent Leafy-stemmed Plants

This group of species grows rooted in the lake bottom or attached to rocks or other solid underwater surfaces. Their stems bear numerous leaves, most of which are beneath the water surface, but some at the stem tip may be floating. The group includes aquatic mosses and flowering plants. Flowering and fruiting occur underwater, at the water surface, or on short emergent stem tips.

AQUATIC MOSSES

In our work in lakes, ponds, and streams we have observed several aquatic mosses. They are included here because they might be confused with vascular plants on first glance. However, mosses differ from vascular plants in that they do not have specialized water conducting tissues. They are differentiated into stems and tiny leaves; however, close examination will reveal that although the leaves may have a midrib, there is no network of veins.

Reproduction is by spores; flowers and fruits are not formed. Fragmentation is another way the mosses can spread or disperse.

The three genera discussed here represent the aquatic mosses we have encountered most frequently; however, others may also be present. Additional references should be consulted for more specific information on this group of plants.

Water moss (*Fontinalis sullivantii*) grows attached to rocks or logs below the water surface, its long leafy stems trailing with the water current. The leaves, which are arranged in 3 ranks along the stems, lack a midrib. Water moss is dark green and appears almost black when viewed from above. It is frequently present in lakes and can be quite abundant, especially in low nutrient glacial lakes.

Amblystegium tenax is similar to water moss in general habit, but the leaves have a midrib. It grows attached to rocks in flowing water of rivers and streams.

Sphagnum sp. ×1

Sphagnum, the common moss of bogs, also has aquatic forms. Sphagnum moss is light green and tufted. The cellular structure of sphagnums absorbs large amounts of water which can easily be squeezed out. Aquatic sphagnum moss may be present in shallow water along the edges of glacial lakes with low pH and low nutrient status.

Fontinalis sullivantii ×1

Fontinalis sullivantii—leaves ×2, note lack of midrib

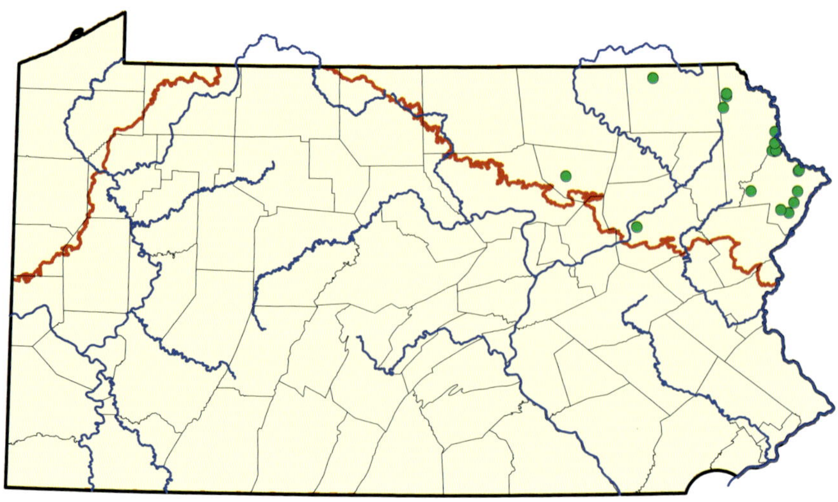

Fontinalis sullivantii

HYDRILLA

Frog's-bit Family—Hydrocharitaceae

Hydrilla *Hydrilla verticillata* (L.f.) Royle

Introduced
Rooted, submergent perennial

Hydrilla is very similar to waterweed (*Elodea canadensis* and *E. nuttallii*). The best characters for distinguishing it are the distinctly toothed leaf margin (visible without magnification) and the presence of four or more leaves per whorl. In addition, whitish tubers are produced on lower portions of the stems buried in the substrate. Late in the season dark green winter buds (turions) may be found in the leaf axils. Hydrilla grows rapidly and spreads via stem fragments that become rooted. Flowering and seed production are rare.

An introduced plant, Hydrilla is native to Eurasia and perhaps elsewhere. It was brought to North America as an aquarium plant. In southeastern Pennsylvania it is abundant in the Delaware and Schuylkill Rivers and at scattered sites elsewhere.

Hydrilla is designated as a federal noxious weed because of its rapid growth and ability to clog waterways and displace native aquatic vegetation.

Hydrilla verticillata ×1/2—note stolon and tuber

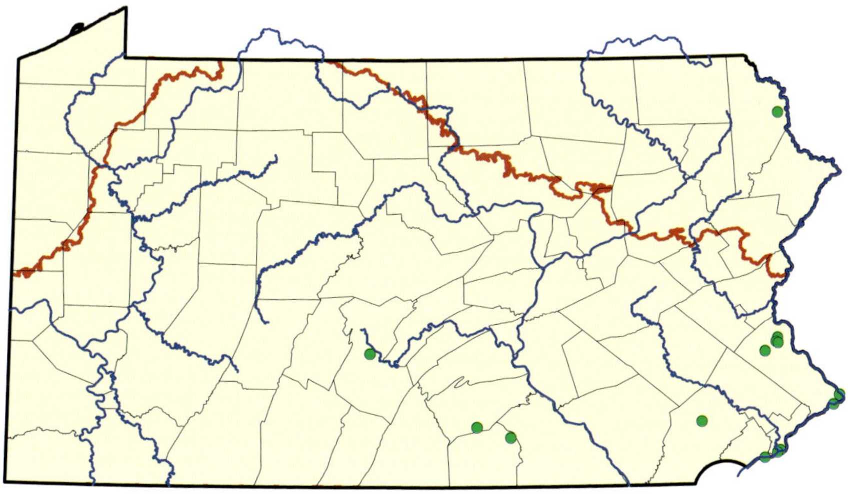

Hydrilla verticillata

PONDWEEDS

POTAMOGETON L. AND *ZANNICHELLIA* L.

Pondweed Family—Potamogetonaceae

The pondweeds all belong to the Pondweed Family (Potamogetonaceae), our largest group of aquatic plants. They are rooted, mostly submergent aquatics, although some species also have floating leaves. When present, floating leaves are broader and firmer than the underwater foliage. Stipules that form a sheath around the stem are a prominent feature; in some pondweed species the stipular sheath is fused to the base of the leaf for part of its length with only the tips free, in others the stipular sheath is completely free from the leaf. Flowers are borne in terminal or axillary spikes, which may extend above the water surface, or be completely submersed. Pollination is by wind or water. Thirty species of pondweeds, segregated in 2 genera (*Potamogeton* and *Zannichellia*), are present in Pennsylvania.

Key to the Genera of Pondweeds

A. submersed leaves opposite, fruits 2–3 times longer than wide and distinctly curved .... *Zannichellia palustris*

A. submersed leaves alternate; fruits rounded to obovate, less than 2 times as long as wide .... *Potamogeton*

***Potamogeton* L.**

The pondweeds in the genus *Potamogeton* have leaves of two types. Submersed leaves are alternate, parallel-veined, usually limp or flaccid (except for a few species), and often have lacunae (lighter colored bands) along each side of the midvein. Many species also have floating leaves, which are opposite, firmer, and elliptical. Tubers and/or turions (winter buds) are formed by some species. Turions are borne at stem tips or in leaf axils; they consist of a condensed stem with reduced leaves and often have a thickened or hardened texture. Flowers are perfect and consist of 4 tepals, 4 stamens, and 4 superior ovaries; the fruit is a cluster of achenes.

Potamogeton is our largest genus of aquatic plants; 29 species are known in Pennsylvania. All are native with the exception of curly pondweed. Fourteen species are classified as endangered, threatened, or rare by the Pennsylvania Natural Heritage Program (PNHP 2010).

While the pondweeds as a group are distinctive, they can be difficult to identify to species if mature fruits are not present. Accurate interpretation of stipule characteristics, leaf apex shapes, and whether or not glands are present at the nodes are all important. The size of fruits and whether or not keels are present are also critical characteristics. As with many aquatic plants, stem length and leaf characteristics, including whether floating leaves are present, vary with water depth. Identification of pondweeds is further complicated by their propensity for hybridization (Kaplan et al. 2009).

Native Americans used the dried stem fibers of pondweeds to make nets and cords. Foliage, fruits, and tubers are important foods for waterfowl and other wildlife.

Key to the Species of *Potamogeton*

A. stipular sheaths fused to the bases of the submersed leaves with the tips extending as a ligule
  B. stipular sheath fused to the base of the leaf blade for ½ or more of its length
    C. base of stipular sheath inflated; achenes not beaked .... *Potamogeton filiformis*
    C. base of stipular sheath not inflated; fruits beaked .... *Potamogeton pectinatus*
  B. stipular sheath fused to the base of the leaf blade for ½ or less of its length
    D. leaves stiff, conspicuously 2-ranked, auriculate at the base; leaf margins finely toothed .... *Potamogeton robbinsii*
    D. leaves not conspicuously 2-ranked, not auriculate; margins smooth
      E. submersed leaves obtuse to acute, floating leaves with rounded tips
        F. free portion of the stipular sheath much shorter than the fused part; achenes mostly greater than 1.5 mm, with a single dorsal wing and smooth sides .... *Potamogeton spirillus*
        F. free portion of the stipular sheath longer than the fused part; achenes mostly less than 1.5 mm, with a prominent dorsal wing and often a suggestion of lateral wings or bumps .... *Potamogeton diversifolius*
      E. submersed leaves acute to acuminate; floating leaves with pointed tips
        G. submersed leaves 0.2–1 (–2) mm wide, with obvious lacunae; floating leaves with 9–23 veins........................ *Potamogeton tennesseensis*
        G. submersed leaves 0.1–0.4 (–0.6) mm wide, without obvious lacunae; floating leaves with 3–7 veins........................ *Potamogeton bicupulatus*
A. stipular sheaths of the submersed leaves completely free from the leaf bases
  H. leaves distinctly wavy-margined and finely toothed........................ *Potamogeton crispus*
  H. leaves not wavy-margined or toothed
    I. submersed leaf blades ovate, lanceolate, or elliptic
      J. submersed leaves with a clasping base, floating leaves not produced
        K. stipules deciduous, leaves ovate to orbicular .... *Potamogeton perfoliatus*

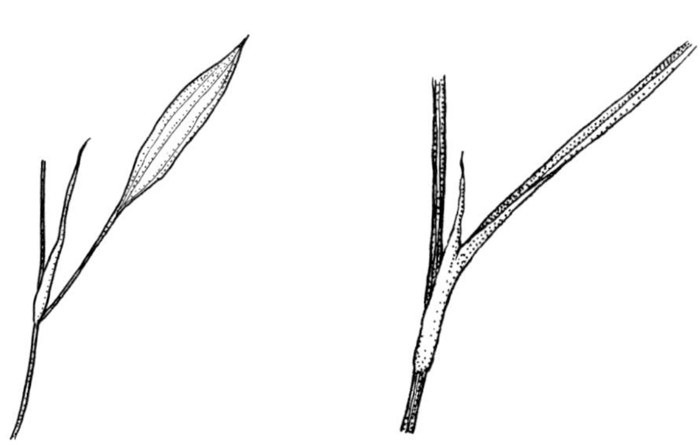

Figure 6.1. Stipular sheath of pondweeds (*Potamogeton* spp.): free (left), fused to the petiole base for part of its length (right). (Not to scale.)

K. stipules persistent at least as shredded fibers, leaves ovate-lanceolate to lanceolate
  L. leaf tip hooded .... *Potamogeton praelongus*
  L. leaf tip flat .... *Potamogeton richardsonii*

J. submersed leaves not clasping, floating leaves present or absent
  M. submersed leaves folded and curved along the midrib, 23–37-veined .... *Potamogeton amplifolius*
  M. submersed leaves not folded and curved, veins 5–19
    N. stems black-spotted .... *Potamogeton pulcher*
    N. stems not black-spotted
      O. submersed leaves sessile
        P. stipular sheath 4–10 cm long; submersed leaves with 7–19 veins and an acute-mucronate tip............................ .... *Potamogeton illinoensis*
        P. stipular sheath 1–3 cm long, submersed leaves with 3–9 veins and an acuminate tip .... *Potamogeton gramineus*
      O. submersed leaves with a definite petiole
        Q. petioles 2–13 cm long; larger submersed leaves acute, but without a sharp awl-like tip; fruits reddish-brown .... *Potamogeton nodosus*
        Q. petioles less than 4 cm; larger submersed leaves acuminate and often mucronate; fruits greenish .... *Potamogeton illinoensis*

I. submersed leaves ribbon-like or thread-like with parallel sides
  R. floating leaves usually present
    S. blades of floating leaves 0.6–1.5 cm long; submersed leaves 1(3)-veined.. *Potamogeton vaseyi*
    S. blades of floating leaves more than 2 cm long
      T. submersed leaves 2–10 mm wide with prominent lacunar bands .... *Potamogeton epihydrus*
      T. submersed leaves 0.3–2 mm wide with 3–5 obscure veins, lacunae present or absent
        U. petiole of floating leaves with a 1–2 cm-long zone at the apex that is lighter colored and more flexible than the rest; base of floating leaves cordate or rounded .... *Potamogeton natans*
        U. petiole of floating leaves continuous in color and texture; base of floating leaves rounded to tapering .... *Potamogeton oakesianus*
  R. floating leaves not produced
    V. submersed leaves very slender and thread-like, 0.1–0.3 mm wide with a single vein; peduncles 10–25 cm .... *Potamogeton confervoides*

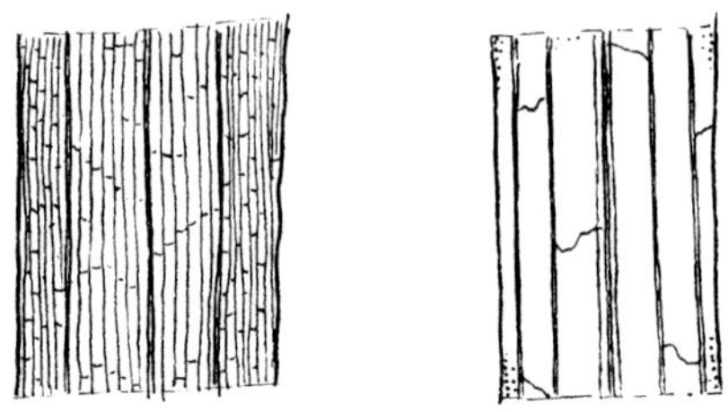

Figure 6.2. Leaf sections of pondweeds (*Potamogeton* spp.) showing a wide lacunar band (left) and a narrow lacunar band (right). (Not to scale.)

V. leaves not thread-like, 0.1–5 mm wide with more than 1 vein; peduncles shorter
  W. stems conspicuously flattened............*Potamogeton zosteriformis*
  W. stems terete or only slightly flattened
    X. leaves bristle-tipped
      Y. leaves firm and rigid; nodal glands present; fruits not keeled..*Potamogeton strictifolius*
      Y. leaves lax, not holding their shape out of water; nodal glands rarely present; fruits with 3 keels ....*Potamogeton hillii*
    X. leaves not bristle-tipped
      Z. fruits keeled
        AA. keel undulate ....*Potamogeton foliosus*
        AA. keel not undulate................*Potamogeton obtusifolius*
      Z. fruits not keeled
        BB. leaf tip apiculate to acute; stipules white and fibrous, 0.5–2 cm long.....................*Potamogeton friesii*
        BB. leaf tip acute to obtuse; stipules brown to green or whitish, less than 1 cm long......*Potamogeton pusillus*

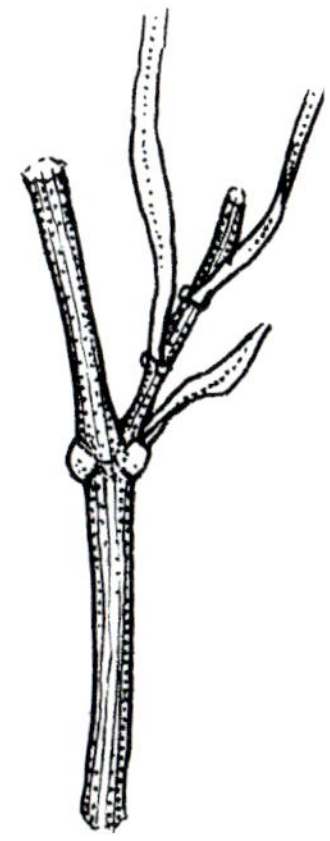

Figure 6.3. Pondweed (*Potamogeton* sp.) stem with glands at the nodes (not to scale).

Figure 6.4. Pondweed (*Potamogeton* spp.) achenes: beaked and with a rounded edge (left), beaked with a distinct keel (center), keeled but without a beak (right). (Not to scale.)

Bigleaf pondweed — *Potamogeton amplifolius* Tuck.

Native

Rooted, submergent and floating-leaf perennial

Bigleaf pondweed is our largest pondweed; its mostly unbranched stems, which grow from rhizomes rooted in the lake bottom, may be 2–3 m or more in length and grow in water up to 3 m deep. Both underwater and floating leaves are produced. The large submersed leaves are 8–20 cm long and are typically curved and folded along the midrib. Floating leaves are elliptic and 5–10 cm long. Stipules are free from the leaf bases. Turions are not produced.

The dense, cylindrical flowering spikes typically extend above the water surface and are associated with floating leaves; however, in deep water flowering may occur beneath the water surface and without the presence of floating leaves. Bigleaf pondweed flowers in early to midsummer; the nearly 6 mm-long achenes mature by late summer or early fall.

This species occurs from Nova Scotia and Quebec west to Minnesota and south to Alabama. In addition, it is found at scattered locations from British Columbia south to California. In Pennsylvania bigleaf pondweed grows in lakes, ponds, and streams across the state, but is most common in the glaciated areas. It is found most often in areas with soft bottom sediments.

The only species that bigleaf pondweed might be confused with is white-stem pondweed (*Potamogeton praelongus*), which in Pennsylvania is known only from the northwestern corner of the state. Both have unbranched stems that can reach 1.5 m or more in length. White-stem pondweed never has floating leaves, but its submersed leaves can be similar in appearance to those of bigleaf pondweed. The stipules of white-stem pondweed are white, clasp the stem tightly and become conspicuously shredded at the tips. Bigleaf pondweed stipules, on the other hand, are light brown or greenish and taper to a slender point.

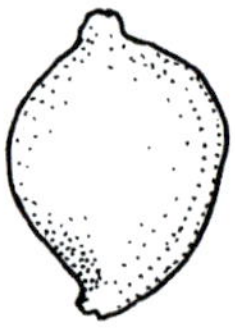

Potamogeton amplifolius—achene ×5

Potamogeton amplifolius ×1/4

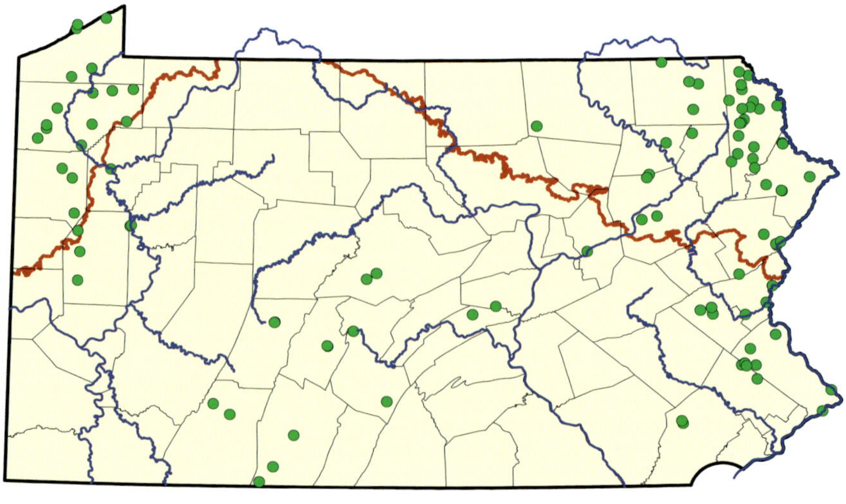

Potamogeton amplifolius

Pondweed

Potamogeton bicupulatus Fernald [synonym: Potamogeton diversifolius Raf. var. trichophyllus Morong]

Native
Rooted, submergent and floating-leaf perennial

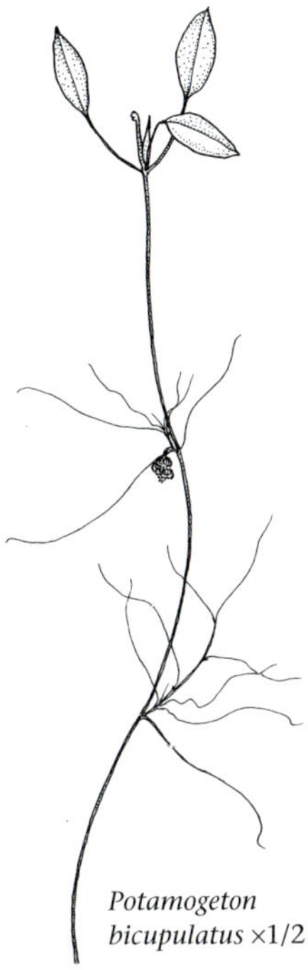

Potamogeton bicupulatus ×1/2

This highly branched pondweed has very slender underwater leaves that are less than 0.5 mm wide, have only a single vein, and lack lacunar strips. The stipules are fused to the leaf bases for less than half their length with the tips protruding. Glands are not present. Floating leaves, when present, are 0.6–2.3 cm long with 3–7 veins, a pointed tip, and tapering base. The stems, which can reach 1 m in length, arise from rhizomes. Turions are not produced.

Inflorescences that extend above the water surface are short-cylindrical; those that form under water are globular in shape. The achenes have one central and two lateral keels and a knobby appearance due to the points on the keels.

This delicate pondweed occurs from Ontario and New England south to Pennsylvania and west to Michigan. In Pennsylvania it is found in lakes and ponds in the glaciated region of the northeast and scattered sites elsewhere.

Potamogeton bicupulatus can be confused with *P. diversifolius*, of which it has been considered a variety by some botanists. The acute to acuminate leaf tips on both underwater and floating leaves, and lack of lacunar strips are useful distinguishing features.

Potamogeton bicupulatus—
achene ×5

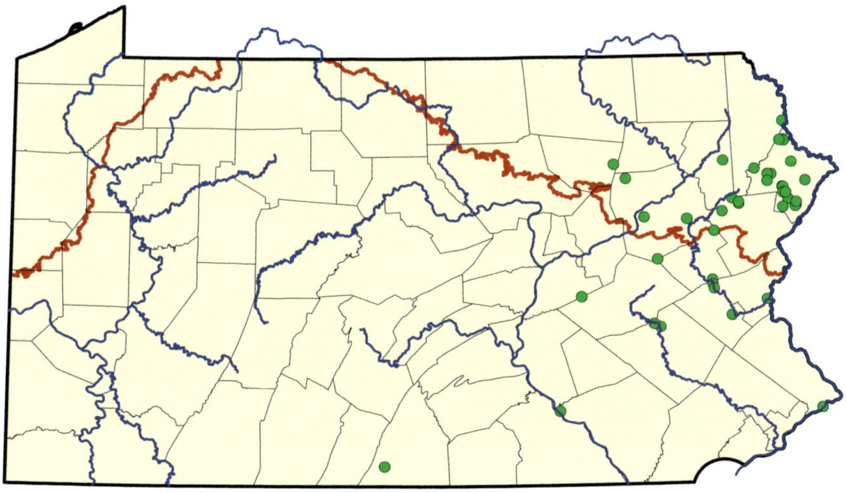

Potamogeton bicupulatus

Tuckerman's pondweed — *Potamogeton confervoides* Reichenb.

PNHP
Native
Rooted, submergent perennial

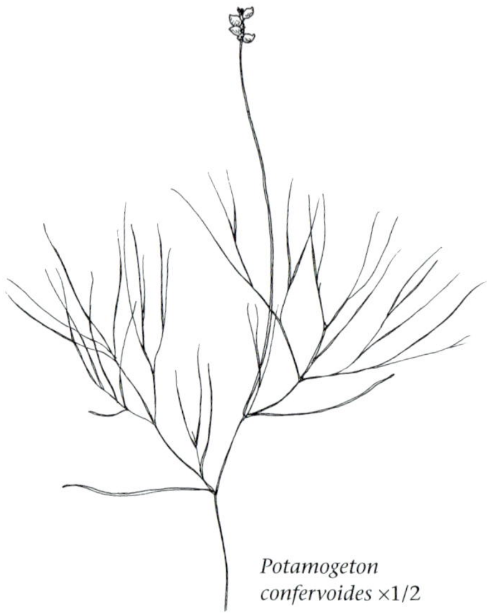

Potamogeton confervoides ×1/2

Tuckerman's pondweed is a rare plant of lakes and boggy ponds in northeastern Pennsylvania. It grows completely under water, and has highly branched stems to 0.8 m long with fine thread-like leaves. The stipules are free from the leaf base; nodal glands are not present. Turions are produced in the axils of older leaves.

Flowers are in dense cylindrical spikes on 10–20 cm-long peduncles. Fruits are 2–3 mm, with a winged keel and a distinct beak.

Ranging from Newfoundland and Labrador south to northern Wisconsin, Tuckerman's pondweed is a northern species at its southern limit of range in Pennsylvania, where it occurs in lakes and ponds of the glaciated northeast and adjacent areas.

Potamogeton confervoides—achene ×5

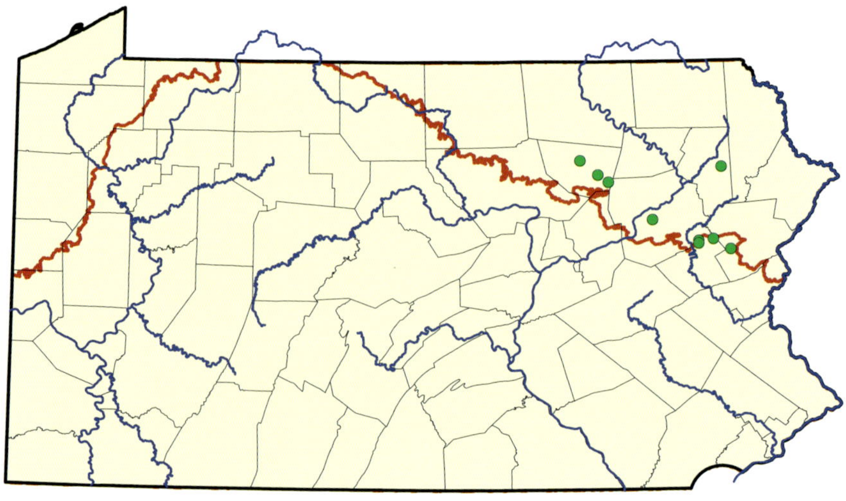

Potamogeton confervoides

Curly pondweed *Potamogeton crispus* L.

Introduced
Rooted, submergent perennial

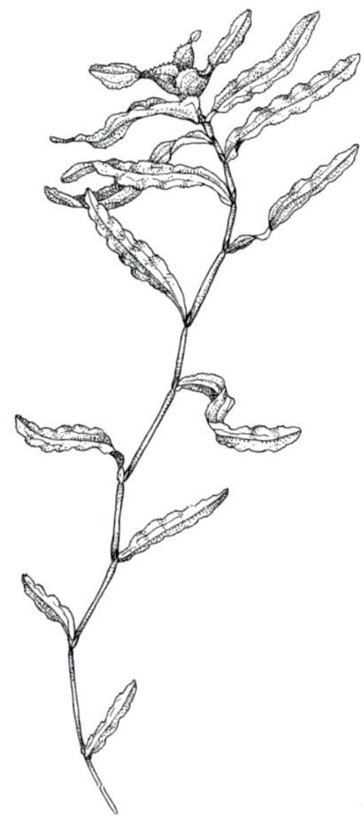

Potamogeton crispus ×1/2

Curly pondweed is one of the most common of the non-native aquatic plants which have become naturalized in Pennsylvania (Table 1.3, p. 18). It has only underwater leaves; floating leaves are not produced. The leaves are ribbon-shaped with wavy, finely toothed edges; they are 2-ranked and stiff enough to retain their shape when the plant is removed from the water. Stipules are free from the leaf bases. The stems, which arise from a rhizome, are only slightly branched.

Flowering occurs from April to June when inflorescences form at the stem tips. Achenes lack keels but have a prominent beak. Turions, dormant stem tips with stiffened and rolled leaves, are formed at branch tips and leaf axils in late spring or early summer.

Curly pondweed is found in lakes, ponds, rivers, and streams; primarily in areas with a soft substrate. Active growth begins in early spring; flowering occurs early in the season. After forming turions, the plants become dormant in midsummer when the water is warmest. Active growth resumes in the fall when the water temperature drops. Plants can remain vegetatively active through the winter and continue to photosynthesize even when lakes are covered with ice and snow.

Potamogeton crispus—achene ×3

Native to Central and South America, Eurasia, and Australia, curly pondweed was introduced in North America sometime before the mid-1800s; the earliest confirmed specimens were collected near Wilmington, Delaware, in 1859 (Stuckey 1979). Early collections from Pennsylvania were obtained from the Lehigh and Juniata Rivers, and in the vicinity of Lancaster; the species has subsequently been found throughout the state.

The only other pondweed in our flora that has toothed leaf margins is flat-leaved pondweed (*Potamogeton robbinsii*). The wavy edges of the leaves and

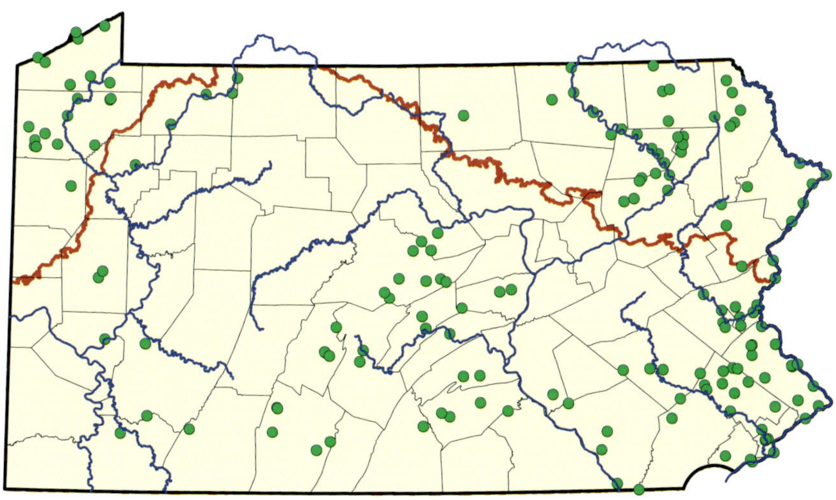

Potamogeton crispus

tapered or rounded leaf base of curly pondweed distinguish it from flat-leaved pondweed. In addition curly pondweed is usually bright yellowish-green, compared to the very dull, dark green color of flat-leaved pondweed.

Snailseed pondweed *Potamogeton diversifolius* Raf.

Native

Rooted, submergent and floating-leaf perennial

Potamogeton diversifolius—achene ×5

Potamogeton diversifolius ×1

Potamogeton diversifolius has both underwater and floating leaves, although the latter may not always be present. The stems arise from rhizomes; turions are not produced. Submersed leaves are linear, 0.1–0.5 mm wide with acute tips and narrow lacunar strips. Stipules are fused to the leaf bases for less than half their length; glands are not present. Floating leaves (when present) are 0.8–1.6 cm long, 3–8.5 mm wide, with 3–7 parallel veins and a mostly rounded tip.

Flowering spikes are produced both above and below the water surface. Fruits are about 1 mm in diameter with a tiny beak, a prominent abaxial keel, and sometimes a suggestion of lateral keels.

Potamogeton diversifolius is a plant of eastern North America from Pennsylvania and Nebraska south to Florida and Texas with scattered locations farther west. In Pennsylvania it is commonly found in shallow, quiet water of lakes and ponds. It is most common in the eastern and southwestern parts of the state.

Potamogeton diversifolius is very similar to *P. spirillus*. The two are best distinguished by the relative length of the free and fused parts of the stipules: in *P. diversifolius* the free part of the stipulular sheath is longer than the portion fused to the leaf. The fruits of *P. diversifolius* are less than 1.5 mm in diameter, whereas those of *P. spirillus* are usually larger.

Potamogeton diversifolius

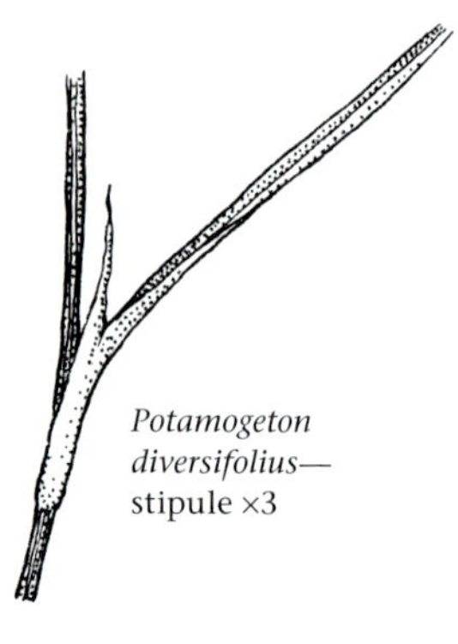

Potamogeton diversifolius—stipule ×3

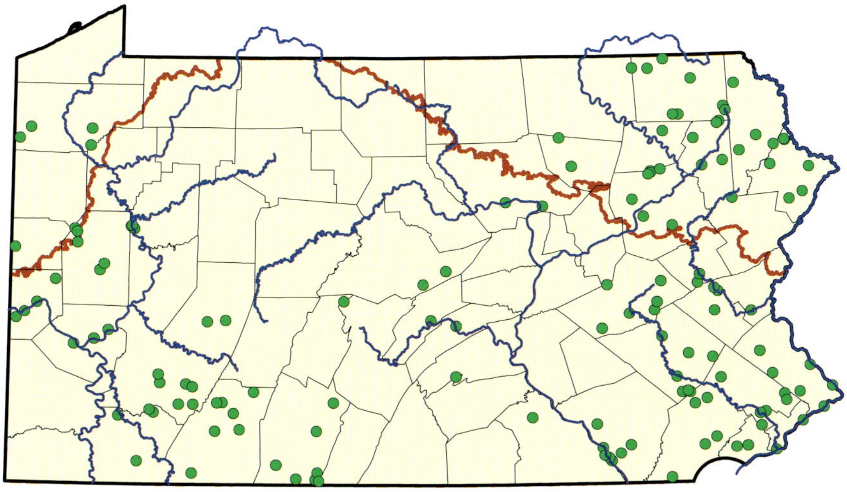

Potamogeton diversifolius

Ribbonleaf pondweed *Potamogeton epihydrus* Raf.

Native

Rooted, submergent and floating-leaf perennial

Ribbonleaf pondweed is the most common pondweed (*Potamogeton*) species in Pennsylvania. The slender, slightly flattened and branched stems of this plant can reach 2 m in length. It has both underwater and floating leaves. The submersed leaves are 8–20 cm long and about 5 mm wide; they are sessile, linear, and flat with prominent bands of translucent cells forming a distinct lacunar band on each side of the midrib. Floating leaves are elliptic, 2.5–9 cm long and up to 2 cm wide, and narrowed to a slender petiole. The stipules are free from the leaf base. Vegetative shoots of ribbonleaf pondweed have a flat appearance under water and often appear brownish-orange.

Flowers are borne in dense cylindrical spikes, about 3 cm in length, which project above the water surface. Flowers are always accompanied by floating leaves in this

Potamogeton epihydrus ×1/2

species. Fruits are greenish-brown, 2.5–4.5 mm long, and have abaxial and lateral keels.

Ribbonleaf pondweed occurs from Newfoundland to southern Alaska, south to Georgia, Louisiana, and California. In Pennsylvania it is common in lakes, ponds, and slow-moving rivers and streams throughout the state.

Potamogeton epihydrus

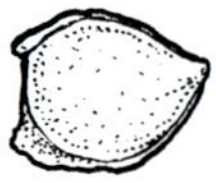

Potamogeton epihydrus—achene ×5

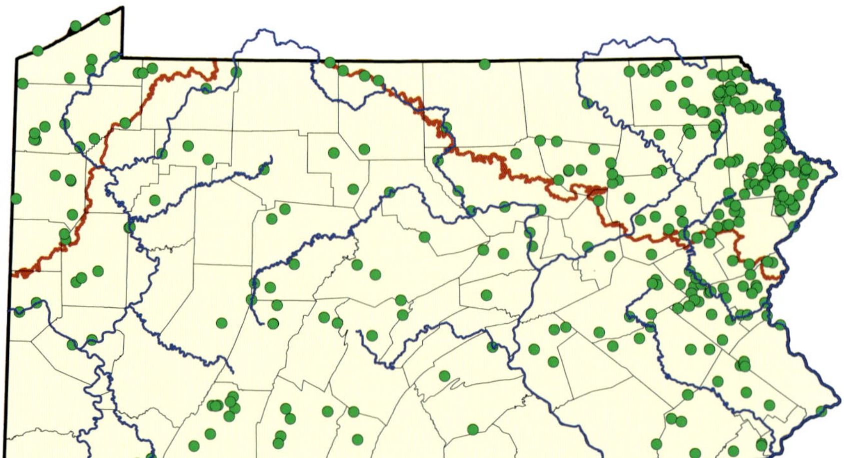

Potamogeton epihydrus

Threadleaf pondweed — *Potamogeton filiformis* Persoon [synonym: *Stuckenia filiformis* (Persoon) Böerner]

PNHP
Native
Rooted, submergent perennial

Threadleaf pondweed is a freely branched, submersed aquatic with alternate, linear leaves to 15 cm long and 0.2–2 mm wide. The stipular sheaths are inflated at the base and adnate to the bases of the leaf blades for two-thirds or more of their length. Inflorescences develop completely under water and consist of 2–6 whorls of flowers/fruits.

This species occurs across the northern United States, Canada, and Asia. It is at its southern limit of range in Pennsylvania and New Jersey. In our state, historically it was a rare plant of calcareous ponds and streams at widely scattered locations. It is thought that it may be extirpated.

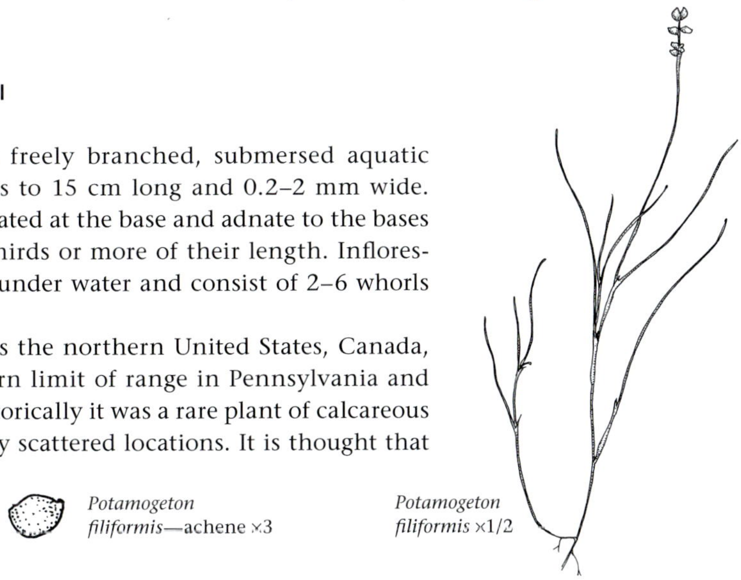

Potamogeton filiformis—achene ×3

Potamogeton filiformis ×1/2

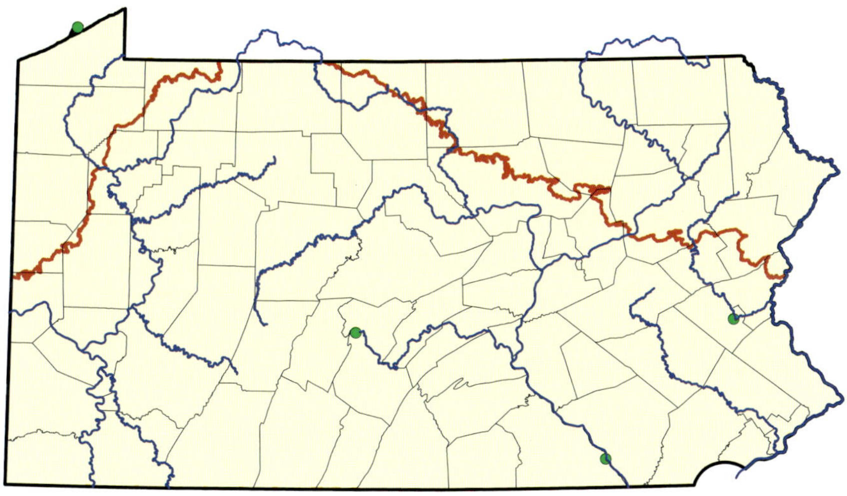

Potamogeton filiformis

Leafy pondweed — *Potamogeton foliosus* Raf.

Native
Rooted, submergent perennial

Leafy pondweed has freely branched and slightly compressed stems to 0.7 m long; rhizomes are not produced. Leaves are all submersed, linear, 0.3–2.3 mm wide with an acute or apiculate apex, 1–5 veins, and lacunar strips that are narrow or lacking. Stipules are free, brown or whitish, veiny, and 0.5–2 cm long; glands are occasionally present at the nodes. Turions are uncommon;

if present, they are lateral or terminal with the inner leaves rolled into a hardened structure.

Flower spikes extend above the water surface; fruits are 1.4–2.7 mm, pale green to brownish, with a single undulate keel and a prominent beak.

Leafy pondweed occurs throughout North America from southern Canada south with the exception of Florida and Texas; it is also known from Central America. In Pennsylvania it occurs in lakes and streams throughout. Ours is *Potamogeton foliosus* ssp. *foliosus*.

The undulate keel on the fruit distinguishes this species from all the other linear-leaved pondweeds in our region.

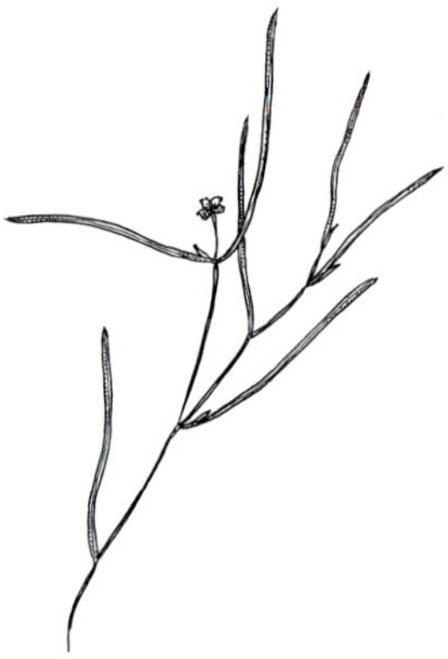

Potamogeton foliosus ×1/2

Potamogeton foliosus—achene ×5

Potamogeton foliosus—leaf tip ×3

Potamogeton foliosus—stipule ×1.5

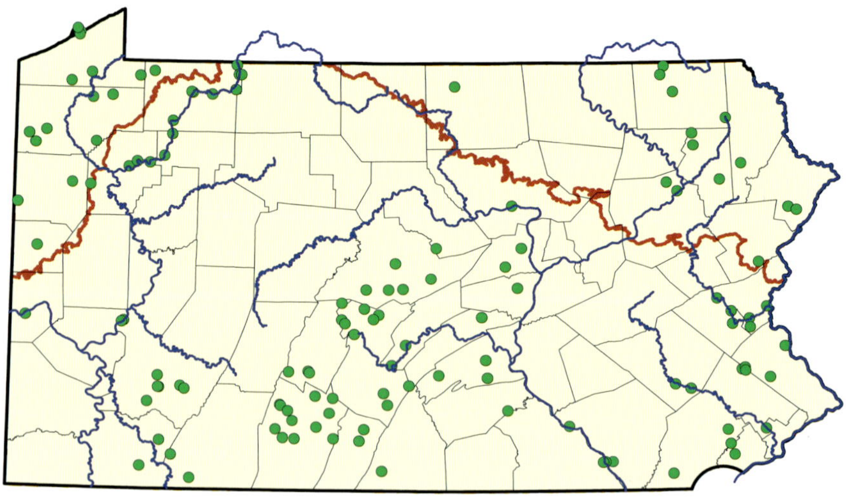

Potamogeton foliosus

Fries' pondweed — *Potamogeton friesii* Rupr.

PNHP
Native
Rooted, submergent perennial

Fries' pondweed has slightly compressed, branched stems to 1 m; rhizomes are not produced. Only submersed leaves are present; they are linear, 1.5–3 mm wide, with a narrow lacunar strip, and an apiculate to acute apex. Glands are present at the nodes. Stipules are free, fibrous, and white; however, they are not persistent. Turions, with a fan-shaped arrangement of the inner leaves, are common in leaf axils or at stem tips.

Flower spikes are slender with 3–5 well-spaced whorls of flowers; achenes are 1.8–2.5 mm and lack a keel.

The total range of this northern species extends from Nova Scotia, Newfoundland, and Labrador to Alaska, and south to northern Iowa, Ohio, and Pennsylvania. It is a rare plant of calcareous lakes in northwestern Pennsylvania and several other scattered sites across the state.

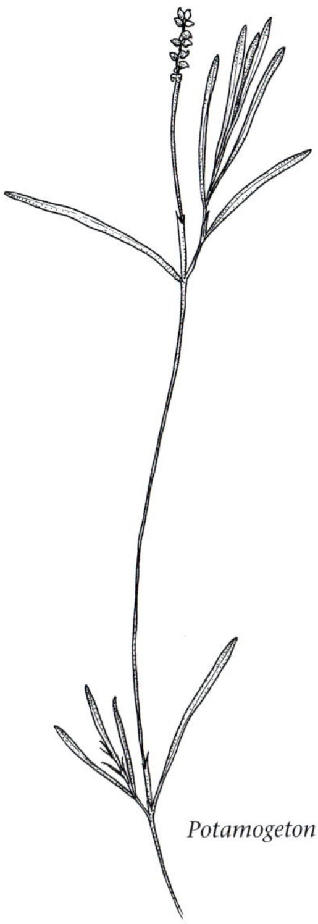

Potamogeton friesii ×1/2

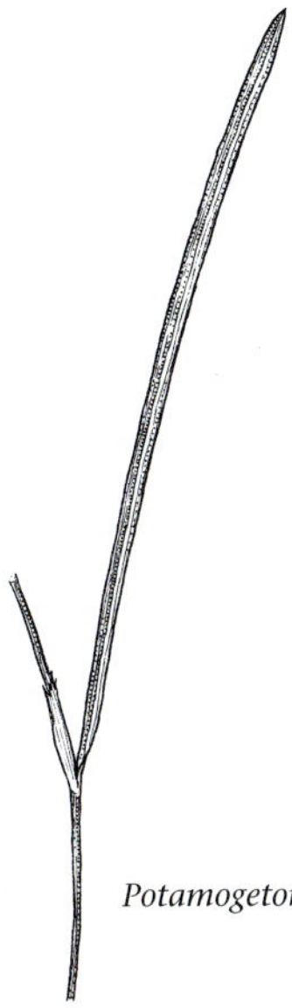

Potamogeton friesii—stipule ×1

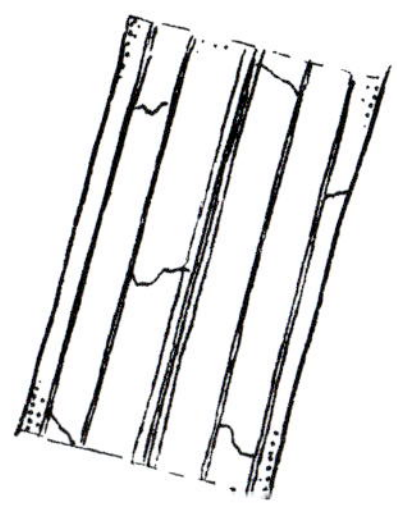

Potamogeton friesii—leaf section showing narrow lacunar strips ×10

Potamogeton friesii—leaf tip ×5

Potamogeton friesii—achene ×3

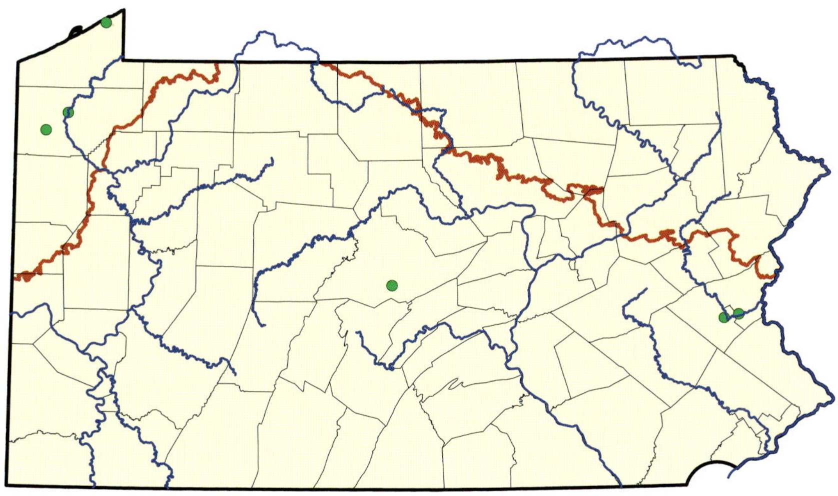

Potamogeton friesii

Grassy or variable pondweed *Potamogeton gramineus* L.

PNHP

Native

Rooted, submergent and floating-leaf perennial

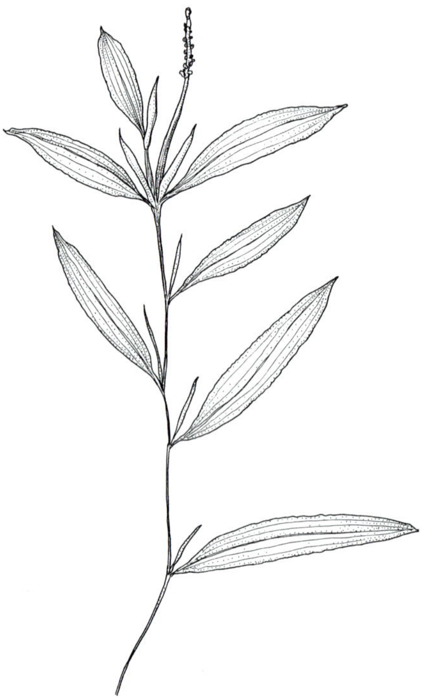

Potamogeton gramineus ×1/2

Grassy pondweed has elongate, branched stems to 1.5 m long that arise from a rhizome. Underwater leaves are present and sometimes also floating leaves; turions (dormant buds) are not produced. Submersed leaves are alternately arranged, lack petioles, and are elliptical in shape and about 10 cm long. Floating leaves are similar in shape to the underwater ones, but have distinct stalks. Stipules are distinct from the leaf and membranous in texture.

Flowering and fruiting occur on stem tips that extend above the water surface. The achenes have an abaxial keel, lateral ridges, and an erect beak.

Grassy or variable pondweed is found throughout the boreal regions of the northern hemisphere. In North America it extends south as far as California, Iowa, and Pennsylvania, where it is found in lakes, ponds, and slow-moving streams in widely scattered locations.

Grassy pondweed could be confused with Illinois pondweed (*Potamogeton illinoensis*) or longleaf pondweed (*P. nodosus*); however Illinois pondweed and longleaf pondweed both have larger submersed leaves with 7–19 veins and a distinct petiole. Underwater leaves of grassy pondweed have 3–9 veins and lack a petiole.

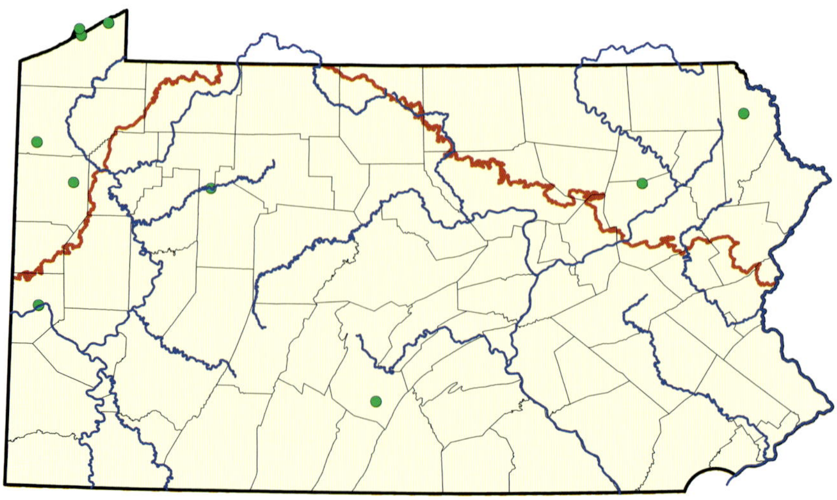

Potamogeton gramineus

Hill's pondweed *Potamogeton hillii* Morong

PNHP
Native
Rooted, submergent perennial

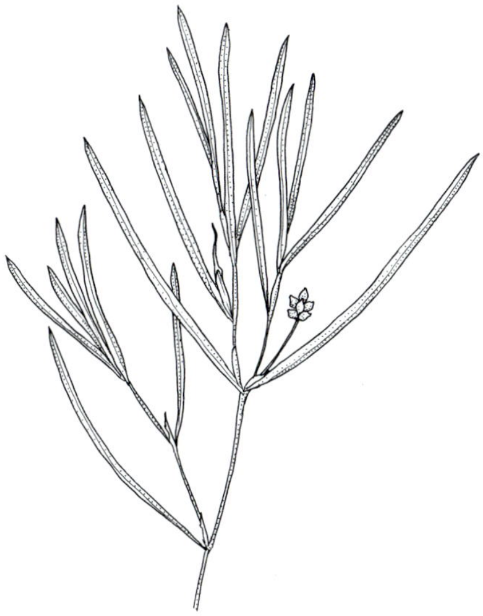

Potamogeton hillii ×1/2

Hill's pondweed has very slender, freely branched stems to 1 m in length; rhizomes are not produced. Leaves are all submersed, linear, 0.6–2.5 mm wide, with 3 veins, and narrow lacunar strips; the leaf apex is bristle-tipped. Stipules are free, white to light brown, and somewhat fibrous; nodal glands are rare. Turions are rarely produced.

Hill's pondweed flowers and fruits freely; inflorescences are emersed and few-flowered. The achenes are greenish brown, flattened, 3.5–4 mm wide, and have 3 keels.

This species has a very limited geographic range extending from southern Ontario to Virginia and west to Wisconsin. It is a plant of alkaline waters. In Pennsylvania, Hill's pondweed occurs as an early successional species in beaver ponds in the northwestern region of the state and at scattered locations elsewhere.

Hill's pondweed can be distinguished from other submergent pondweeds with linear leaves by its bristle-tipped leaves with 5 or fewer veins, and an almost total lack of nodal glands.

Potamogeton hillii—achene ×3

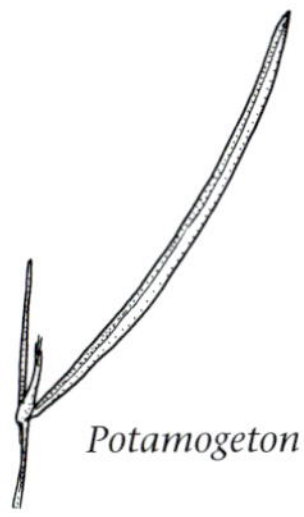

Potamogeton hillii—stipule ×1/2

Potamogeton hillii—leaf tip ×3, note bristle-tipped apex

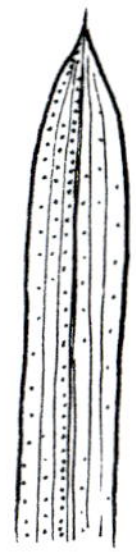

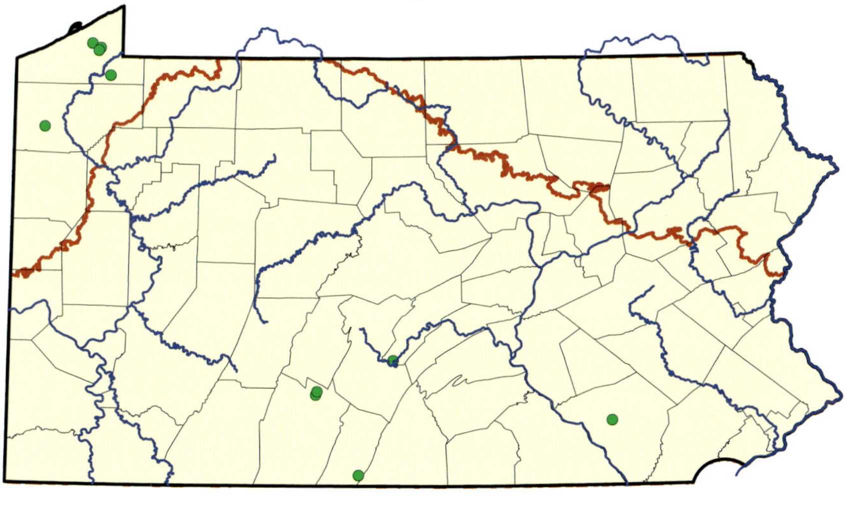

Potamogeton hillii

Illinois pondweed *Potamogeton illinoensis* Morong

Native

Rooted, submergent and floating-leaf perennial

Potamogeton illinoensis—achene ×3

The stems of Illinois pondweed arise from rhizomes and reach a length of 1 m or more. No glands are present and turions are not produced. Submersed and floating leaves are produced, but floating leaves are not always present. Submersed leaves are sessile or petioled, elliptic to lanceolate, 5–20 cm long and 2–5 mm wide with 2–5 rows of lacunar cells on each side of the midrib. The leaf apex is flat and acute-mucronate. The stipules are free from the base of the leaf, 4–10 cm long, and persistent. Floating leaves are oblong elliptic, 4–19 cm long and 2–6.5 cm wide, with 13–29 veins.

The 4–30 cm-long inflorescences are produced above the water surface; fruits are 2.5–3.5 mm in diameter, with a well-developed abaxial keel and lateral ridges.

Illinois pondweed ranges across the northeastern United States from New Hampshire to Wisconsin and south to West Virginia; it is scattered elsewhere. It also occurs in Mexico, the West Indies, Central America, and South America. In Pennsylvania it is a rare plant of lakes and rivers, particularly in the Susquehanna drainage and the upper Delaware.

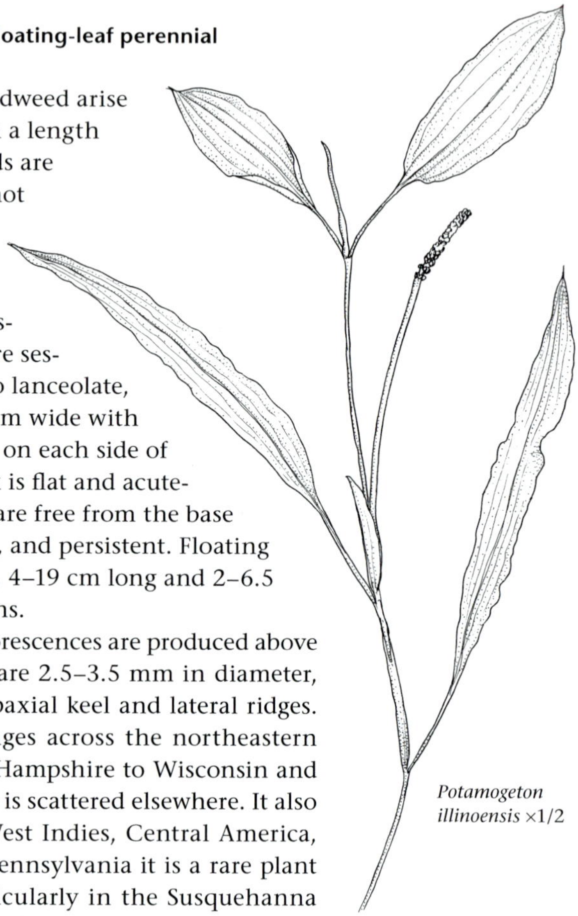

Potamogeton illinoensis ×1/2

See description of *Potamogeton gramineus* for comments on distinguishing these two species.

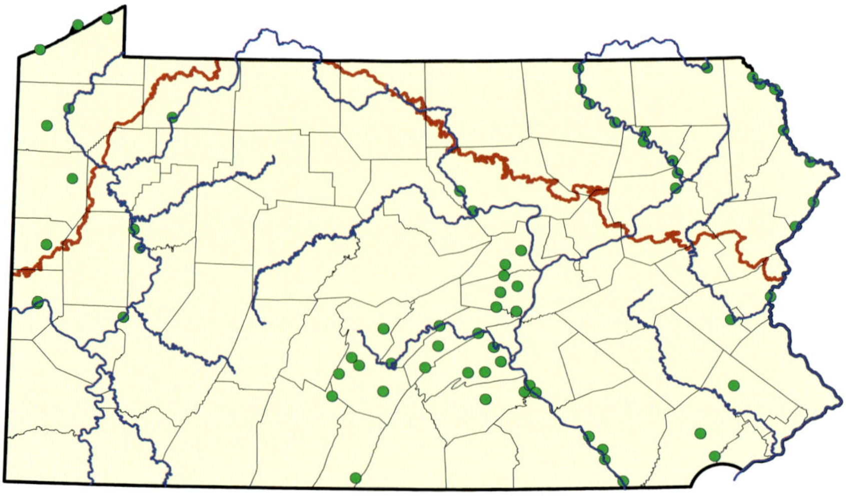

Potamogeton illinoensis

Floating-leaf pondweed *Potamogeton natans* L.

Native
Rooted, submergent and floating-leaf perennial

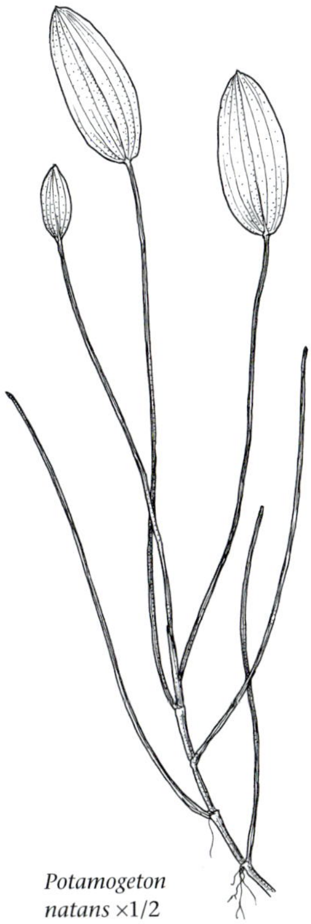

Potamogeton natans ×1/2

Floating-leaf pondweed has both submersed and floating leaves, but the former are very narrow and appear to consist only of an elongate petiole or midrib without a blade. Stipules are free, whitish, and persistent. Blades of floating leaves are elliptic to ovate, 3.5–11 cm long, with 17–37 veins, a cordate or rounded base, and an acute to rounded apex.

The flowering spikes extend above the water surface. They are cylindrical and 10–15 mm long. Achenes are reddish-brown, 3.5–5 mm long, and have an erect or recurved beak; they are not keeled.

Floating-leaf pondweed occurs throughout the boreal regions of the northern hemisphere. In northeastern North America it grows from Newfoundland and Labrador to southern Quebec and south to New Jersey and Iowa. It is also present in Manitoba and Saskatchewan and from Alaska to southern California. In Pennsylvania it is a species of lakes and streams, mostly north of the glacial boundary.

Very narrow, rigid underwater leaves are a good characteristic for distinguishing *Potamogeton natans* from several similar-appearing species. *Potamogeton oakesianus* also has very narrow submersed leaves, but they are lax rather than rigid; in addition, the fruits are keeled. *Potamogeton natans* can also be confused with *P. nodosus*, because the underwater leaves of *P. nodosus* often deteriorate leaving only petioles. However, floating leaves of *P. nodosus* are usually tapered at the base and the petioles lack a light-colored zone at their apex.

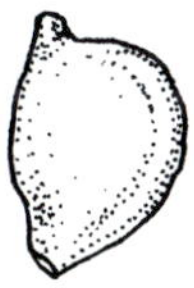

Potamogeton natans—achene ×5

Potamogeton natans—stipule ×1/2

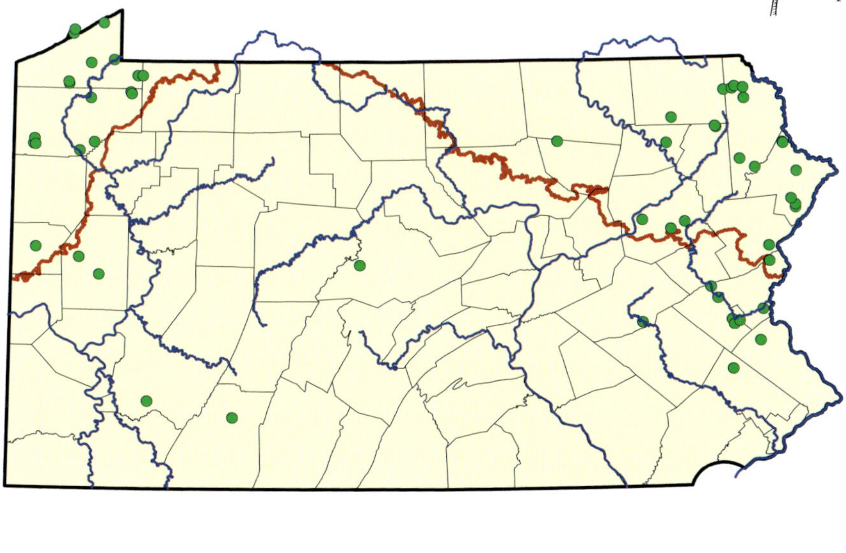

Potamogeton natans

Floating leaves of *Potamogeton natans* in a stream in Bucks County, Pennsylvania

Longleaf pondweed *Potamogeton nodosus* Poir.

Native

Rooted, submergent and floating-leaf perennial

Potamogeton nodosus ×1/2

Longleaf pondweed has stems to 2 m long arising from whitish rhizomes; nodal glands are not present. Both submersed and floating leaves are produced. Submersed leaves are petiolate, lance-elliptic, 9–20 cm long and 1–3.5 cm wide, with conspicuous lacunar bands and an acute apex. The conspicuous stipules are free from the leaf blade and 4–10 cm long. Floating leaves are elliptic, 5–13 cm long, and 15–45 mm wide, with 9–21 veins.

Flowering spikes are all emersed and 2–7 cm long. Fruits are 3–4 mm in diameter with a well-developed abaxial keel, lateral ridges, and an erect beak.

Longleaf pondweed occurs throughout the eastern two-thirds of the United States, plus California, Oregon, and Washington. Its range also includes Mexico, the West Indies, Central America, South America, and Eurasia. In Pennsylvania it is frequent in lakes, ponds, and streams across the state.

Underwater leaves of long-leaf pondweed frequently senesce before fruits are produced, complicating identification. See description of *Potamogeton natans* above for comments on distinguishing this pondweed from similar-appearing species.

Potamogeton nodosus—stipule ×1/2

Potamogeton nodosus—achene ×3

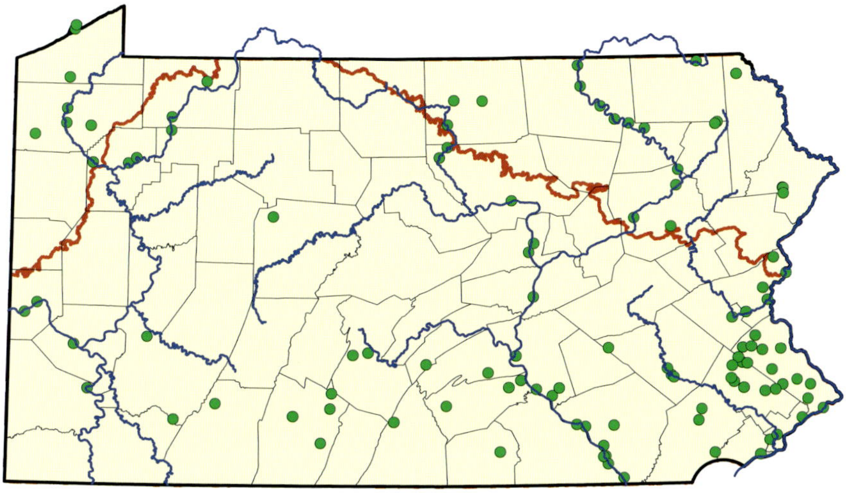

Potamogeton nodosus

Oakes' pondweed — *Potamogeton oakesianus* J.W. Robbins

PNHP
Native
Rooted, submergent and floating-leaf perennial

Potamogeton oakesianus ×1/2

Oakes' pondweed has freely branched stems to 1 m; they are rounded, and red-spotted. Rhizomes are produced, but nodal glands and turions are both absent. Both submersed and floating leaves are present. Submersed leaves are linear, 0.3–1 mm wide, and essentially without an expanded blade; they are often completely deteriorated by flowering time. Stipules are free and 2.5–4 cm long. Floating leaves are lanceolate to elliptic or ovate, 3.2–7.5 cm long, with 9–19 veins.

Flowering spikes extend above the water surface; they are cylindrical and 10–35 mm long. The fruits are greenish brown, 2.5–3.5 mm, and have both abaxial and lateral keels.

Oakes' pondweed grows from Newfoundland and Labrador to Virginia and west to Minnesota and southern Quebec. In Pennsylvania it is a rare plant of acidic ponds and lakes at widely scattered locations.

Like *Potamogeton natans*, this species has underwater leaves that are reduced to an elongated petiole without an expanded blade. However, the submersed leaves of *P. oakesianus* are lax, not rigid as in *P. natans*, and the fruits of *P. oakesianus* are smaller and keeled.

Potamogeton oakesianus—achene ×3

Potamogeton oakesianus—stipule ×1/2

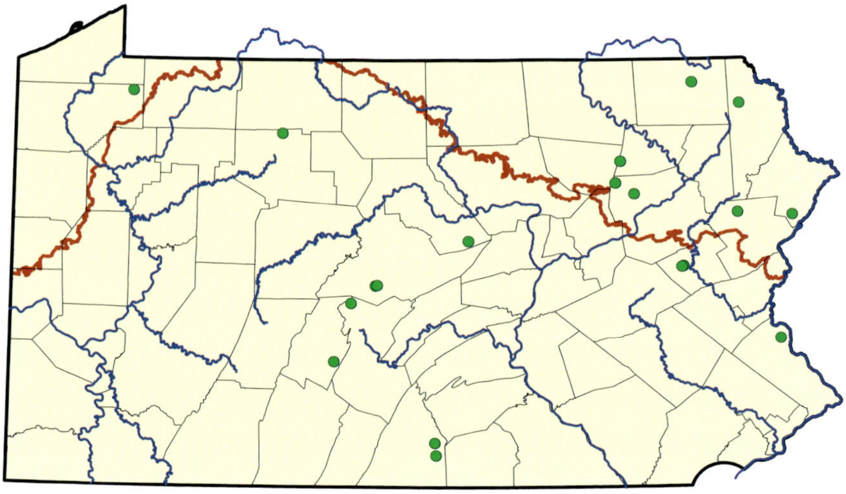

Potamogeton oakesianus

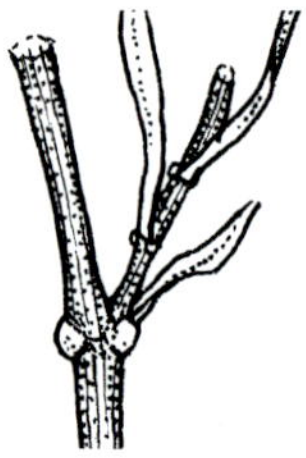

Potamogeton obtusifolius—node with glands ×2

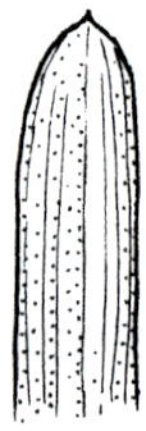

Potamogeton obtusifolius—round-apiculate leaf tip ×3

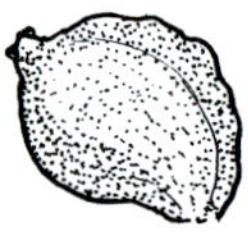

Potamogeton obtusifolius—achene ×5

Blunt-leaved pondweed — *Potamogeton obtusifolius* Mert. & Koch

PNHP
Native
Rooted, submergent perennial

Stems slender, slightly compressed, to 1 m with glands at the nodes; rhizomes are not produced. Turions may be present at the stem tips. Leaves are all submersed, linear, 1–3.5 mm wide, with 3 veins and narrow lacunar strips; the apex is obtuse or round-apiculate. Stipules are free, 1–2 cm long, white, and membranous.

Flowering spikes are emersed, 1.5–2 cm long with 3–5 whorls of flowers; achenes 2.5–3.6 mm long, smoothly rounded on the sides and with an abaxial keel and a distinct beak. Flowering and fruiting occur in summer and fall.

Blunt-leaved pondweed has a limited distribution from Nova Scotia to Michigan and south to New Jersey and Pennsylvania. It is also present at scattered locations farther west. In Pennsylvania it is a rare species of boggy ponds and lakes in the glaciated northeast.

Blunt-leaved pondweed resembles several other linear-leaved species, especially if flowers and fruits are not present. The rounded leaf tips and conspicuous nodal glands are good distinguishing characters. In addition, blunt-leaved pondweed often has a reddish appearance.

Potamogeton obtusifolius ×1/2

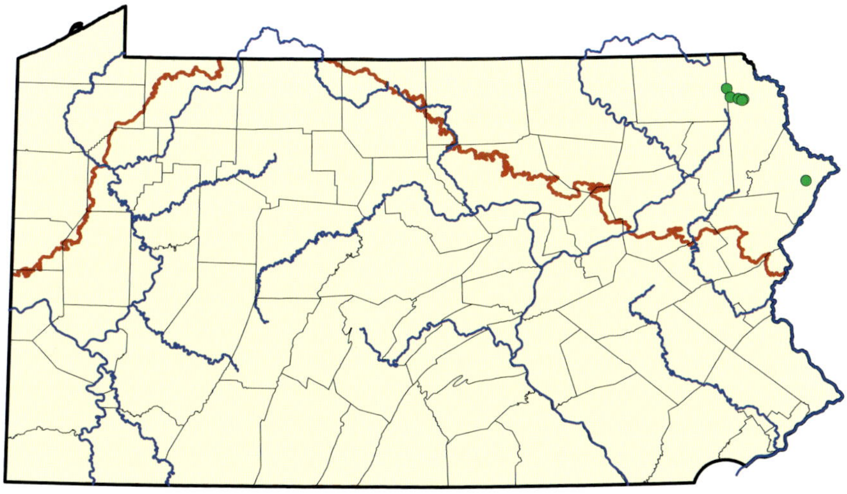

Potamogeton obtusifolius

Sago pondweed — *Potamogeton pectinatus* L. [synonym: *Stuckenia pectinata* (L.) Böerner]

Potamogeton pectinatus—stipule ×1

Native
Rooted, submergent perennial

Sago pondweed is a submergent aquatic with branched stems and alternate narrow leaves 5–10 cm long and up to 1 mm in width. The stipular sheath is fused to the base of the leaf for most of its length. The plants do not produce turions, but underground tubers are formed. Flower spikes are terminal or axillary and contain 3–5 whorls of flowers/fruits.

Sago pondweed is a worldwide species; in North America its range includes all of the United States (except a cluster of states in the southeast), southern Canada, Yukon, and Alaska. In Pennsylvania it can be found in shallow brackish or calcareous waters of lakes and rivers.

The tubers of sago pondweed are an important food source for ducks and other waterfowl, which also spread the plant from site to site.

Potamogeton pectinatus ×1/2

Potamogeton pectinatus—achene ×3

Potamogeton pectinatus—fruiting inflorescence ×1

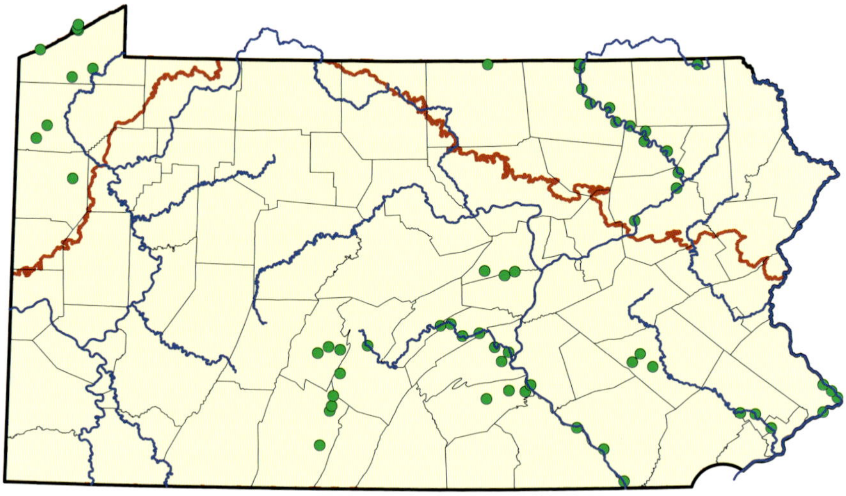

Potamogeton pectinatus

Perfoliate pondweed, redhead-grass — *Potamogeton perfoliatus* L.

Native

Rooted, submergent perennial

Perfoliate pondweed is completely submergent except for the flower spikes. It has round branched stems that arise from a rooted rhizome; nodal glands are not present and turions are not produced. The leaves are ovate to orbicular, and rounded and clasping to perfoliate at the base. Stipules are free from the leaf base and soon disappear. The inflorescences are 1–7 cm long. The fruits are 1.6–3 mm in diameter with an erect beak; they are not keeled.

The range of perfoliate pondweed extends from Newfoundland and Labrador west to Ohio and south to Delaware. It is also known from Louisiana, Florida, Costa Rica, Eurasia, Africa, and Australia. In Pennsylvania it grows in lakes, ponds, rivers, and streams throughout.

Potamogeton perfoliatus is very similar to *Potamogeton richardsonii*, from which it is distinguished by shorter leaves (ovate to orbicular as opposed to ovate-lanceolate to narrowly lanceolate), and deciduous stipules.

Potamogeton perfoliatus ×1/2

Potamogeton perfoliatus—achene ×3

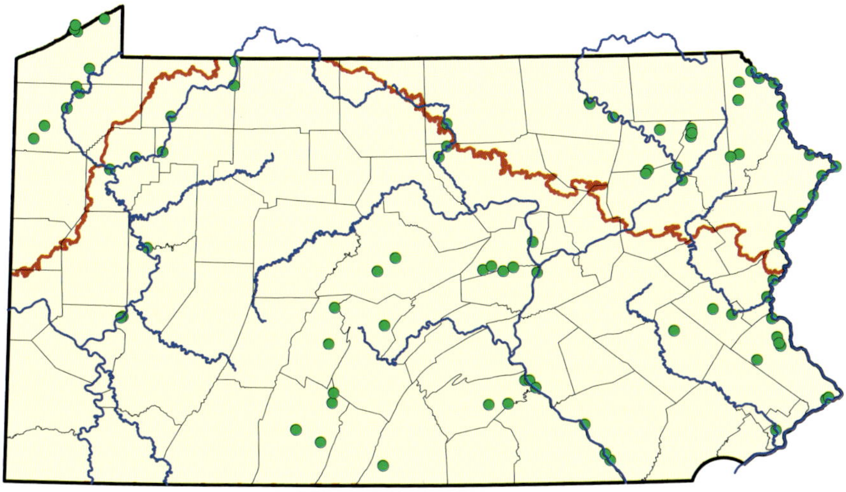

Potamogeton perfoliatus

White-stem pondweed *Potamogeton praelongus* Wulfen

PNHP
Native
Rooted, submergent perennial

White-stem pondweed is completely submergent except for the inflorescences. Its stems, which arise from rhizomes, may be up to 3 m long, freely branched, and somewhat zigzag in appearance. Leaves are lanceolate to lance-linear and sessile to clasping; the tips are hooded. The stipules are free from the leaves, persistent, and soon become shredded and white at the tips. Achenes are 4–5.7 mm, keeled, and have an erect beak.

This distinctive pondweed can easily be recognized by the zigzag stem, persistent, whitish stipules, and hooded leaf tips.

White-stem pondweed is known from only a few locations in Erie and Crawford Counties in northwestern Pennsylvania. Its total range includes the northeastern United States and adjacent areas of Canada plus scattered sites to the north and west.

Potamogeton praelongus ×1/2

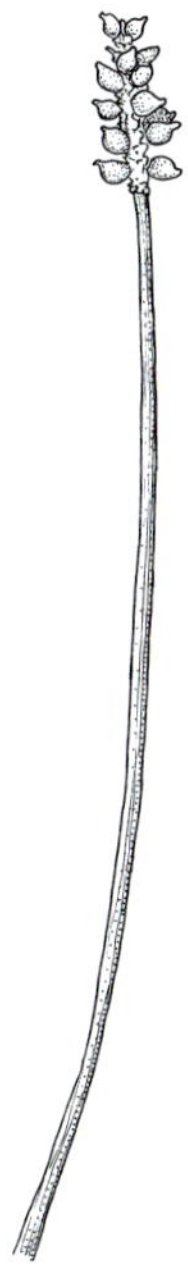

Potamogeton praelongus—inflorescence ×1/2

Potamogeton praelongus—achene ×3

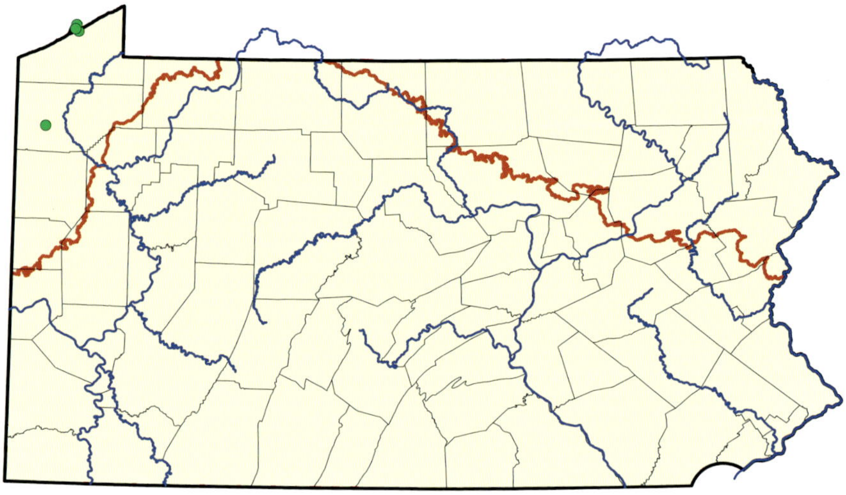

Potamogeton praelongus

Heartleaf or spotted pondweed *Potamogeton pulcher* Tuck.

PNHP
Native
Rooted, submergent and floating-leaf perennial

Potamogeton pulcher ×1/2

Spotted pondweed has unbranched stems up to 0.6 m long arising from rhizomes; the stems and leaf petioles are conspicuously black-spotted. Nodal glands are absent. Submersed leaves are broadly lanceolate, and 1–3 cm wide; floating leaves are broadly elliptic, obtuse or subcordate at the base with black-spotted petioles. The stipules are free approximately 6 mm long, and deciduous.

The dense flower spikes are 2–4 cm long. Achenes are approximately 5 mm in diameter, with a well-developed abaxial keel and lateral ridges.

Spotted pondweed is a plant of the southeastern United States north to Massachusetts and western New York with outliers in Nova Scotia. In Pennsylvania it is a rare plant of streams, swamps, and muddy shores at widely scattered sites from the Atlantic Coastal Plain to the northwest corner of the state.

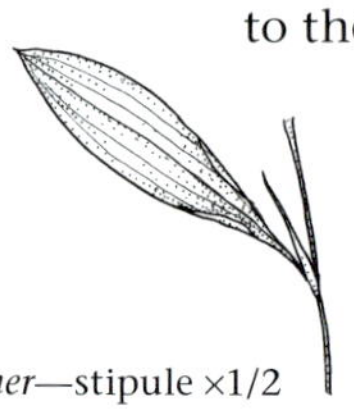

Potamogeton pulcher—stipule ×1/2

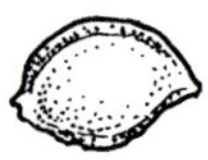

Potamogeton pulcher—achene ×3

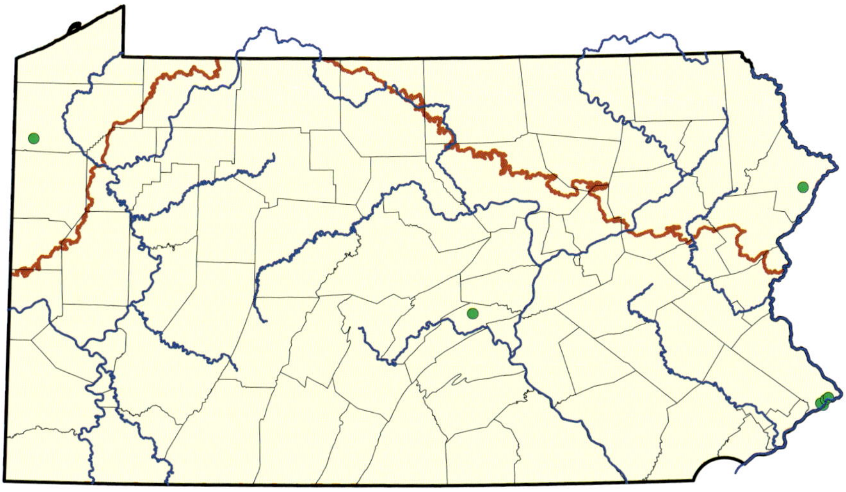

Potamogeton pulcher

Pondweed *Potamogeton pusillus* L.

Native
Rooted, submergent perennial

Potamogeton pusillus has slender, much branched stems to 1.5 m; rhizomes are absent. Leaves are all submersed, linear, 0.2–2.5 mm wide, 1–3 veined, with or without lacunar strips, and with acute to obtuse leaf apices. Stipules are free, delicate, and up to 1 cm long. Glands are present at at least some of the nodes.

Turions are common and either axillary or terminal; they consist of a hardened, fusiform center flanked by several stiff, 2-ranked leaves.

Flowering spikes are submersed or emersed, 2–10 mm long, with 3–5 well-spaced whorls of flowers. Achenes are green to brown, 1.5–2.2 mm and rounded; they lack abaxial or lateral keels.

Potamogeton pusillus has a worldwide distribution. In Pennsylvania it is common in lakes and ponds throughout, and can be locally abundant. Two subspecies are present in our region. *Potamogeton pusillus* ssp. *pusillus* has obovoid fruits with concave sides and 1–3 peduncles per plant. *Potamogeton pusillus* ssp. *tenuissimus* has ovoid fruits with rounded sides and more than 3 peduncles per plant.

Potamogeton pusillus ×1/2

Potamogeton pusillus—turion ×1/2

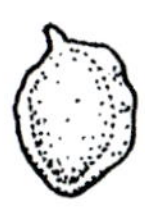

Potamogeton pusillus—achene ×5

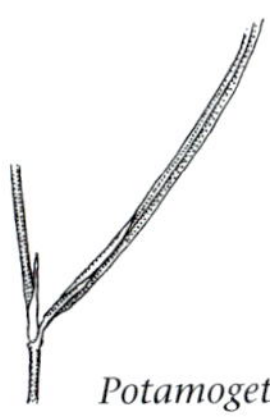

Potamogeton pusillus—stipule ×1.5

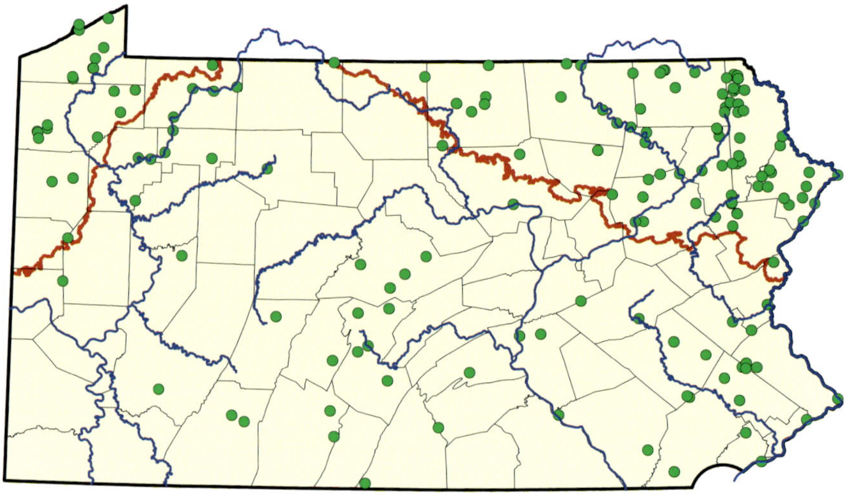

Potamogeton pusillus

Redhead pondweed, Richardson's pondweed

Potamogeton richardsonii (A. Bennett) Rydberg

PNHP

Native

Rooted, submergent perennial

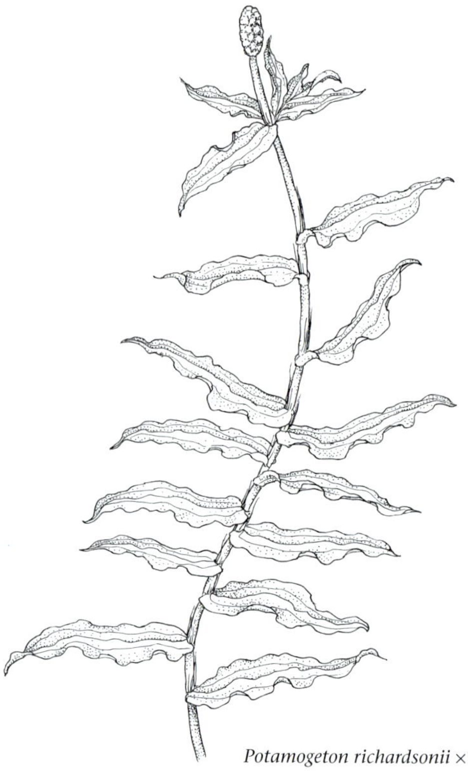

Potamogeton richardsonii ×1/2

Redhead pondweed is completely submergent; it lacks floating leaves. Stems arise from a rhizome and bear ovate-lanceolate to narrowly lanceolate leaves with clasping bases. Stipules are free from the leaves and soon disintegrate leaving persistent fibers. The cylindrical inflorescences are emergent; fruits have an erect beak and are rarely keeled.

Redhead pondweed is a plant of calcareous waters. Its range extends across Canada and the northern United States. In Pennsylvania it is a rare (but occasionally locally abundant) plant found only in Erie County in the northwestern corner of the state.

This species is similar to the closely related *Potamogeton perfoliatus*, from which it differs by the longer leaves and persistent fibrous stipules. The two hybridize to form *Potamogeton* x *absconditus*, a fertile hybrid that has been identified in Maine and Vermont (Kaplan et al. 2009).

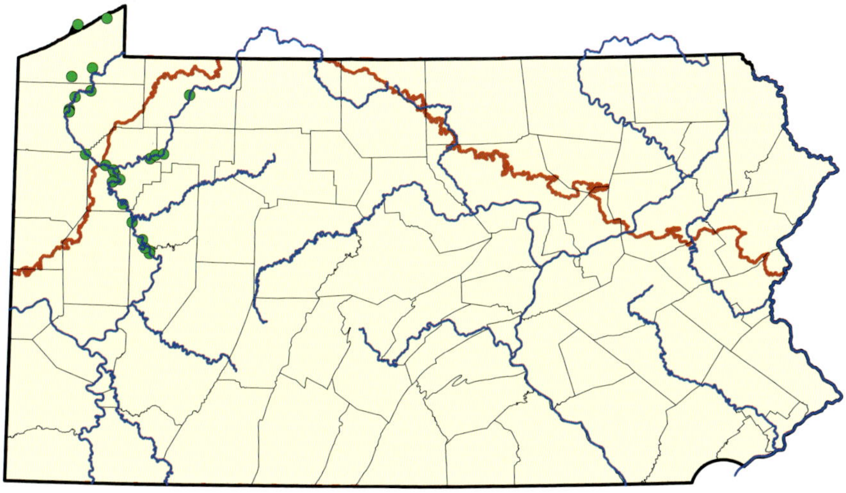

Potamogeton richardsonii

Flat-leaved or fern pondweed

Potamogeton robbinsii Oakes

Native

Rooted, submergent perennial

This species is one of the most easily recognized pondweeds in our flora. It is entirely submergent, and often forms a dense stand on the lake or pond bottom. The lightly branched stems are up to 1 m long and arise from rhizomes. Leaves are stiff and closely 2-ranked, producing a flat, frond-like form that holds its shape out of water; the leaves have small rounded lobes at the base and finely toothed margins. Stipules are fused to the leaf base with the extended tip becoming fibrous. Turions are produced.

Flat-leaved pondweed rarely flowers, but when it does another distinguishing characteristic is its branched inflorescence. Fruits are 3–4 mm with abaxial and lateral keels.

Flat-leaved pondweed is primarily a northern species; its range extends from Nova Scotia and Newfoundland to Quebec, south to New Jersey, Pennsylvania, Ohio, and Michigan. It is also found at scattered locations farther west. In Pennsylvania it grows in lakes, ponds, and rivers in water from 1–2 m deep; it is found primarily in the glaciated regions of the state.

The only other pondweed in our flora that has toothed leaf margins is curly pondweed (*Potamogeton crispus*). The wavy edges of the leaves and tapered or rounded leaf base distinguish curly pondweed from *Potamogeton*

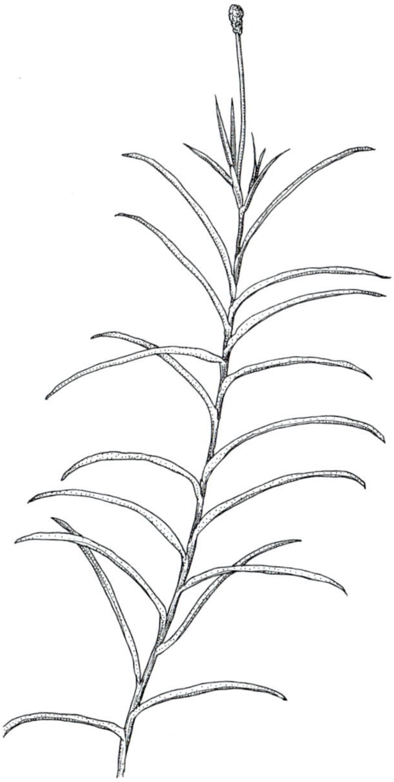

Potamogeton robbinsii ×1/2

robbinsii. In addition, curly pondweed is usually bright yellowish-green compared to the very dull, dark green color of *P. robbinsii*.

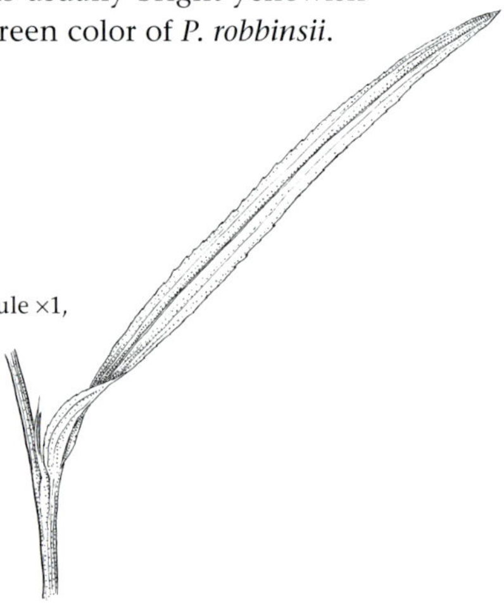

Potamogeton robbinsii—leaf and stipule ×1, note toothed margin of leaf

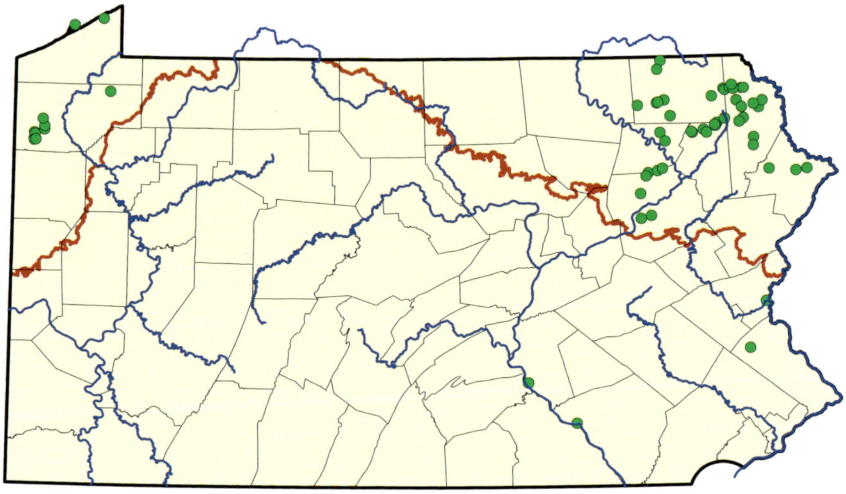

Potamogeton robbinsii

Snailseed pondweed *Potamogeton spirillus* Tuck.

Native

Rooted, submergent and floating-leaf perennial

Potamogeton spirillus is a rhizomatous species with compressed stems that lack nodal glands. Turions are not produced. It has both underwater and floating leaves, although floating leaves are not always present. Underwater leaves are linear, 0.8–8 cm long, and 0.5–2 mm wide; a broad lacunar band is present on each side of the midvein. Stipules are fused to the base of the leaf blade for more than half their length. Floating leaves are 0.7–3.5 cm long, 2–13 mm wide, and have 5–15 veins and an obtuse or rounded tip.

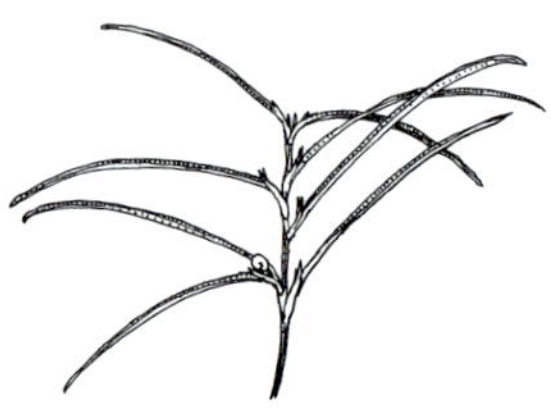

Potamogeton spirillus ×1/2

Flowering spikes are submergent or emergent; those underwater are globular with about 8 flowers, whereas those above the water surface are longer with up to 35 flowers. Fruits are (1.3)1.5–2.4 mm in diameter with an abaxial keel but usually lack lateral keels.

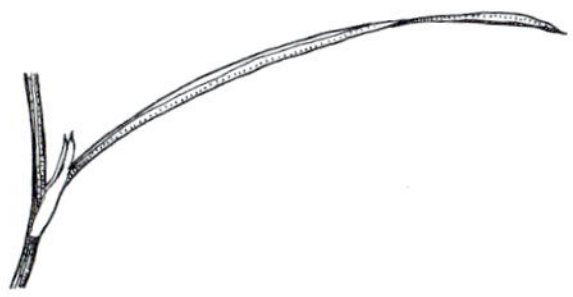

Potamogeton spirillus—stipule ×1.5

Potamogeton spirillus is a northern species, ranging from Newfoundland and Labrador west to southern Ontario and south to Pennsylvania and Michigan. In Pennsylvania it occurs primarily in the northeastern corner of the state, north of the glacial boundary, where it grows in quiet water of lakes, ponds, and rivers.

Potamogeton spirillus is very similar to *P. diversifolius*. The two are best distinguished by the relative length of the free part of the stipular sheath: in *P. spirillus* the free part of the stipular sheath is shorter than the portion fused to the leaf; and by fruit size; fruits of *P. spirillus* are mostly more than 1.5 mm in diameter whereas those of *P. diversifolius* are smaller.

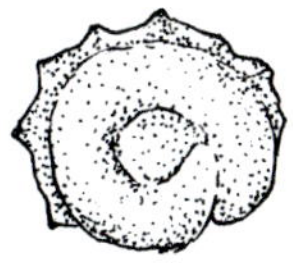

Potamogeton spirillus—achene ×10, note prominent keel

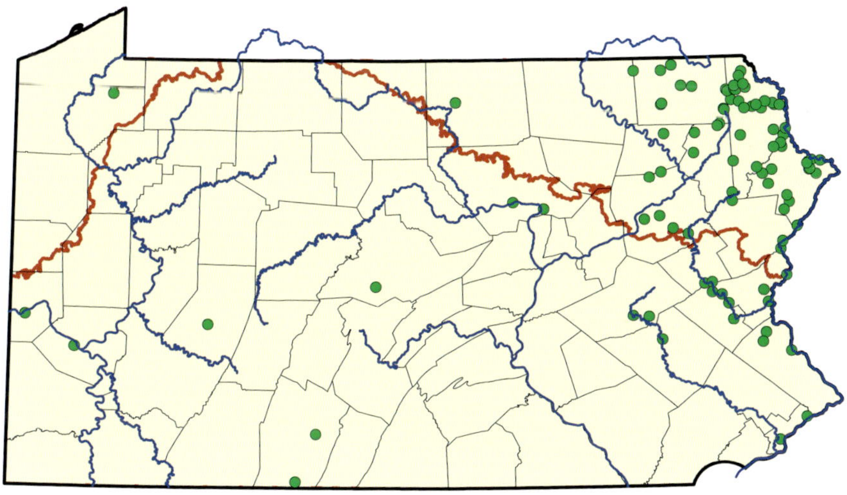

Potamogeton spirillus

Narrow-leaved or straight-leaved pondweed — *Potamogeton strictifolius* A. Benn.

PNHP
Native
Rooted, submergent perennial

Narrow-leaved pondweed has slender, rounded, branched stems to 1 m; rhizomes are not produced. Nodal glands are present, and turions are common in either axillary or terminal positions. The leaves are all submersed, linear, rigid, 0.6–2 mm wide, with 3–7 veins and an acute or bristle-tipped apex. Lacunar bands are absent. Stipules are free, white, 7–15 mm long; with time they become fibrous and shredded at the tips.

Flowering spikes are mostly terminal, slender, 10–15 mm long, with 3–4 well-spaced whorls of flowers. Fruits are 2–3 mm in diameter, greenish-brown, and lack keels.

Narrow-leaved pondweed grows in Canada from New Brunswick west to Alberta and south to Pennsylvania, Ohio, and South Dakota. In Pennsylvania it is a very rare plant of calcareous ponds and streams in widely scattered locations.

The rigid leaves of this species hold their shape when removed from water. Only a few other pondweed species in our flora share that characteristic: *P. robbinsii*, *P. crispus*, and *P. zosteriformis*.

Potamogeton strictifolius ×1/2

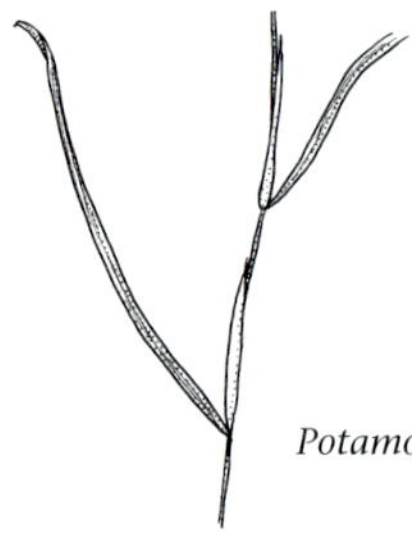

Potamogeton strictifolius—stipule ×1.5

Potamogeton strictifolius—achene ×3

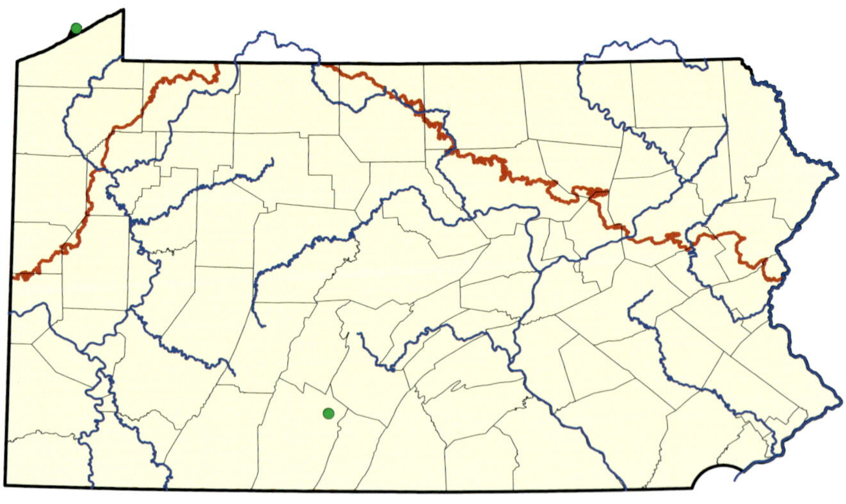

Potamogeton strictifolius

Tennessee pondweed — *Potamogeton tennesseensis* Fernald

PNHP
Native
Rooted, submergent and floating-leaf perennial

Tennessee pondweed is a rhizomatous species with rounded stems that lack glands at the nodes. Both submersed and floating leaves are produced, but floating leaves are not always present. Submersed leaves are linear, 2.5–12 cm long and 0.2–1 mm wide with a broad lacunar band. Stipules are fused to the leaf bases for 25 percent or less of their length. Floating leaves are 2–4 cm long and 5–13 mm wide with 9–23 veins. Turions are not produced.

Flowering spikes are emersed, 3–8 cm long; fruits are about 2.5 mm in diameter with an abaxial keel, lateral ridges, and an erect beak.

Tennessee pondweed has a very limited range extending from Pennsylvania south in the Appalachian Mountains to Tennessee and Kentucky. In Pennsylvania it is a rare plant of ponds and streams in the southwestern corner of the state.

Potamogeton tennessensis—achene ×3

Potamogeton tennessensis—submersed leaf with stipule ×1/2

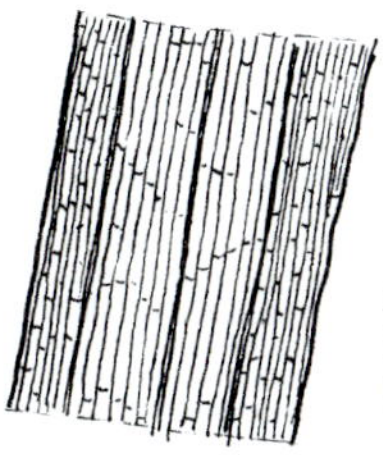

Potamogeton tennessensis—leaf section showing wide lacunar band ×10

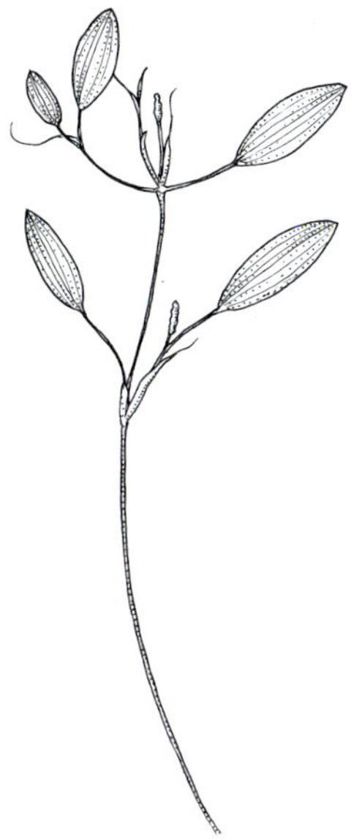

Potamogeton tennessensis ×1/2

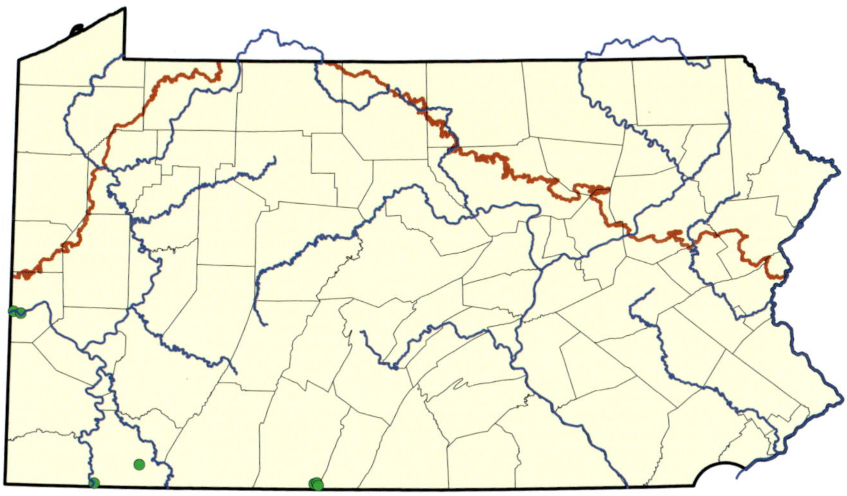

Potamogeton tennessensis

Vasey's pondweed — *Potamogeton vaseyi* J.W. Robbins

PNHP
Native
Rooted, submergent and floating-leaf perennial

Like many of the pondweeds this species has an elongate branching stem bearing alternately arranged underwater leaves; broader floating leaves are formed near the stem tip. Vasey's pondweed is a delicate plant to 1 m long; its submersed leaves are less than 1 mm wide and 2.5–5 cm long. The blades

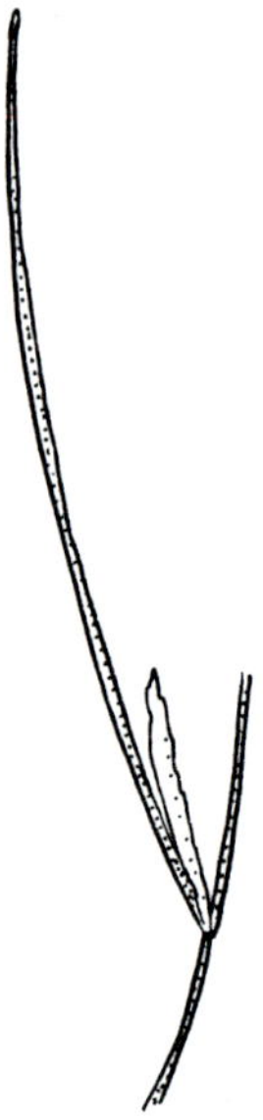

Potamogeton vaseyi—stipule ×3

of the floating leaves are at most 1.5 cm long and 1 cm wide. Stipules are completely free from the leaf bases and fibrous; a pair of small glands is present on the stem at the base of each underwater leaf.

Flowering spikes are associated with the floating leaves and bear 1–4 whorls of flowers or fruits. Fruits are green to brown, 1.5–2.5 mm with an abaxial keel and an erect beak. Flowering and fruiting occur in early summer, and the plant senesces by late July.

Although sometimes classified as an annual, Vasey's pondweed, like many other species in the genus, forms turions. They appear as axillary shoots with rolled and hardened leaves.

Vasey's pondweed grows from New Brunswick and southern Quebec to Minnesota and south to Pennsylvania, Ohio, and Illinois; thus Pennsylvania is at the southern limit of its natural range. Here it is found in lakes, ponds, and other slow-moving waters at a handful of widely scattered locations. Although it is known from only a few scattered sites, the fact that this species senesces early in the season may have resulted in its being missed in some surveys over the years.

Plants without floating leaves or inflorescences could easily be confused with *Potamogeton pusillus*.

Potamogeton vaseyi ×1

Potamogeton vaseyi—achene ×3

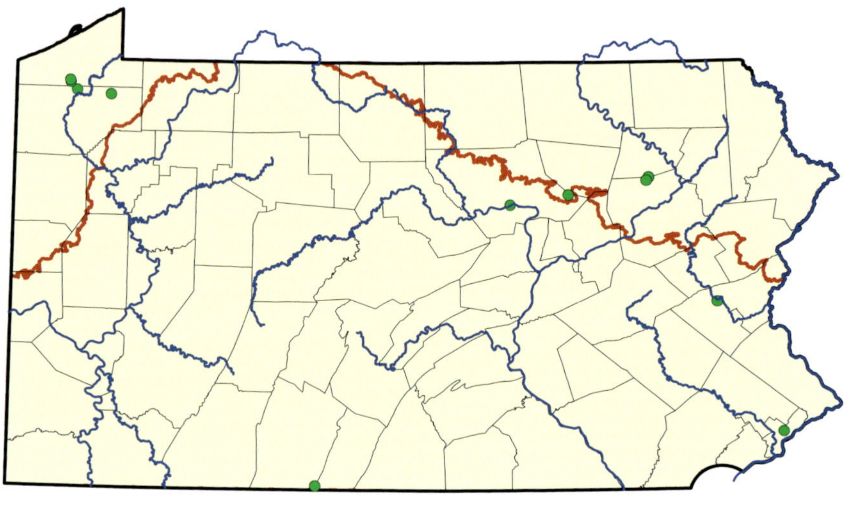

Potamogeton vaseyi

Flat-stem pondweed *Potamogeton zosteriformis* Fernald

PNHP
Native
Rooted, submergent perennial

Flat-stem pondweed has conspicuously flattened stems, the only *Potamogeton* species in our region with this characteristic. Rhizomes are not produced; nodal glands are absent or very rarely present, but turions are common, either terminal or lateral. Leaves are all submersed, 2-ranked, and rigid. The blades are linear, 2–5 mm wide, with 15–35 veins and a blunt, acuminate, or bristle-tipped apex. Lacunar bands are not present. Stipules are free from the leaf blade, and white and shredding at the tips.

The flowering spikes of flat-stem pondweed extend above the water surface; they are cylindrical and 2–5 cm long. Fruits are light green to olive green, 4–5 mm, with a toothed abaxial keel.

Flat-stem pondweed grows from Newfoundland and Labrador south to New Jersey and west to Alaska and California. In Pennsylvania it is a rare plant of ponds and slow-moving streams, mostly in the northwestern corner of the state.

Potamogeton zosteriformis ×1/2

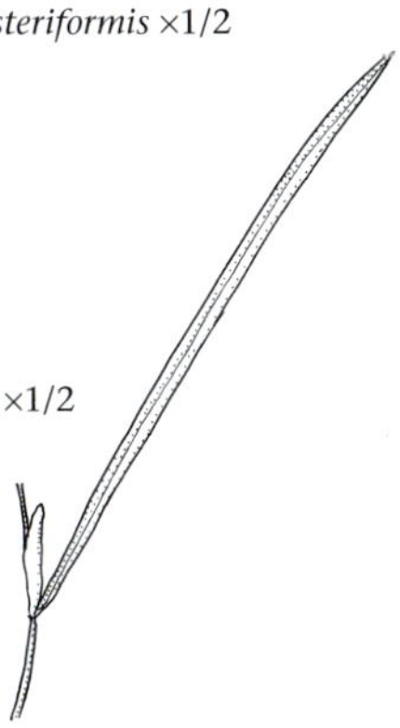

Potamogeton zosteriformis—stipule ×1/2

Potamogeton zosteriformis—achene ×3

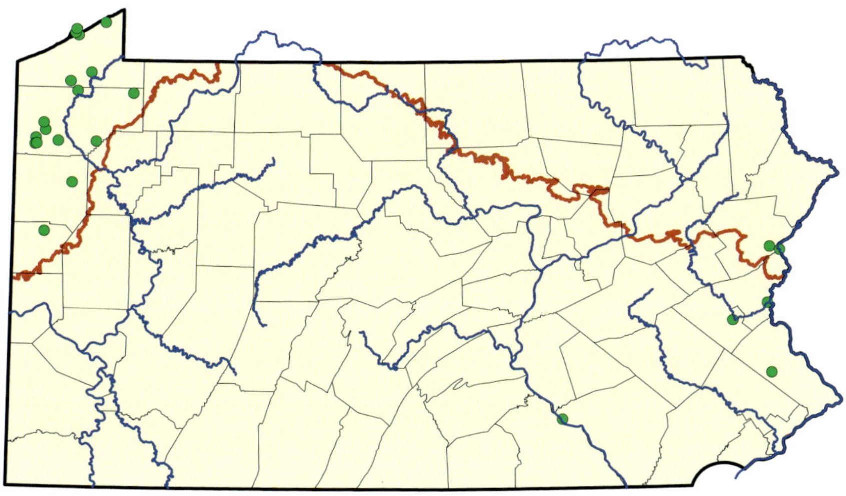

Potamogeton zosteriformis

Horned pondweed *Zannichellia palustris* L.

Native
Rooted, submergent annual

Horned pondweed is a rooted submergent annual with slender branching stems that arise from slender rhizomes. Leaves are opposite, linear, 0.2–1 mm wide, and have sheathing stipules. Flowers are unisexual, but male and female flowers are present on the same plant, making the species monoecious. Flowers are located in the leaf axils; pollination occurs under water. Fruits consist of a cluster of stalked achenes, each 2–3 mm long and curved. A single species occurs in our area.

Horned pondweed occurs throughout North America from southern Canada to Mexico and in South America, Eurasia, Africa, and Australia. In Pennsylvania it is a plant of lakes, springs, and tidal mudflats in brackish or alkaline waters at scattered locations across the state. The fruits and vegetative stems of horned pondweed are a highly preferred food of ducks.

Zannichellia palustris—achene ×5

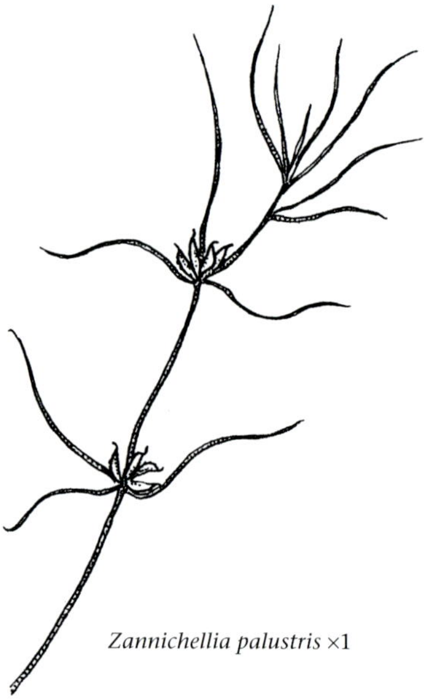

Zannichellia palustris ×1

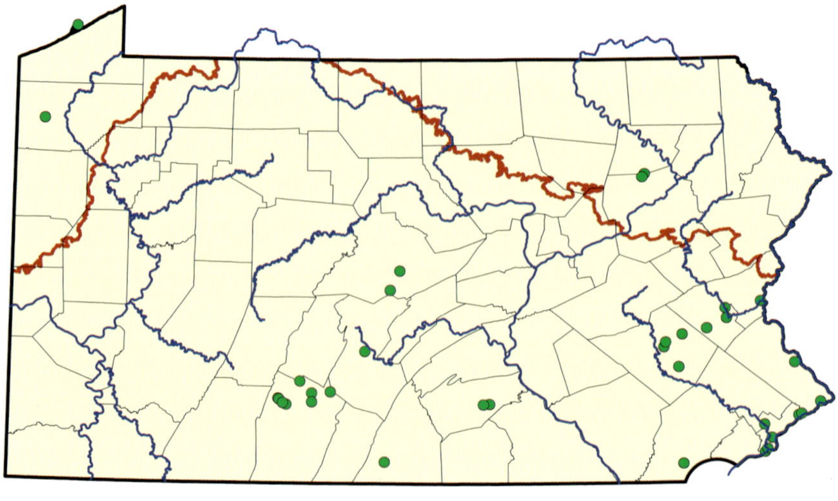

Zannichellia palustris

WATER STAR-GRASS
HETERANTHERA RUIZ & PAVON
Pickerelweed Family—Pontederiaceae

Water star-grass ***Heteranthera dubia* (Jacq.) MacM.**
[synonym: *Zosterella dubia* (Jacq.) Small]

Native
Rooted, submergent perennial

Water star-grass is a submergent, branched perennial with elongate stems up to 2 m long that extend out in the direction of water flow. The alternate, linear leaves have neither a stalk nor a midrib. Solitary yellow flowers are produced at the water surface; flowering occurs from June to September.

The range of water star-grass extends from New Brunswick to Wisconsin and south to Texas in the east; it also includes Washington, California, and Arizona and extends south to Cuba and Central America. In Pennsylvania, water star-grass is a plant of rivers, lakes, and streams throughout. In clear water it can grow at depths of 3 m or more. During periods of low water the plants may persist on exposed mud flats.

When not in flower this species could be confused with several of the pondweeds (*Potamogeton* spp.). However, the lack of stipules is a reliable distinguishing characteristic.

Heteranthera dubia ×1/2

Heteranthera dubia

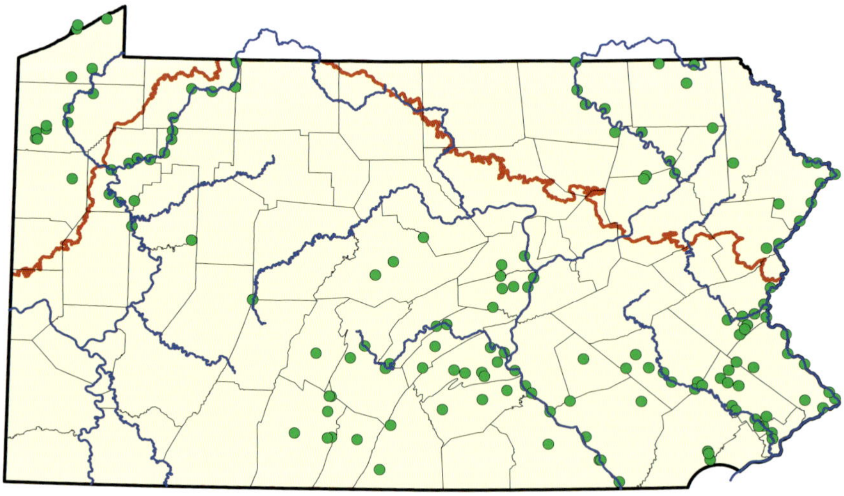

Heteranthera dubia

WATERNYMPH, NAIAD

Najas L.

Frog's-bit Family—Hydrocharitaceae

The waternymphs are rooted, submergent annuals with opposite leaves that have a linear blade and sheathing base. In clear water they can grow at depths of 3 m or more. Flowers and fruits are located individually in the leaf axils; the flowers are unisexual, but both male and female flowers are present on each plant, making the species monoecious. Pollination occurs underwater. The fruit is a fusiform achene tapering at the top to the persistent style.

There are five species of waternymph in Pennsylvania; four are native, and one is introduced. *Najas* species are an important food source for ducks; both the fruits and the vegetative parts of the plants are consumed.

Key to the Species of Waternymph

A. leaf teeth coarse, visible to the naked eye; stem and underside of the leaf midvein prickly .... *Najas marina*

A. leaf teeth finer, in some species visible only with magnification; leaves and stem not prickly

- B. margins of the leaf blades minutely spinulose (visible at 20×); tops of the leaf sheaths gradually tapered to the base of the blade
  - C. leaves 0.2–0.6 mm wide, up to 3 cm long, with an acute tip; fruit obovoid, 2.5–3.7 mm long including the 1.5–1.7 mm persistent style .... *Najas flexilis*
  - C. leaves 0.2–2.1 mm wide, up to 1.8 cm long, with an acute tip; fruit 1.2–3.8 mm including the 0.3–1.5 mm persistent style .... *Najas guadalupensis*
- B. margins of the leaf blades denticulate; tops of the leaf sheaths lobe-like or truncate
  - D. leaves recurved, with 7–15 teeth per side; lobes of the sheath fan-shaped and finely toothed .... *Najas minor*
  - D. leaves not recurved, thread-like with 13–17 teeth per side; lobes of the sheath wedge-shaped and coarsely jagged .... *Najas gracillima*

Northern waternymph — *Najas flexilis* (Willd.) Rostk. & Schnidt

Native
Rooted, submergent annual

Northern waternymph is our most common *Najas* species. Its stems are 25–50 cm tall and profusely branched above. Leaf blades are 0.2–0.6 mm wide and up to 3 cm long; they taper to a long slender tip and slope gradually to the sheathing base. Plants are variable and can range in color from green to reddish. The fruits are 2.5–3.7 mm long including the 1.5–1.7 mm-long persistent style.

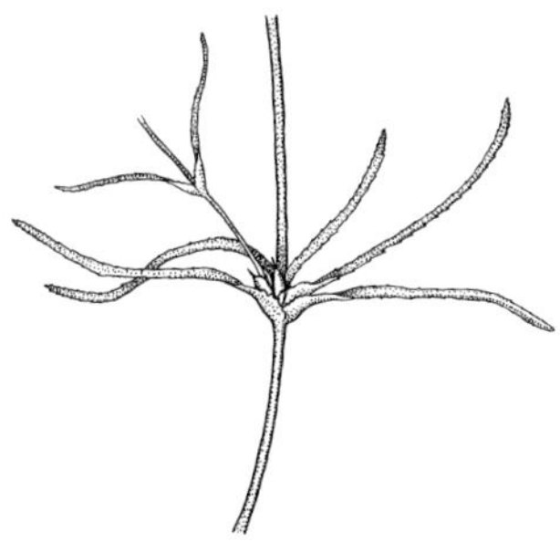

Najas flexilis—node showing leaf bases and fruits in leaf axils ×1

The range of northern waternymph includes the northeastern United States and adjacent areas of southern Canada plus scattered sites farther west. In Pennsylvania it is common in lakes, ponds, streams, and canals throughout.

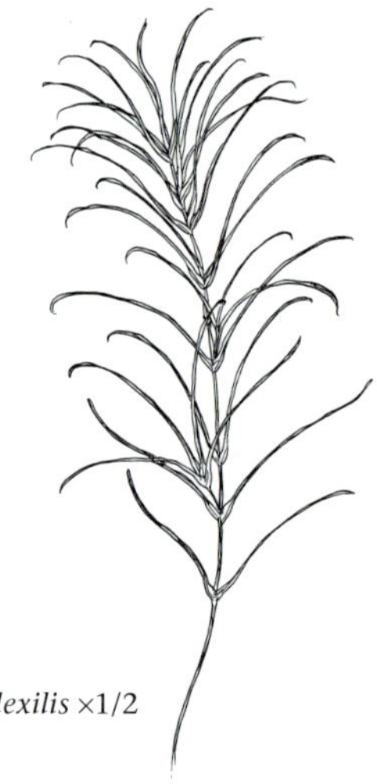

Najas flexilis ×1/2

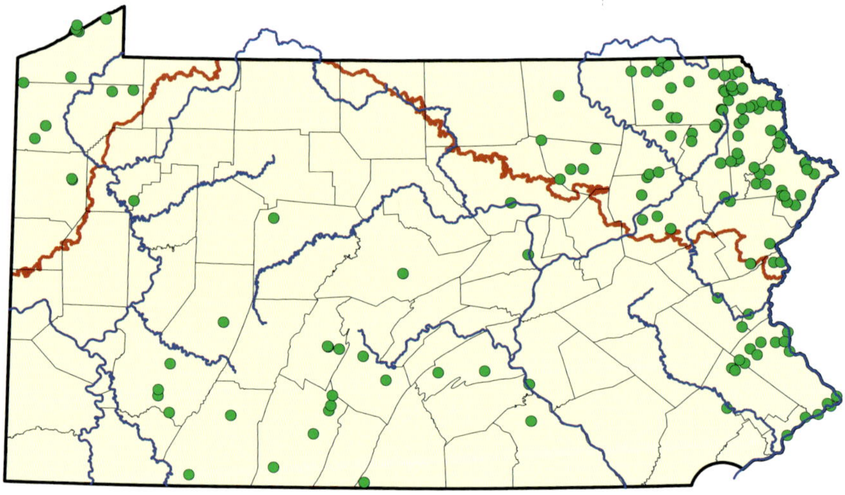

Najas flexilis

Slender waternymph, bushy naiad — *Najas gracillima* (A. Braun) Magnus

Native

Rooted, submergent annual

Slender waternymph is a bushy, rooted plant with sparingly branched stems that are 25–50 cm tall. Leaf blades are 0.1–0.5 mm wide, and spreading. The sheath is abruptly expanded and minutely toothed across its lobe-like summit. Seeds are light brown, 2.5–3.5 mm long with a slightly asymmetrical apex.

The range of slender waternymph extends from New Brunswick to North Carolina and west to Wisconsin; it also grows in Eurasia. In Pennsylvania this species is found in shallow water of lakes, ponds, and reservoirs throughout.

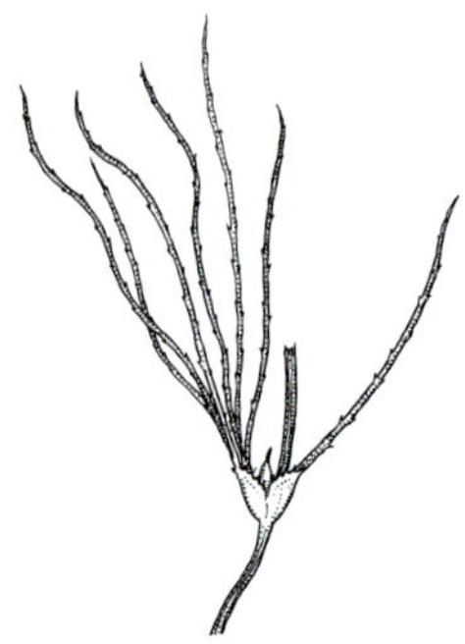

Najas gracillima—node showing leaf bases and a fruit in the leaf axil ×1

Slender waternymph is more delicate and has narrower leaves than any of our other *Najas* species; the leaves and stems are brittle and easily broken during handling.

Najas gracillima ×1/2

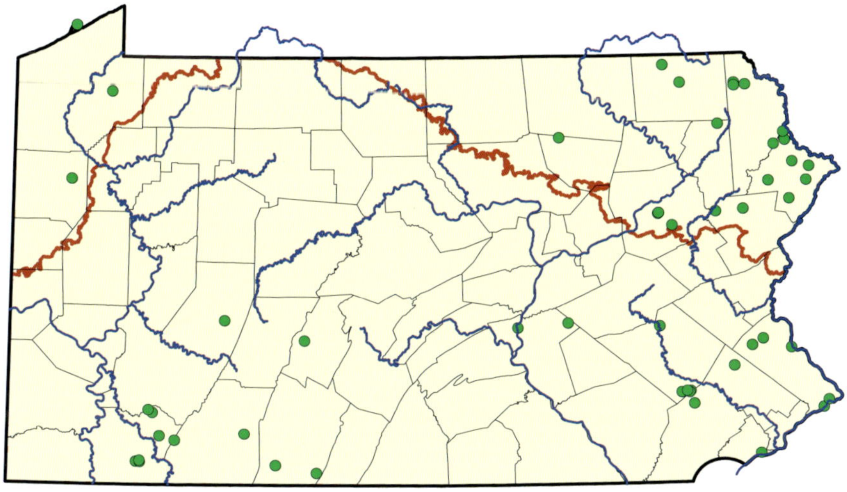

Najas gracillima

Southern waternymph, southern naiad — ***Najas guadalupensis* (Spreng) Magnus**

Native

Rooted, submergent annual

Stems of southern waternymph are up to 1 m long and lightly to profusely branched. Stems and leaves are bright green, with the pairs of leaves well separated along the stems. Leaf blades are 0.2–2 mm wide, up to 1.8 cm long and very finely toothed; they have an acute tip and taper to the expanded sheathing base. The fruits are 1.2–3.8 mm long including the 0.3–1.5 mm persistent style.

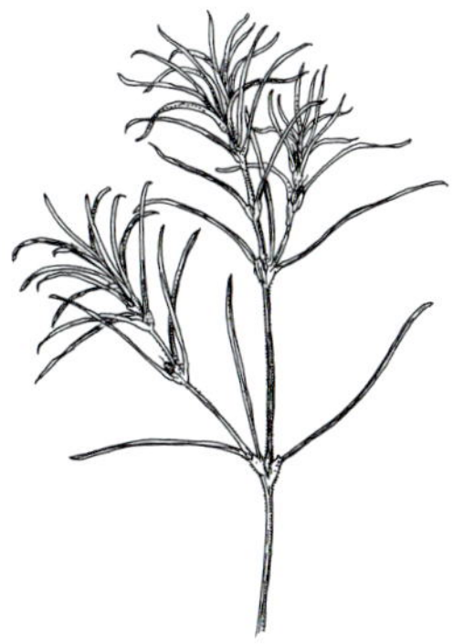

Najas guadalupensis ×1/2

Southern waternymph occurs throughout the eastern United States and at scattered locations farther west. Its range also extends into Mexico, the West Indies, Central America, and South America. In Pennsylvania it is known from scattered sites including lakes, ponds, reservoirs, and tidal flats.

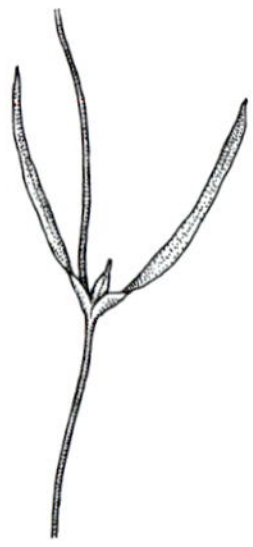

Najas guadalupensis—node with a fruit in the leaf axil ×1

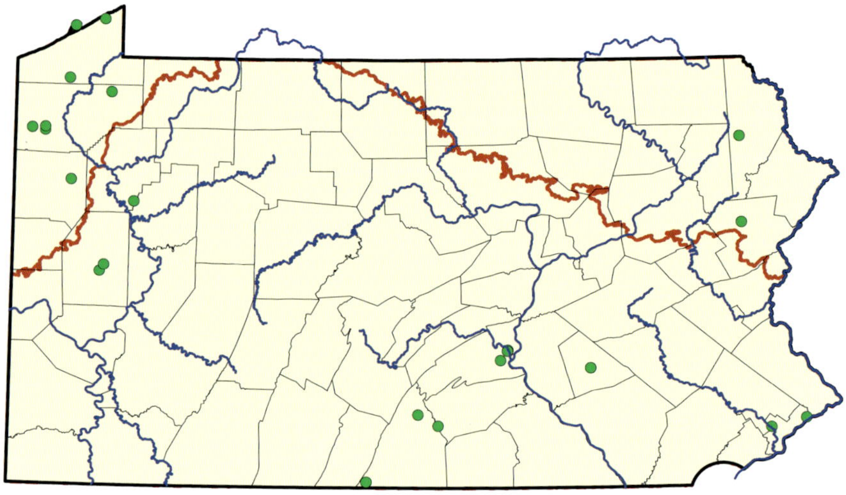

Najas guadalupensis

Holly-leaved naiad *Najas marina* L.

Native

Rooted, submergent annual

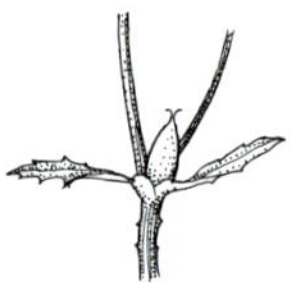

Najas marina—node with a fruit in the leaf axil ×1

Holly-leaved naiad has stems 5–45 cm long that are often prickly; leaves are coarsely toothed with prickles on the midrib beneath. Stems and leaves are bright or pale green or sometimes reddish, and brittle. The reddish-brown fruits are 4–7.5 mm long.

A plant of alkaline or brackish water, holly-leaved naiad is known from Pennsylvania, New York, the Great Lakes region, and scattered locations across the southern and southwestern United States. It also grows in Mexico, Central America, South America, Europe, and Asia. The current distribution of this species is interpreted as remnants of an earlier circumpolar distribution that was fragmented by continental glaciation (Stuckey 1985). In Pennsylvania the species is known from a single site at a quarry in Blair County.

The coarse leaf teeth and prickles make this species the easiest of our *Najas* species to identify.

Najas marina ×3/4

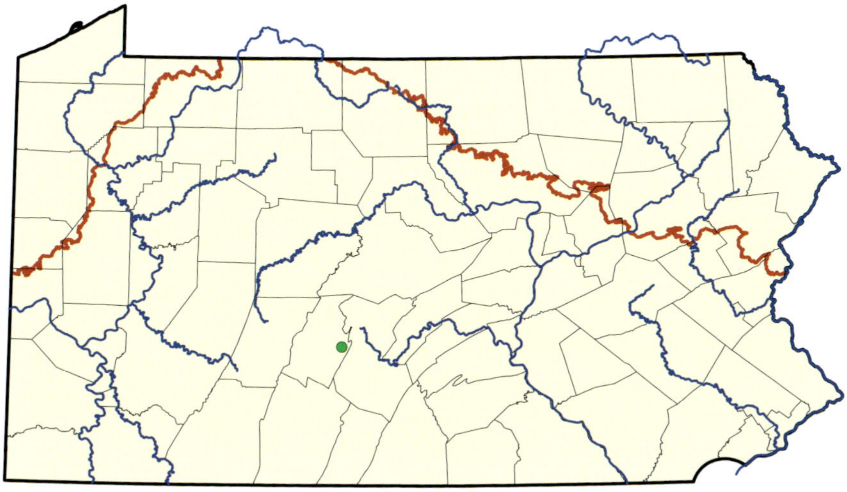

Najas marina

Waternymph, naiad — *Najas minor* All.

Introduced
Rooted, submergent annual

Najas minor has profusely branched stems 10–20 cm long; the leaves become stiff and recurved with age. Leaf margins are conspicuously toothed, but unlike *N. marina*, there are no prickles on the midrib or stem. The apex of the leaf sheath is truncate with coarse, jagged teeth. The seeds are purplish and slightly recurved.

Najas minor is native to Eurasia and Africa. In Pennsylvania it is naturalized in lakes, ponds, and reservoirs scattered across the state.

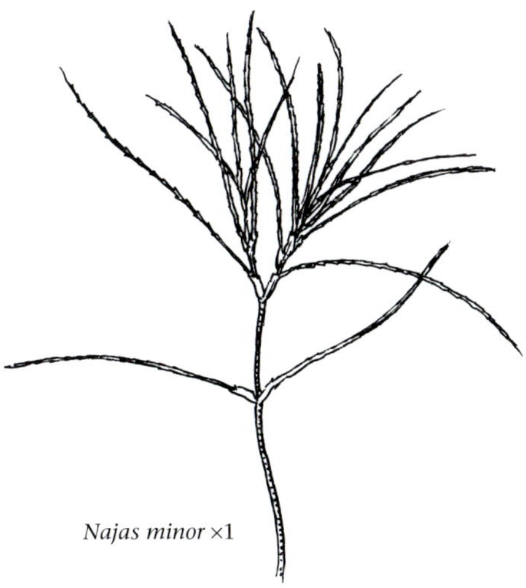

Najas minor ×1

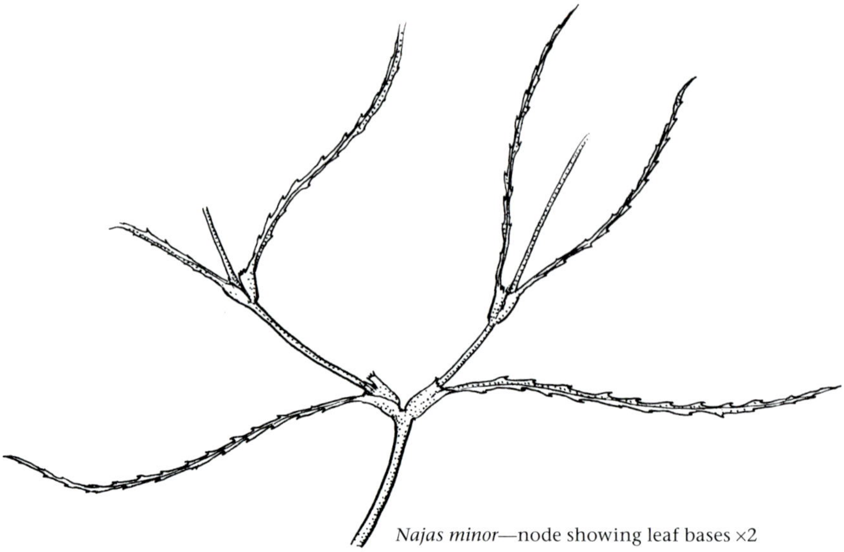

Najas minor—node showing leaf bases ×2

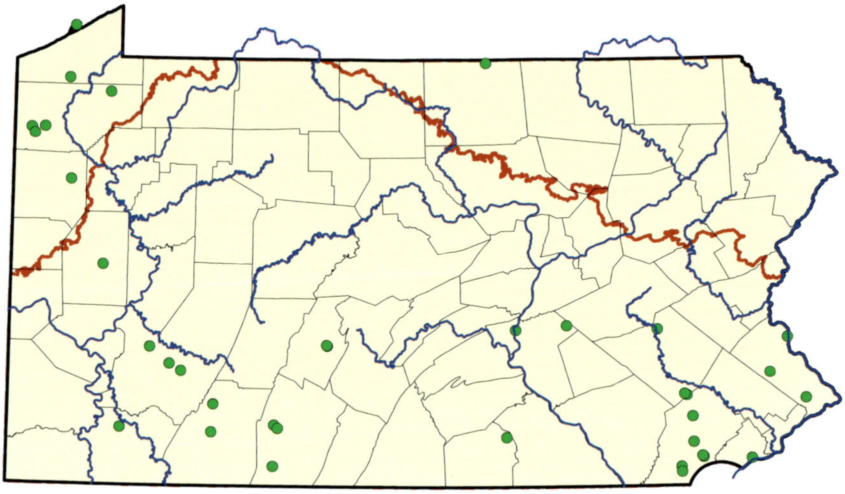

Najas minor

WATER-STARWORT
CALLITRICHE L.
Plantain Family—Plantaginaceae

The water-starworts are small, branched annuals of aquatic or moist sites. Leaves may be submersed, floating, or emersed; they are opposite and lack stipules. Flowers are axillary, much reduced, unisexual, and mostly submersed. Pistillate flowers contain a flattened, 2-lobed ovary; male flowers consist of 1–3 stamens. Pollination takes place beneath the surface of the water. The fruits separate into 4 nutlets at maturity.

Because of the variability of leaf shape, fruit characteristics are essential for accurate identification of the three native and one introduced species of water-starwort present in Pennsylvania.

Key to the Species of Water-starwort

A. plants mostly terrestrial; fruit with a short pedicel ....*Callitriche terrestris*
A. plants mostly submersed; fruits sessile
  B. fruits shallowly grooved, margins rounded .... *Callitriche heterophylla*
  B. fruits distinctly grooved, margins winged
    C. leaves to 3 mm wide and 0.5–1.5 cm long; fruit obovate, longer than wide, 1–1.4 mm long ....*Callitriche palustris*
    C. leaves 5–8 mm wide and up to 2 cm long; fruit suborbicular, 1.5–2 mm long .... *Callitriche stagnalis*

Water-starwort — ***Callitriche heterophylla* Pursh emend Darby**

Callitriche heterophylla ×1/2

Native
Rooted, submergent annual

Callitriche heterophylla is the most common species of water-starwort in Pennsylvania. It is typically mostly submersed with a few floating leaves clustered at the stem tip. Submersed leaves are linear; floating leaves are more nearly round. Fruits are 1 mm or less, wider above the middle and very shallowly grooved with rounded edges.

Callitriche heterophylla occurs throughout North and South America; in Pennsylvania it is a common plant of low-nutrient lakes, ponds, slow-moving streams, and muddy shores throughout.

Callitriche heterophylla—fruit ×10

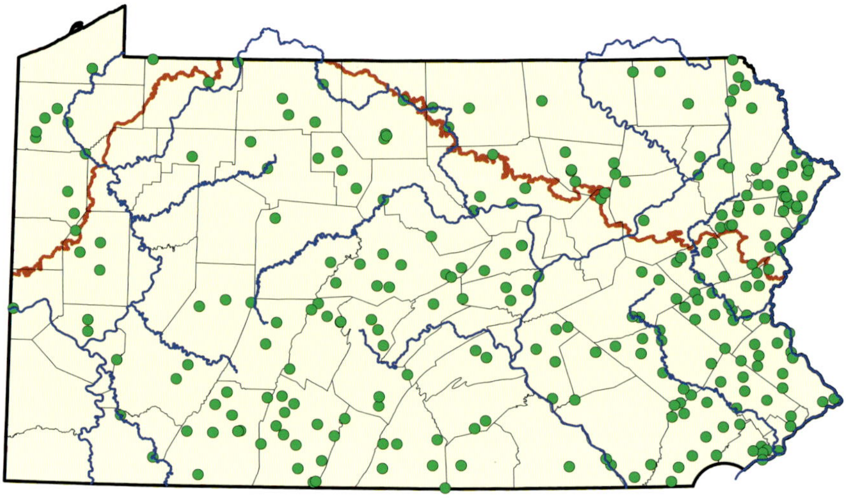

Callitriche heterophylla

Water-starwort — *Callitriche palustris* L.

Native

Rooted, submergent annual

Callitriche palustris is a submergent plant with only the stem tips floating or emergent. Stems are loosely clustered and frequently produce roots at the lower nodes. Submersed leaves are linear; floating or emergent ones are more rounded and up to 3 mm wide with 3 nerves. The fruits are 1–1.4 mm long and widest above the middle; they are compressed, grooved, and narrowly winged.

Callitriche palustris occurs throughout the boreal regions of the northern hemisphere. Pennsylvania populations are near the southern limit of the species in eastern North America. It is occasional in ponds, streams, swamps, springs, and muddy shores.

Callitriche palustris—fruit ×10

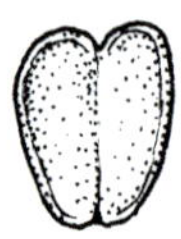

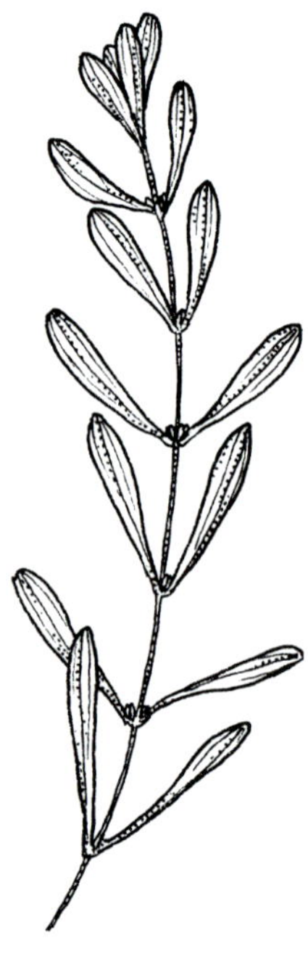

Callitriche palustris ×1

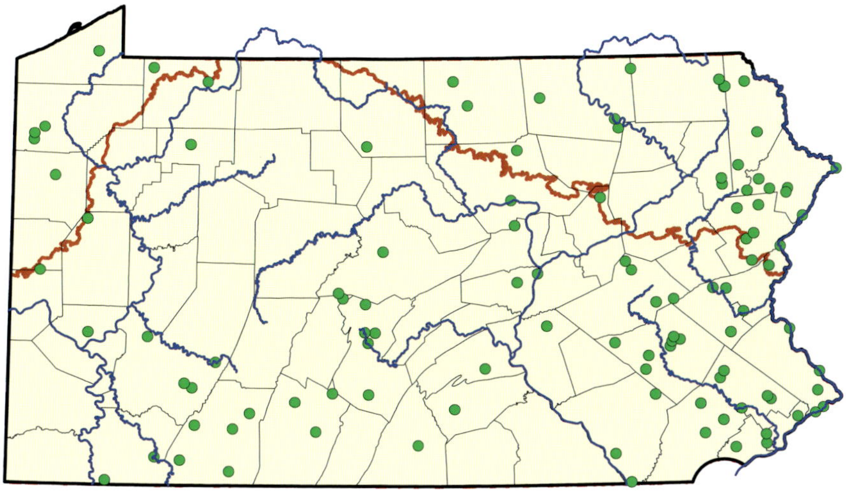

Callitriche palustris

Water-starwort, water-chickweed

***Callitriche stagnalis* Scop.**

Introduced
Rooted, submergent annual

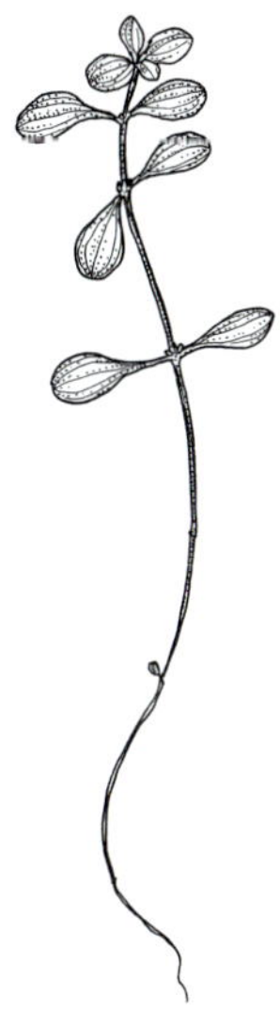

Callitriche stagnalis ×1/2

Callitriche stagnalis—fruit ×4

Callitriche stagnalis is similar to *Callitriche palustris*, but larger. It can be distinguished by having leaves 5–8 mm wide and up to 2 cm long, and by the fruits, which are 1.5–2 mm in diameter, nearly round, and distinctly winged on the edges.

Callitriche stagnalis is native to Europe; it the only non-native species of water-starwort that occurs in our area. In Pennsylvania it has become naturalized in slow-moving waters of rivers, streams, swamps, and ditches. It is extremely hardy, remaining green all winter in southeastern Pennsylvania where most of the populations are concentrated. Flowering and fruiting appear to occur in the spring; late season plants are almost always sterile.

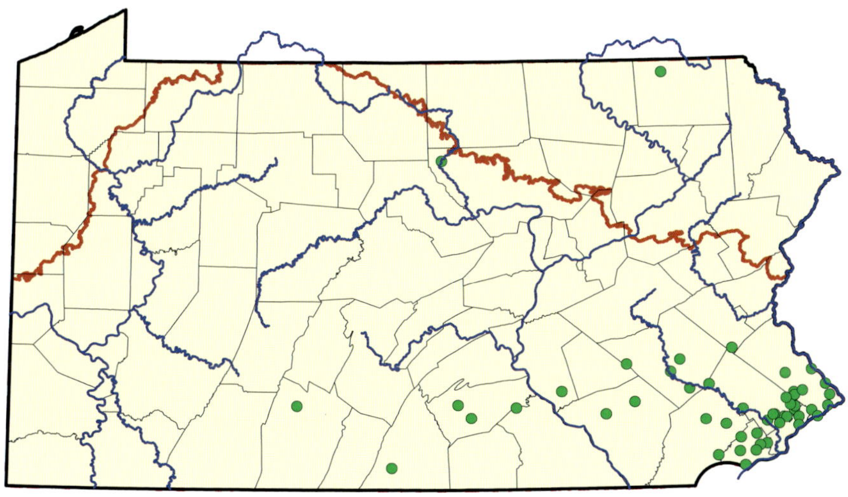

Callitriche stagnalis

Callitriche stagnalis in a spring seep in Bucks County, Pennsylvania, in January

Water-starwort ***Callitriche terrestris*** **Raf. emend Torrey**

Native

Rooted, emergent annual

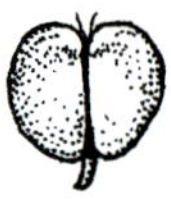

Callitriche terrestris—fruit ×10

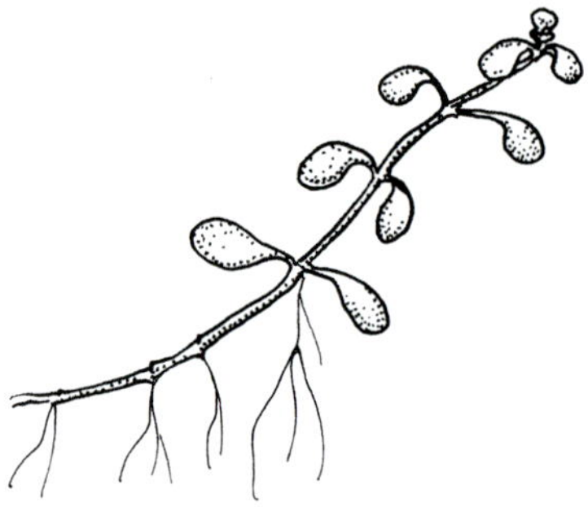

Callitriche terrestris ×1

Callitriche terrestris is a tiny, creeping plant with leaves that are 2–5 mm long and widest near the tip. The fruits are stalked, approximately 1 mm wide, strongly flattened and deeply grooved, forming 4 disk-like segments. The fruit is notched at both ends and has a short persistent style.

The range of *Callitriche terrestris* extends from Massachusetts to Ohio and south to Florida and Texas. In Pennsylvania it is found mostly in the southern half of the state, where it is an occasional plant of moist shaded pond edges, stream banks, and even moist soil.

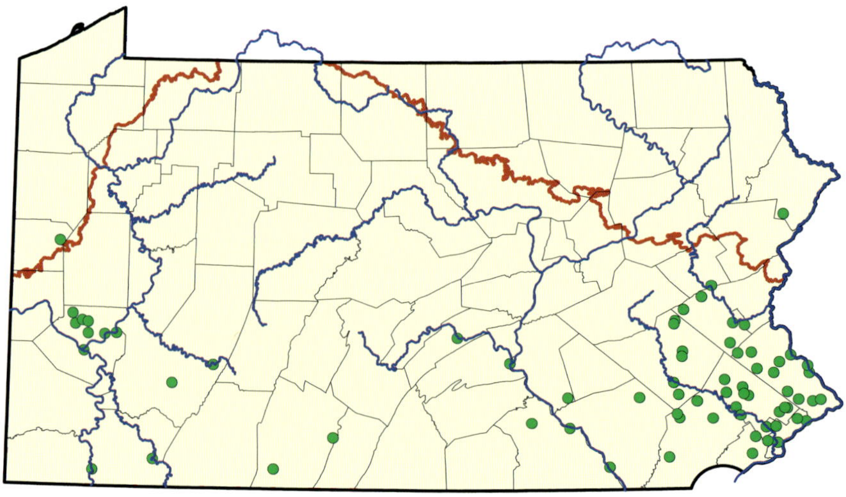

Callitriche terrestris

WATERWEED

Elodea Michaux and *Egeria* Planch.

Frog's-bit Family—Hydrocharitaceae

Waterweed will be familiar to anyone who has kept an aquarium, as it is often sold in pet stores for that purpose. The flowers are unisexual, with male and female flowers borne on separate plants (dioecious). The ephemeral, delicate flowers are white with 3 petals and 3 sepals, borne on long slender stalks that allow the flowers to reach the water surface. Male flowers release their pollen to float on the water surface; when the floating pollen grains contact a female flower, pollination takes place. After pollination a leathery fruiting capsule develops below the water surface. Turions, consisting of condensed stem tips, are also formed.

There are two native species of waterweed in Pennsylvania and one naturalized species. The report of a fourth species, *Elodea schweinitzii*, characterized by perfect flowers, which was based on a few collections from a small area in Northampton County in the early 1800s, is now considered questionable.

The leafy stems of waterweed are eaten by ducks and other waterfowl.

Key to the Species of Waterweed

A. leaves in whorls of 4–8, crowded, those from one node overlapping the next; leaf margins finely toothed...*Egeria densa*

A. leaves in whorls of 3 (occasionally 2 or 4)

B. most leaves more than 1.8 mm wide, flat; male flowers remaining attached to elongate peduncles until just before pollen is shed.........*Elodea canadensis*

B. most leaves less than 1.7 mm wide, folded along the midrib; male flowers detaching and floating to the surface...................................... *Elodea nuttallii*

Brazilian waterweed — *Egeria densa* Planchon

Introduced

Rooted, submergent perennial

Egeria densa ×1/2

Brazilian waterweed consists of elongate stems densely covered with overlapping whorls of 4–8 linear leaves that are 2–4 cm long and 3–6 mm wide. White flowers, approximately 6 mm in diameter, appear in late summer or fall. Male and female flowers, which are produced on separate plants, float on the surface attached to slender stalks.

Brazilian waterweed is an introduced plant that has become naturalized at several sites in southeastern Pennsylvania. It is frequently sold for use in aquariums and water gardens. It has a larger, bushier appearance than the native waterweeds (*Elodea canadensis* and *E. nuttallii*), which it otherwise resembles.

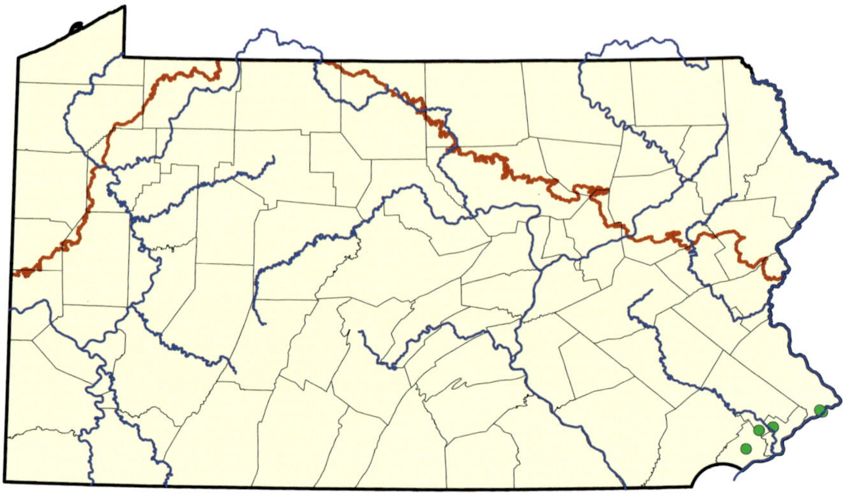

Egeria densa

Elodea canadensis ×1/2

Common waterweed, ditch-moss — *Elodea canadensis* Michaux

Native
Rooted, submergent perennial

Elodea canadensis has its leaves mostly in whorls of 3; they are linear, oblong, or ovate, 6.2–10 mm long and 1.3–2.7 mm wide with an acute to obtuse apex. Those near the stem apex are imbricate; farther down the stem they are flat and spreading or even recurved. Male flowers remain attached to their long stalks until just before pollen is shed. However, male plants are rare in the state; most known populations are female.

The range of *Elodea canadensis* extends across North America from New England and southern Canada south to North Carolina and west to Califor-

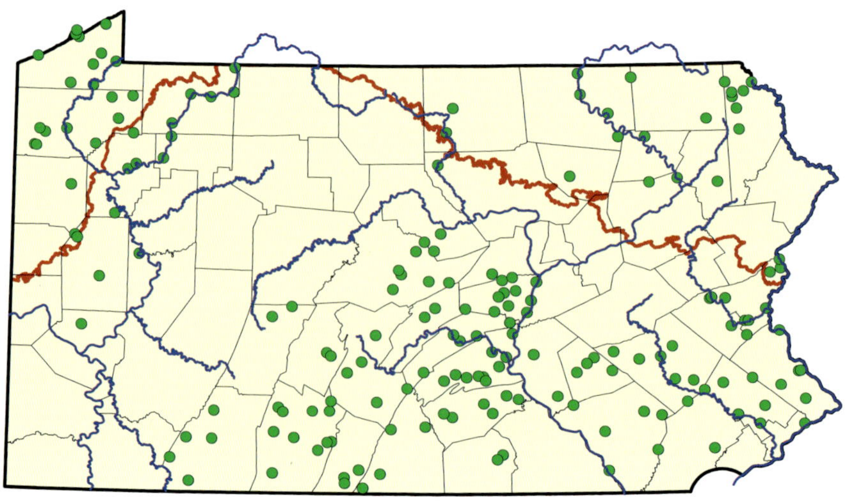

Elodea canadensis

nia, Oregon, and Washington, with the exception of the Great Plains and the Rocky Mountains. In Pennsylvania it is found in limestone-influenced waters of lakes, rivers, and streams throughout.

Elodea canadensis and *Elodea nuttallii* are difficult to tell apart with certainty unless flowers are present. Overall, *E. canadensis* leaves are larger than those of *E. nuttallii*; however, leaf dimensions are variable within each species; season, water depth, and the sex of the plant all have an impact.

Elodea canadensis—flower ×1

Waterweed — *Elodea nuttallii* (Planchon) H. St. John

Native
Rooted, submergent perennial

Elodea nuttallii is the most common waterweed in Pennsylvania. Its leaves are linear or lanceolate, 5–9 mm long and 0.9–1.8 mm wide with an acuminate apex; they are frequently recurved and slightly folded along the midrib. The middle and upper leaves are typically whorled, although some may be opposite. Male flowers are not stalked, but rather are released to float freely to the water surface, where pollen is shed. Flowering occurs late June to early August.

The range of *Elodea nuttallii* extends from southern Maine, Quebec, and Ontario south to Missouri and North Carolina. It is also known from parts of Montana, Oregon, and California. In Pennsylvania it is common in shal-

Flowering stems of *Elodea nuttallii* in a stream in Bucks County, Pennsylvania

low water of rivers, streams, lakes, and ponds throughout. This species is very similar to *Elodea canadensis*, but the leaves of *E. nuttallii* are shorter and narrower with pointed tips, in contrast to the larger size and blunt tips of *Elodea canadensis* leaves. See additional comments above.

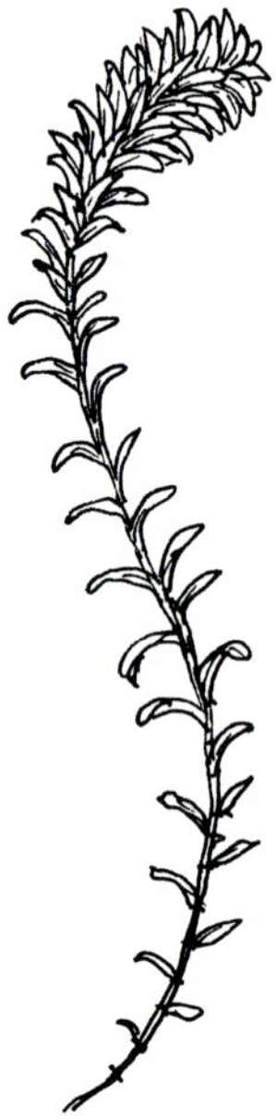

Elodea nuttallii ×1

Elodea nuttallii—flower ×5

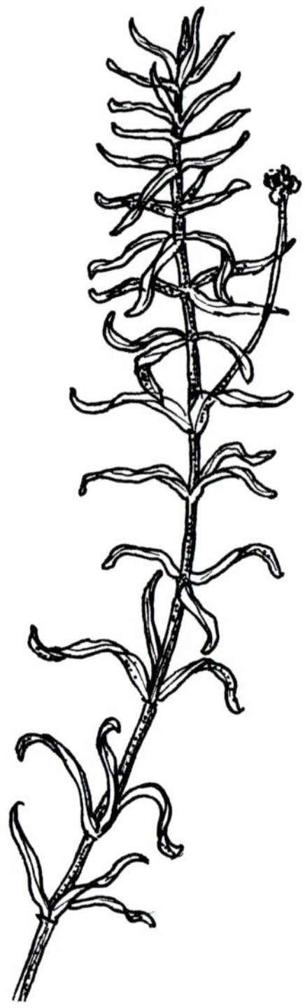

Elodea nuttallii—habit with flower ×1

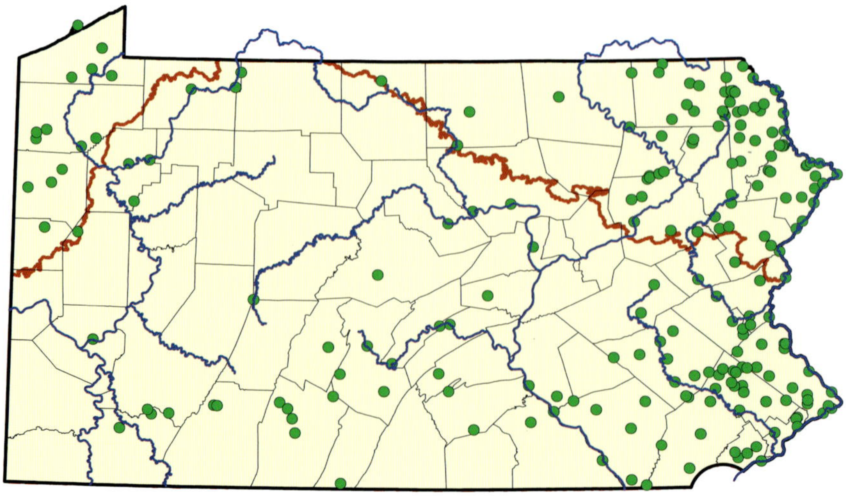

Elodea nuttallii

WATERWORT
ELATINE L.
Waterwort Family—Elatinaceae

The waterworts are tiny rooted annuals that grow completely under water. The plants are only a few centimeters tall at most; they have opposite leaves and tiny flowers that are produced singly in the leaf axils. Roots develop from the lower nodes of the branching stems. Flowers are bisexual; the fruit is a papery capsule containing 4–8 elongate seeds.

Three species of waterwort are known from Pennsylvania; one, *Elatine americana,* is known only from the freshwater tidal marshes of the Delaware River estuary. The other 2 are plants of lakes and ponds. Long-stem waterwort (*Elatine triandra*) has recently spread into Pennsylvania and other areas in the northeast from farther west.

Key to the Species of Waterwort

A. plants holding their shape when removed from the water; leaves rounded; flowers 2-merous; seeds with rounded areolae............................. *Elatine minima*

A. plants limp and flaccid when out of water; leaves longer than wide; flowers 3-merous; seeds with angular areolae

B. leaves obovate or spatulate, rarely notched at the tip.......... *Elatine americana*

B. leaves linear to obovate, often notched.................................... *Elatine triandra*

American waterwort — *Elatine americana* (Pursh) Arn.

PNHP
Native
Rooted, submergent or emergent annual

Long thought to be extirpated in Pennsylvania, American waterwort was recently found on the tidal shore of Little Tinicum Island in the Delaware River after a 50-year absence from the state. Extant populations are also known in New Jersey, Delaware, and Maryland.

American waterwort differs from small waterwort by its flaccid leaves that are longer than wide and the angular areole on the seeds. Like long-stem waterwort its flowers are 3-merous. It is a plant of freshwater tidal mud flats from New Brunswick to Virginia.

Elatine americana ×1

Elatine americana—seed ×40

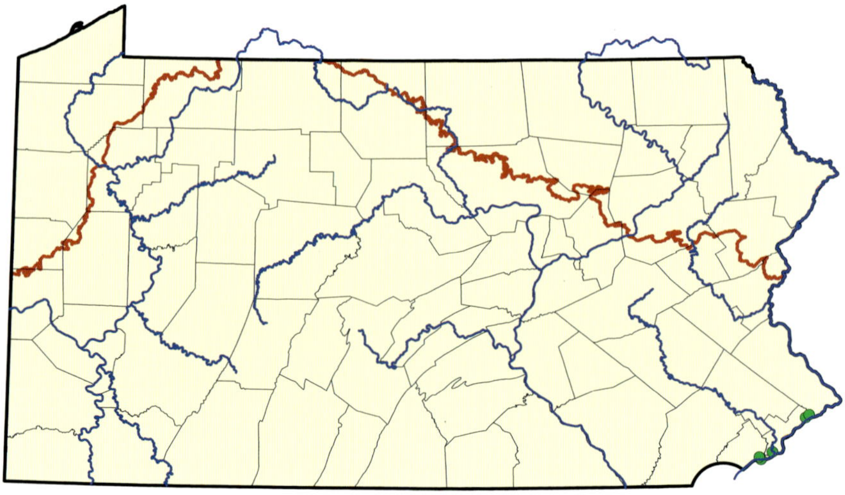

Elatine americana

Small waterwort ***Elatine minima* (Nutt.) Fisch. & C.A. Mey.**

Native
Rooted, submergent annual

Small waterwort is a stiff little plant that retains its shape when removed from the water.

The stems of small waterwort are up to 2.5 cm long and usually branched. Pairs of simple, rounded leaves that are mostly less than 6 mm long are present at the nodes; roots grow from the lower nodes. The 2-merous flowers are produced in the leaf axils, as are the small seed capsules. The seeds are marked on the surface with rows of narrow, rounded areolae.

This northern species grows from Newfoundland west to Ontario and Wisconsin and south along the coast to Maryland and Virginia. In Pennsylvania it is a common native plant of glacial lakes with a sandy bottom, occurring most often in water from a few centimeters to 0.6 m deep.

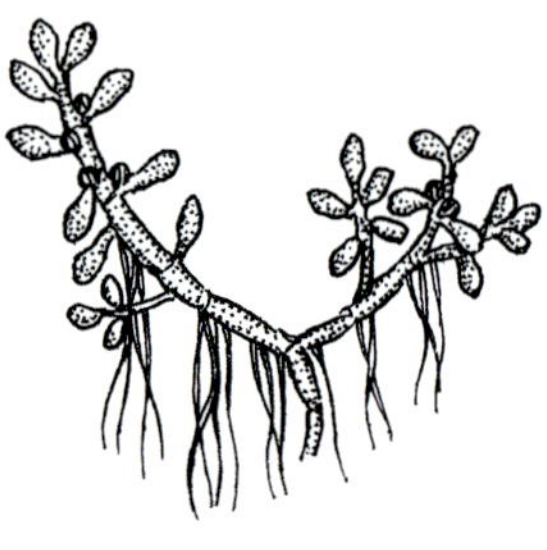

Elatine minima ×1

Elatine minima—seed ×20

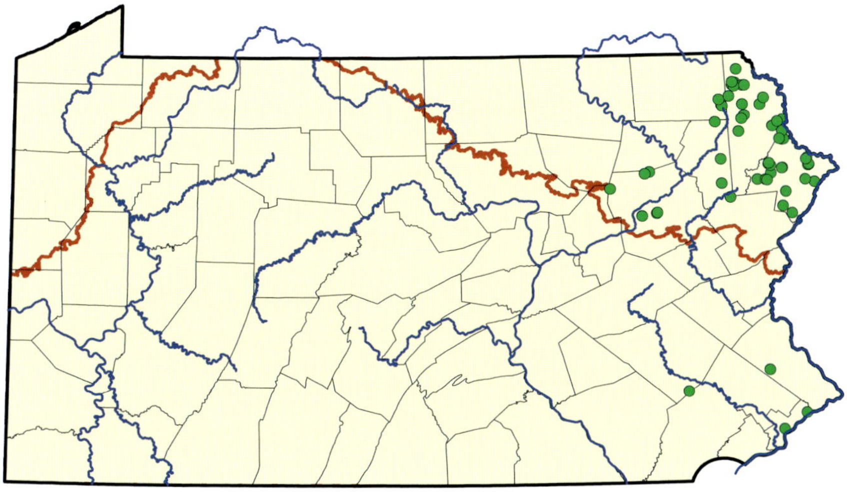

Elatine minima

Long-stem waterwort *Elatine triandra* Schkuhr.

Introduced, or perhaps undergoing a natural range expansion
Rooted, submergent annual

The bright yellow-green plants of long-stem waterwort are conspicuous on the lake bottom where they often grow so densely that they form a carpet of closely spaced individual plants. The branched stems are up to 10 cm long. Linear to spatulate leaves up to 5 mm long with a notched apex occur in pairs at the nodes. The tiny flowers develop in the leaf axils and are followed by papery capsules containing 4–8 seeds. The surface of the elongate seeds is ornamented with parallel rows each containing 16–26 areolae.

Long-stem waterwort grows rooted in the lake bottom in water to about 1 m deep. It grows in soft sediments or a firmer sandy-peaty substrate, often forming a dense carpet. It sometimes occurs in mixed populations with small waterwort.

Long-stem waterwort is native to North America, but has only recently been found in the northeast, where it appears to be expanding its range from Maine to Pennsylvania. We have recorded its presence at more than a dozen

Elatine triandra ×1

lakes in eastern Pennsylvania in the last few years, but a review of herbarium specimens reveals that it was present in at least one Pocono area lake as early as the 1960s. It has long been known from the western states and also from Europe and Asia. It has also recently been found in Wisconsin.

The leaves of long-stem waterwort are longer than wide and limp and flaccid when the plant is removed from the water, in contrast to the stiff form and rounded leaves of *Elatine minima*. In addition, long-stem waterwort flowers are 3-merous, whereas small waterwort has its flower parts in 2s. Long-stem waterwort might also be mistaken for submergent forms of water-starwort (*Callitriche* spp.). Both genera have opposite leaves and axillary flowers and fruits. The fruit shape is entirely different, with a papery, easily ruptured capsule in waterwort compared to the flattened, non-dehiscent 2-lobed nutlet of water-starwort.

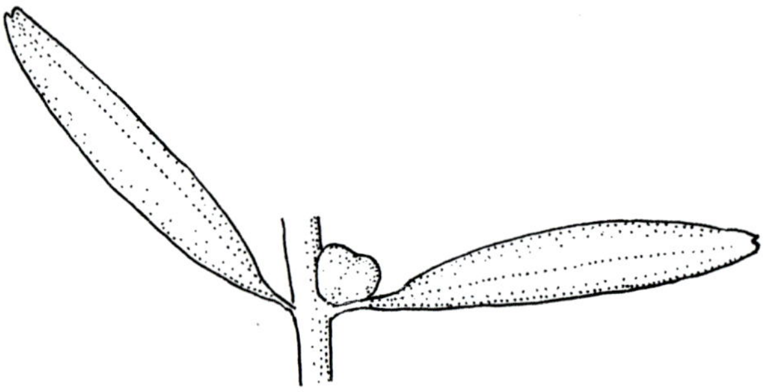

Elatine triandra—node with capsule in leaf axil ×3

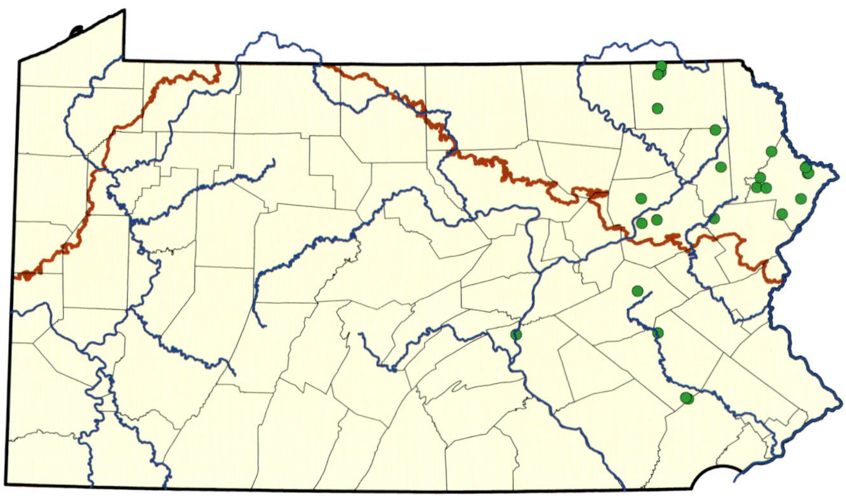

Elatine triandra

Chapter 7. Submergent Rosette-forming Plants

Plants in this group are rooted in the lake bed and form a basal rosette of leaves. The group includes flowering plants and the quillworts, which reproduce by spores. In the flowering species an erect, leafless, flowering stem arises from the basal rosette, and flowers are usually borne above the water surface.

Sterile (non-flowering) rosettes of species other than those treated in this section may be present at some sites. These are very difficult to identify although leaf venation and leaf shape may provide some clues. See additional comments under arrowheads, especially *Sagittaria rigida* and *Sagittaria graminea*.

MUDMAT
Glossostigma L.
Lopseed Family—Phrymaceae

Mudmat — ***Glossostigma cleistanthum* W.R. Barker [*Glossostigma diandrum* (L.) Kuntze of Rhoads and Block 2000]**

Introduced
Rooted, submergent or emergent perennial

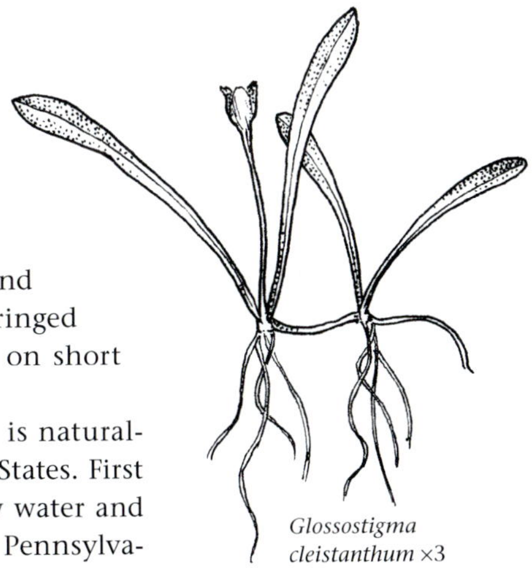

Glossostigma cleistanthum ×3

Mudmat is a tiny submergent or emergent mat-forming, rooted perennial only 1–2 cm tall. Its leaves are strap-shaped and develop in pairs from nodes of slender stolons. Under emergent conditions flowers are produced singly on stalks about as long as the leaves; the 1 mm-wide corolla is pinkish and irregular. The flowers have 2 stamens and a flattened, fringed stigma. Closed (cleistogamous) flowers are produced on short pedicels under submergent conditions.

Mudmat is native to New Zealand and Australia; it is naturalized at scattered locations in the northeastern United States. First detected in Pennsylvania in 1994, it grows in shallow water and on exposed shoreline mudflats of several southeastern Pennsylvania lakes. It has also been found in New Jersey, Connecticut, and Rhode Island (Les et al. 2006).

Available in the aquarium trade, mudmat probably became established after someone emptied a fish tank into a local lake or pond.

Glossostigma cleistanthum—flower ×5

Glossostigma cleistanthum growing in a lake in Bucks County, Pennsylvania

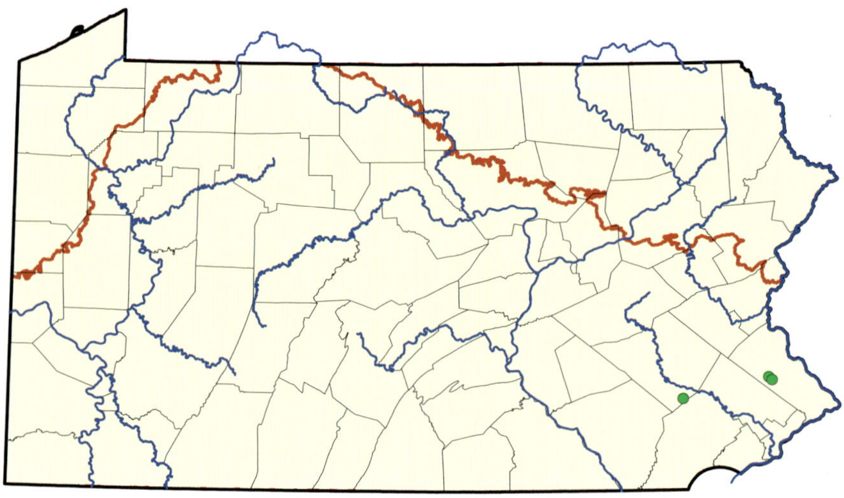

Glossostigma cleistanthum

PIPEWORT
ERIOCAULON L.
Pipewort Family—Eriocaulaceae

Seven-angle pipewort, white-buttons — ***Eriocaulon aquaticum* (Hill) Druce**

Native
Rooted, submergent or emergent perennial

White-buttons is a very descriptive name for this little plant, which can be completely submerged but more often produces its flowering heads on stems that extend above the water surface. The pipeworts are perennial plants consisting of a rosette of narrow pointed leaves and one or more unbranched stems each bearing a dense head of tiny white flowers.

Flowering stems range from a few centimeters tall on saturated peat or muck in bogs to 0.5 m in deeper water. The sterile rosettes, which grow from rhizomes, sometimes form a turf on the lake bottom. The leaves have a distinctive silvery appearance under water, making them recognizable even in the absence of flowering stems. The roots too have a distinctive cross-hatched texture.

Seven-angle pipewort is fairly common in low-nutrient northern lakes in the glaciated portions of Pennsylvania. It is a plant of the northeastern United States and adjacent areas of Canada.

Two other pipewort species found in Pennsylvania in the past are believed to be extirpated. Ten-angle pipewort (*Eriocaulon decangulare* L.) was known from only a single site in northeastern Pennsylvania; Parker's pipewort (*E. parkeri* B.L. Robbins) still grows in freshwater intertidal marshes of the lower Delaware River in New Jersey, but is gone from the Pennsylvania portion of the estuary.

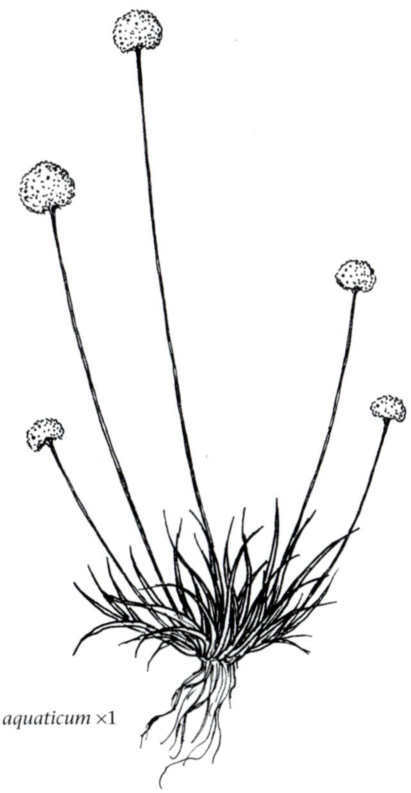

Eriocaulon aquaticum ×1

Eriocaulon aquaticum

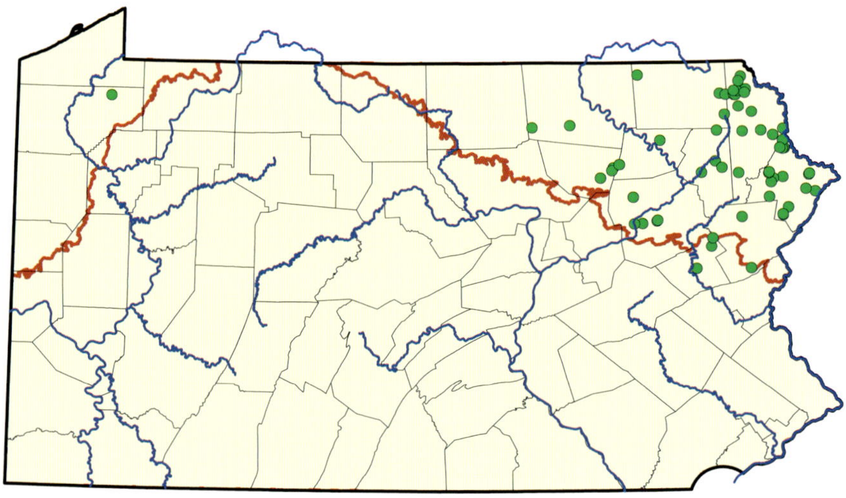

Eriocaulon aquaticum

QUILLWORT
ISOETES L.
Quillwort Family—Isoetaceae

The quillworts are tufted, grass-like plants; however, they do not flower or produce seeds but rather reproduce by spores. Their leaves, which taper to a pointed tip, radiate from a root crown; the spore-bearing structures (sporangia) are enclosed in the expanded bases of the leaves. All of our quillworts have a similar growth habit, and all grow as rooted submergents or occasionally as emergent species along a lake or river margin.

While recognizing a quillwort is not difficult, identifying them to species is another matter. Two types of spores are produced; the larger megaspores are necessary to identify species. Typically megaspores are spherical with 4 distinct ridges. The areas between the ridges are ornamented variously with smaller ridges or spines. A strong hand lens (20X or more) or better yet, a dissecting microscope, is needed to see the sculptural details on surface of the spores. The extent of a flap of tissue known as the velum, that partially covers the sporangium, is also useful in identification (see key).

Hybrids occur frequently where 2 or more species grow in close proximity; they can be recognized by their misshapen or undeveloped spores.

Key to the Aquatic Species of Quillwort

A. megaspores covered with spine-like projections .... *Isoetes echinospora*
A. megaspores covered with ridges
  B. surface of megaspore covered with a reticulate network of ridges
    C. velum covering 10–20 percent of the sporangium
      D. reticulations on the surface of the megaspores regular and smooth-crested; megaspores 0.4–0.5 mm .... *Isoetes engelmannii*
      D. reticulations on the surface of the megaspores broken and ragged; megaspores larger (0.5–0.6 mm) .... *Isoetes appalachiana*
    C. velum covering 50 percent of the sporangium; reticulations on the surface of the megaspores irregular .... *Isoetes valida*
  B. surface of megaspore covered with short, branching ridges....... *Isoetes riparia*

Appalachian quillwort — *Isoetes appalachiana* D.F. Brunton & D.M. Britton [synonym: *Isoetes engelmannii* A. Braun var. *georgiana* Engelmann]

Native
Rooted, submergent or emergent perennial

Isoetes appalachiana was classified as *Isoetes engelmannii* var. *georgiana*. Recent studies have led to recognition of this tetraploid at the species level (Brunton and Britton 1997). The main distinguishing features are characteristics of the megaspores, including larger size (0.5–0.6 mm) and the irregular pattern

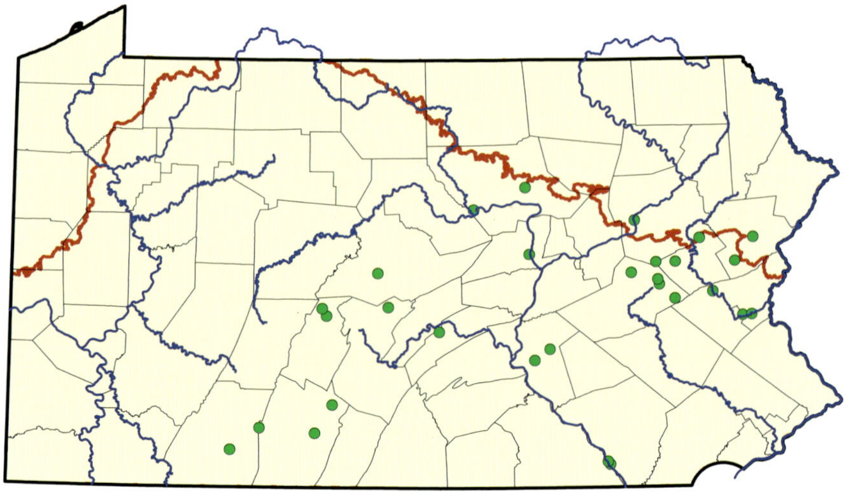

Isoetes appalachiana

of ridges on the surface (see key above). Leaves are erect to loosely spreading, dull green, and white at the base.

Appalachian quillwort is a robust plant of submergent or emergent aquatic habitats; including lakes, stream banks, and boggy woodland pools; it typically grows in acidic substrates. The species ranges from Pennsylvania to Florida with a greater concentration in Pennsylvania than any other state (D. Brunton, personal communication).

Spiny-spored quillwort — *Isoetes echinospora* A. Braun

Native

Rooted, submergent perennial

Spiny-spored quillwort is a rooted submergent perennial of shallow water of cold, slightly acidic lakes, ponds, and slow-moving streams. Its leaves are approximately 15 cm long. The megaspores are 0.4–0.5 mm in diameter and covered with thin spines. The velum covers 20–40 percent of the sporangium.

Spiny-spored quillwort is a circumboreal species that extends across Canada and south to the northern United States. It reaches its southern limit of range in the glaciated portions of Pennsylvania.

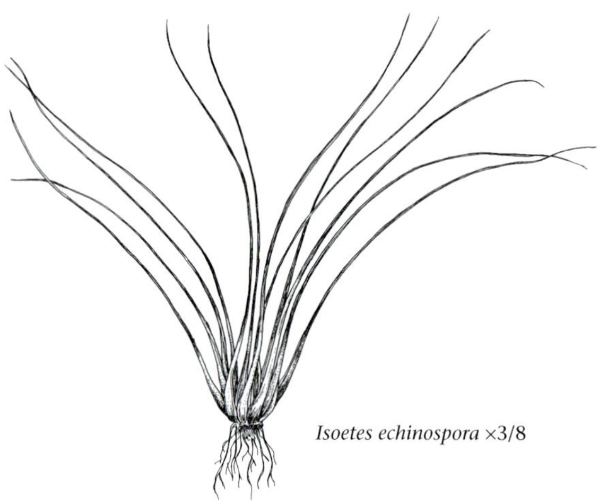

Isoetes echinospora ×3/8

Isoetes echinospora—megaspore ×30

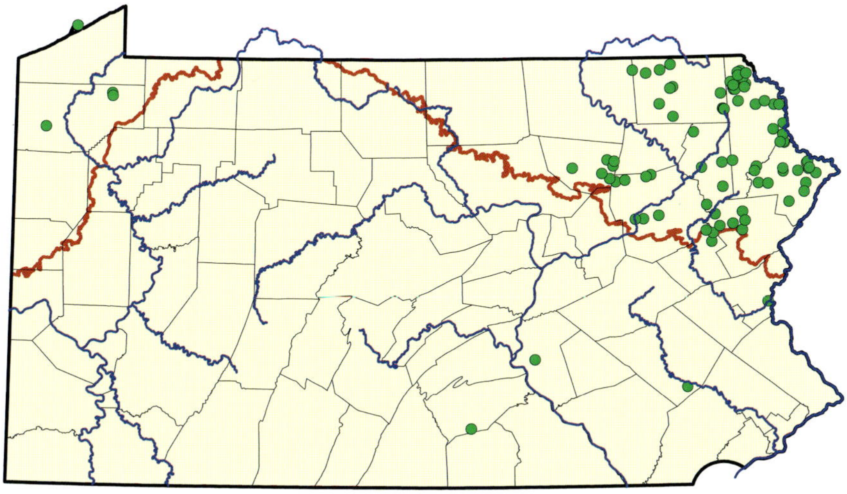

Isoetes echinospora

Engelmann's quillwort *Isoetes engelmannii* A. Braun

Native

Rooted, submergent or emergent perennial

Engelmann's quillwort is a submergent or emergent perennial; its leaves are up to 20 cm long, and loosely spreading to reflexed. Megaspores are 0.40–0.56 mm in diameter, with a reticulate surface ornamented with even ridges. The velum covers 10–20 percent of the sporangium.

The overall range of Engelmann's quillwort extends from New York State to northern Florida and west to Missouri. In Pennsylvania it occurs in ponds and slow-moving streams throughout the eastern and central regions, primarily south of the glacial boundary and east of the Allegheny Front.

Isoetes engelmannii—
megaspore ×30

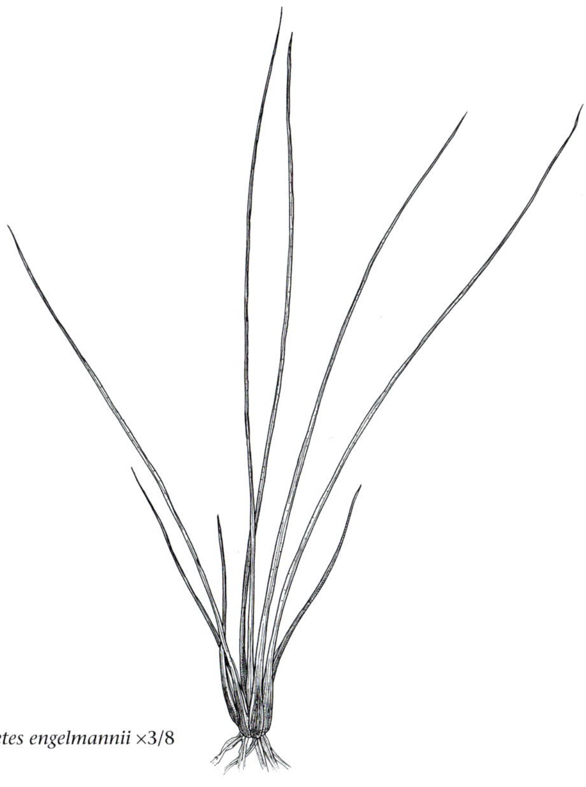

Isoetes engelmannii ×3/8

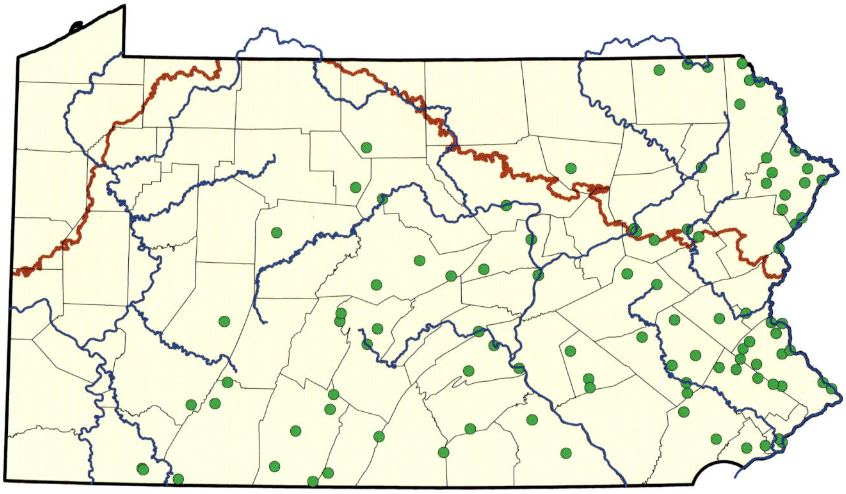

Isoetes engelmannii

Shore quillwort — *Isoetes riparia* A. Braun

Native

Rooted, submergent or emergent perennial

Shore quillwort grows in slow-moving rivers, streams, and tidal mud flats of the Delaware and Susquehanna River systems. It has leaves about 25 cm long. The sporangium wall is brown-spotted or all brown; the velum covers 20–40 percent of the sporangium. Megaspores are 0.45–0.65 mm in diameter with the surface ornamented by short branching ridges.

Shore quillwort is a plant of calcareous to slightly acidic waters in eastern North American near the coast, from Ontario to North Carolina. In Pennsylvania it occurs primarily in the eastern half of the state.

Isoetes riparia—
megaspore ×30

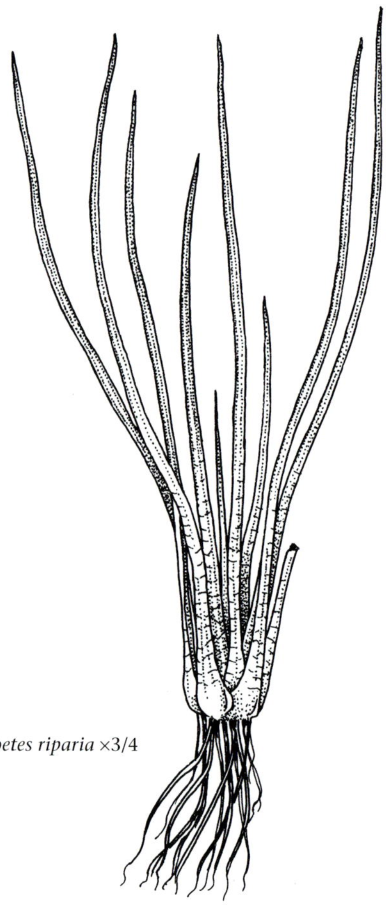

Isoetes riparia ×3/4

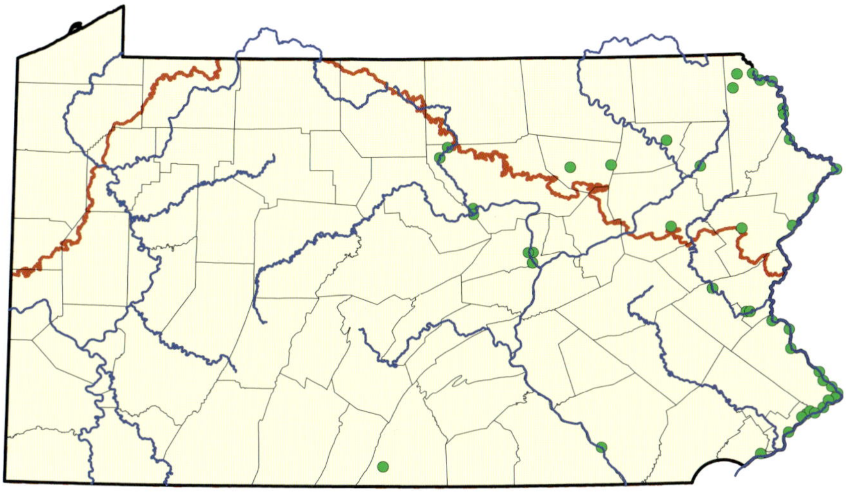

Isoetes riparia

Isoetes riparia growing on the tidal shore of the Delaware River in Bucks County, Pennsylvania

Isoetes valida—
megaspore ×30

Quillwort

***Isoetes valida* (Engelm.) Clute**
[synonyms: *Isoetes engelmannii* A. Braun var. *valida* Engelm.
***Isoetes engelmannii* var. *caroliniana* A.A. Eaton]**

Native
Rooted, emergent perennial

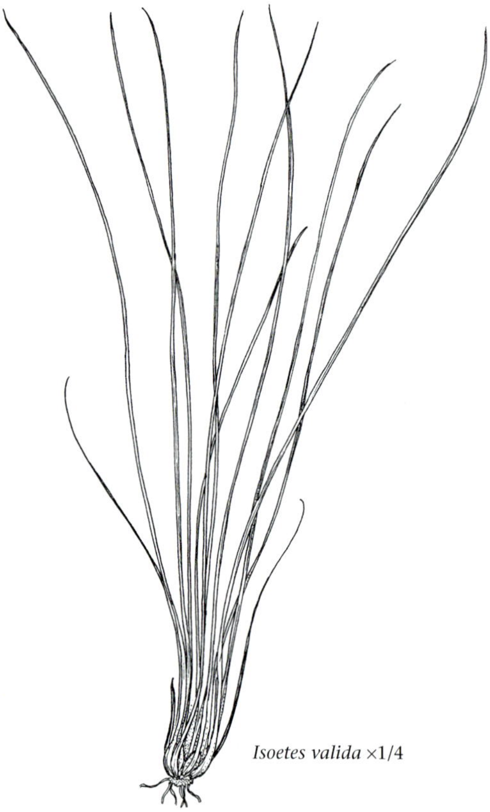

Isoetes valida ×1/4

Isoetes valida was classified as *Isoetes engelmannii* var. *valida* or *Isoetes engelmannii* var. *caroliniana*. Although these two varieties were once considered to be distinct, recent studies have led to the conclusion that they represent members of a single species now designated as *Isoetes valida* (Brunton and Britton 1996).

Isoetes valida is distinguished by its longer and wider leaves, and megaspores with irregular or broken reticulate ridges that contrast with the very even reticulations of *Isoetes engelmannii*. In addition, the velum covers 50 percent of the sporangium. *Isoetes valida* is most conspicuous early in the season (April and May) when the bright green leaves are at their maximum development.

Isoetes valida occurs from southern Pennsylvania and northern Delaware along the Appalachians to northern Alabama and Mississippi. In Pennsylvania it is known primarily from the south central and southeastern counties. It is a plant of small ponds, ditches, and stream beds in forested sites.

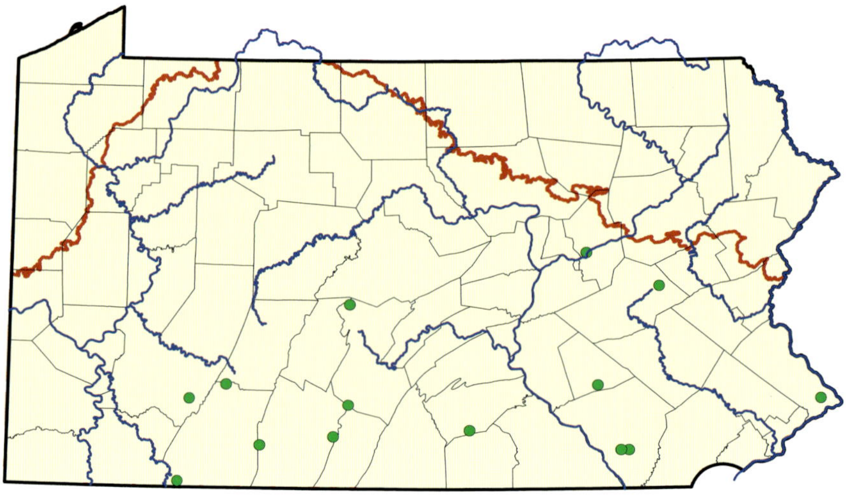

Isoetes valida

WATER LOBELIA

Bellflower Family—Campanulaceae

Water lobelia *Lobelia dortmanna* L.

PNHP
Native
Rooted, submergent perennial

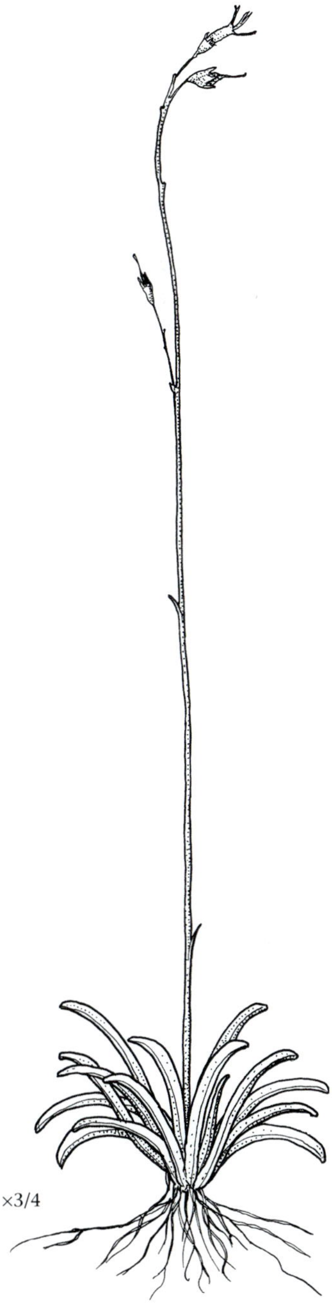

Water lobelia is one of several aquatic species that form a basal rosette of underwater leaves. Its flat, linear leaves have blunt tips that bend abruptly outward, distinguishing it from the quillworts, pipeworts, and arrowheads, which also produce underwater rosettes.

The flowering stems of water lobelia extend above the water surface and bear a raceme of 1 cm-long, white or pale blue flowers. The flowers have an irregular, tubular corolla that is split to the base on one side, 4 anthers, and a 2-lobed stigma that pushes the pollen out as it grows through a tube formed by the anthers.

Water lobelia is a northern species that reaches its southernmost extent in the northeastern corner of Pennsylvania, where it grows in natural glacial lakes. Its range extends from Newfoundland to Minnesota and south to Pennsylvania and northern New Jersey. It also occurs on the Pacific coast and in Europe.

Lobelia dortmanna—flower ×2

Lobelia dortmanna ×3/4

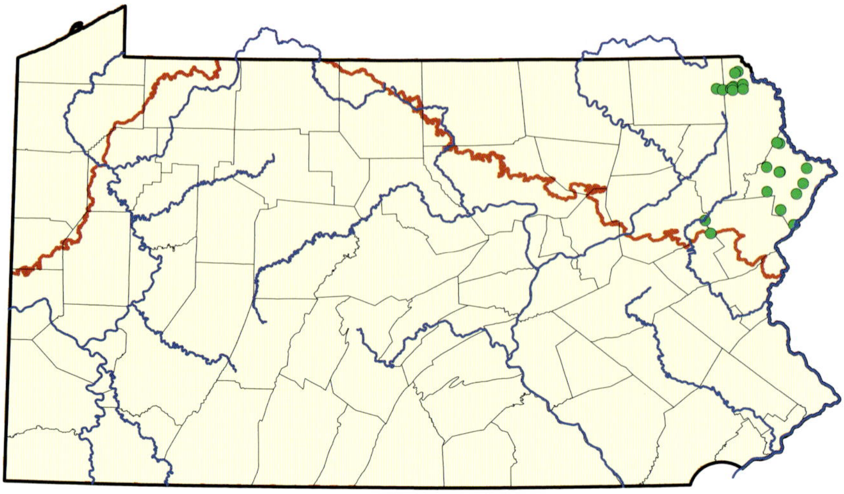

Lobelia dortmanna

Flowering stems of *Lobelia dortmanna* in a natural glacial lake in Wayne County, Pennsylvania

WATER-CELERY, EEL-GRASS

VALLISNERIA L.

Frog's-bit Family—Hydrocharitaceae

Water-celery, eel-grass *Vallisneria americana* Michaux

Native

Rooted, submergent perennial

Water-celery is a rooted submergent perennial with long, ribbon-like leaves that arise in buoyant upright clusters from the base. Because it is stoloniferous, water-celery usually grows in colonies; it is very unusual to find single plants. The leaves are usually bright yellowish-green but may also be reddish in color, especially in shallow water. Leaf length is controlled by water depth and varies from a few cm in shallow water to 1–1.5 m. The tips of the leaves typically reach to just below the water surface; however, occasionally the ends of the leaves are floating.

Water-celery is dioecious, that is, male and female flowers are produced on separate plants. The female flowers have long peduncles that allow the tip of the flower to create a dimple in the water surface. Male flowers, in contrast, are released from the base of the plant to float freely to the surface; there they are drawn by the surface tension of water into the vicinity of a female flower (see Figure 1.2). After pollination occurs, the peduncle of the female flower coils, pulling the flower beneath the water surface where maturation of the fruit proceeds. Flowering occurs July to October.

Water-celery is native to northeastern North America from Nova Scotia to Wisconsin and south to North Carolina and northern Illinois. It is common in lakes, streams, and rivers throughout Pennsylvania. It occurs most often in firm, sandy substrates. It is a valuable food for ducks, which consume the leaves, young shoots, and stolons.

Deep water, non-flowering plants of various bur-reeds (*Sparganium* spp.) can be very similar in appearance to water-celery. Look for the keel (pronounced, raised midrib) on the underside of the leaves of bur-reed; water-celery leaves do not have a keel. In addition, water-celery leaves also have a translucent, veiny quality that contrasts with the opaque green of bur-reed foliage.

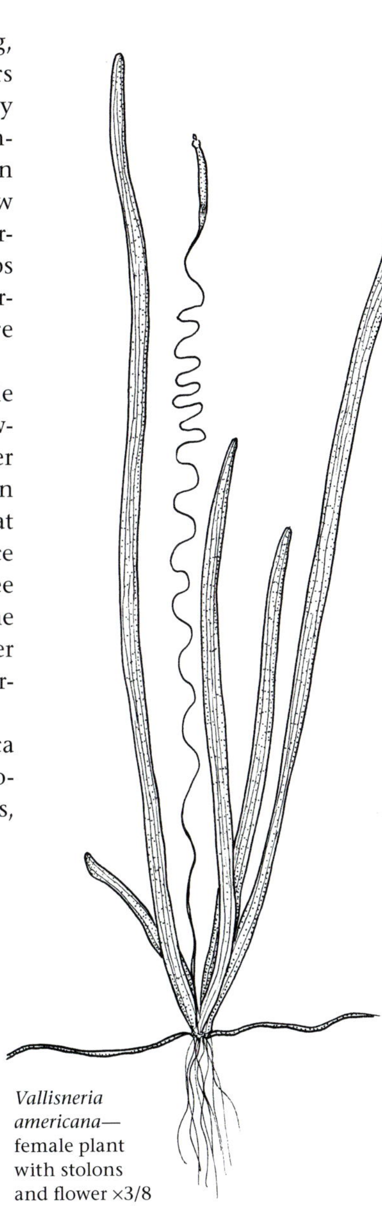

Vallisneria americana—female plant with stolons and flower ×3/8

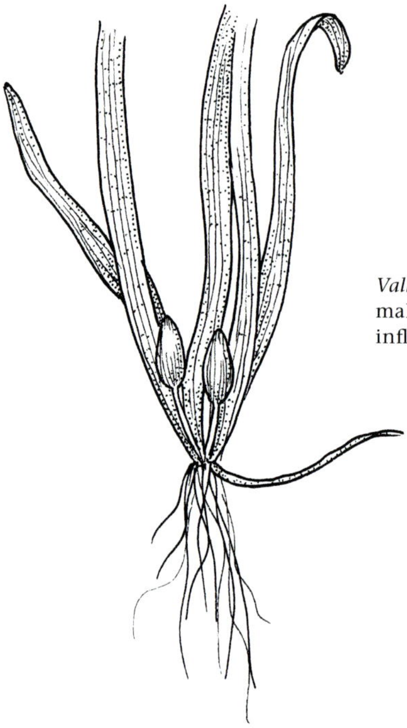

Vallisneria americana—
male plant with
inflorescences ×1/2

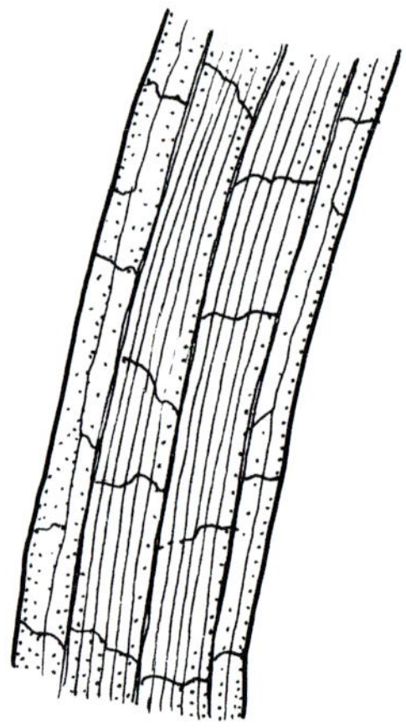

Vallisneria americana—
leaf section showing
venation pattern ×2

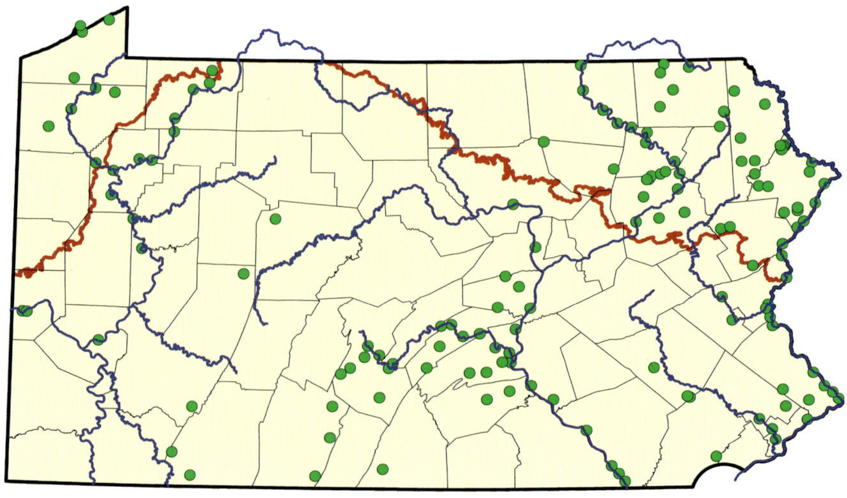

Vallisneria americana

Chapter 8. Submergent Plants with Finely Divided Leaves

The characteristic that defines this group is submergent leaves that are divided into fine, capillary segments. The plants may be either free-floating or rooted. Most are completely submergent, but stem tips or flowering stems of some species extend above the water surface.

BLADDERWORT

Utricularia L.

Bladderwort Family—Lentibulariaceae

Most bladderworts are found in low nutrient environments such as oligotrophic lakes and bogs. These native carnivorous plants trap tiny aquatic invertebrates in small bladders (traps) on their leaves (Figure 8.1). The small sac-like traps have a suction-operated mechanism that is triggered by sensitive hairs at the base of the trap door. When a small aquatic invertebrate comes too close, stimulation of the hairs causes the trap to open, drawing the prey in in a rush of water. Digestion occurs in the trap, releasing nitrogen, phosphorus and other elements that are absorbed by the plant. The trap is later reset when water is actively pumped out, restoring the suction.

Charles Darwin, who was intrigued by bladderworts and other carnivorous plants, noted the similarity in the form of bladderwort traps and some of their micro-invertebrate prey. He suggested that trap appendages which he called "antennae" and "bristles" helped to funnel prey into the traps (Darwin 1898).

Figure 8.1. Bladderwort (*Utricularia* sp.) trap magnified, showing trigger hairs.

Several recent studies have investigated the cost of carnivory, the investment required to build and maintain traps which, themselves, have limited ability to photosynthesize. Kibriya and Jones (2007) determined that fewer traps were produced when the phosphorus content of the surrounding medium was increased. Trap number was not affected by changes in nitrogen levels. This suggests that the acquisition of phosphorus via carnivory is more important that the acquisition of nitrogen.

Other research suggests that there may be more to bladderwort traps than just the capture and digestion of animal prey. Algae have also been found in *Utricularia* traps, where they are killed and digested like other prey (Petroutka et al. 2008).

Sirova et al. (2010) determined that bladderwort plants secrete sugars into newly formed traps. They suggest that the sugars support microbial communities within the traps which in turn facilitate the availability of phosphorus, nitrogen, and possibly other substances through increased decomposition of organic matter. The traps thus become reservoirs of available nutrients that are isolated from competitors in the surrounding water.

This active area of research is sure to yield additional insights into the evolution of structure and function of these fascinating plants in the years to come.

The bladderworts are free-floating at or beneath the water surface, or, in a few cases the leaves (and traps) are imbedded in water-saturated peat or sand. Roots are not present, but some species produce simple root-like structures called rhizoids. The leaves of most species are finely dissected; traps are borne on the tips or margins of leaf segments.

Bladderwort flowers are produced on unbranched leafless stalks (peduncles) that extend above the water surface. The flowers are yellow or purple, bilabiate, and spurred; they are insect-pollinated except for cleistogamous (closed) flowers, which are self-pollinating. The fruit is a capsule that splits open at maturity. Bladderwort seeds and plant fragments are transported from site to site on the feet or feathers of waterfowl.

Most bladderwort species are classified as perennials because of their production of compact winter buds known as turions. Formed late in the season in response to short day length, turions float, or sink to the bottom, and resume growth in the spring when the days lengthen. A few species also form tubers.

All Pennsylvania bladderworts belong to the genus *Utricularia*. Identifying bladderworts to species if flowers are not present can be difficult in some cases. See key and comments below under individual species.

Bladderworts might be mistaken for several other submersed aquatic plants characterized by finely divided leaves with a radiating (palmate) branching pattern: coontail (*Ceratophyllum* species), fanwort (*Cabomba caroliniana*), Beck's water-marigold (*Bidens beckii*), and white water buttercup (*Ranunculus aquatilis*). Close examination should reveal small bladders to be present in the case of bladderwort, at least on the newest growth. See also discussion under fanwort for other distinguishing characteristics.

Eight species of native bladderworts are present in Pennsylvania. In addition, three species which were found in the state in the past, are now believed to be extirpated. Floating bladderwort (*Utricularia radiata*) is very similar to inflated bladderwort (*U. inflata*), but smaller. It is known from southern Bucks County, but was last collected in 1927. Northeastern bladderwort (*U. resupinata*) is similar to horned bladderwort in growth form, but has purple flowers. It was known only from Presque Isle and was last collected in 1879. Slender bladderwort (*U. subulata*) grows with its stolons and leaves mostly imbedded in the substrate; it has very few traps. Known only from a single site in Chester County, this species was last collected in 1909. The three species believed to be extirpated are included in the key below, but they are not described other than a brief comparison under the most similar extant plant.

Several bladderwort species, especially purple bladderwort and inflated bladderwort, have the potential to become very abundant, occasionally interfering with recreational uses such as swimming, boating, and fishing.

Key to the Species of Bladderwort

A. basal portions buried in the substrate; leaves few, linear; traps few, mostly hidden in the substrate
  B. floral bracts tubular; flowers solitary; corolla purple ..... *Utricularia resupinata*
  B. floral bracts not tubular; flowers 1–10; corolla yellow
    C. flowering stems very slender, 3–20 cm; pedicels longer than the bracts
      D. inflorescence axis zigzag; flowers 1–25; pedicels ascending in fruit ..... *Utricularia subulata*
      D. inflorescence axis straight; flowers 2–6; pedicels curved downward in fruit ..... *Utricularia minor*
    C. flowering stems 4–40 cm long, stout; pedicels shorter than the bracts ..... *Utricularia cornuta*
A. plants anchored or suspended; leaves numerous, arising from the nodes of elongate, floating stems, or just above the surface of the submerged substrate, more or less finely, dichotomously divided into capillary (or flattened) segments; traps well-developed
  E. leaves whorled; traps terminal on the leaf segments; flowers purple ..... *Utricularia purpurea*
  E. leaves alternate or crowded at the base; traps lateral on the leaf segments; flowers yellow
    F. peduncle with a whorl of spongy, inflated floats
      G. peduncles 3–5 cm tall, bearing a ring of floats and 1–7 flowers; floats 2–4 mm thick and 4 cm long, with mostly parallel sides ..... *Utricularia radiata*
      G. peduncles 20–50 cm, bearing a ring of floats and 5–8 flowers; floats to 8 mm wide and 3–10 cm long, tapering from the base ..... *Utricularia inflata*
    F. peduncle not bearing a whorl of floats
      H. divisions of the leaves flat, with a visible midrib; floral bracts auriculate (lobed) at the base
        I. traps on separate stems; leaf divisions spine-toothed; pedicels erect in fruit ..... *Utricularia intermedia*
        I. traps on the leaves; leaf divisions not spine-toothed; pedicels curved downward in fruit ..... *Utricularia minor*
      H. divisions of the leaves filiform, without a distinct midrib; bracts not auriculate
        J. peduncles without extra bracts below the lowest flower ..... *Utricularia geminiscapa*

J. peduncles with 1–5 empty bracts below the lowest flower
  K. upper corolla lip larger than the lower; bracts tapering to a clasping base ....*Utricularia gibba*
  K. upper corolla lip smaller than the lower; bracts not clasping..... ....*Utricularia macrorhiza*

Horned bladderwort *Utricularia cornuta* Michx.

PNHP
Native
Anchored, submergent perennial

Horned bladderwort typically grows partially imbedded in a substrate of saturated peat or sand. Bladders are borne on stolons and reduced leaves below the surface of the substrate and on the margins of the erect simple, linear leaves. Flowering stems are 4–40 cm tall and bear 1–6 yellow flowers near the top. The common name reflects the long spur that is part of each flower.

Growing entirely vegetatively in water to 2 m deep in oligotrophic glacial lakes, horned bladderwort can form a thin turf on the lake bottom, where it may be mixed in with needle spike-rush (*Eleocharis acicularis*). Flowering occurs on plants growing in shallow water or on exposed saturated peat of bog margins.

The range of horned bladderwort extends from Newfoundland, Quebec, northern Michigan, and Wisconsin south to Florida, Texas, the Bahamas, and Cuba. In Pennsylvania it occurs mainly in natural glacial lakes in the northeastern corner of the state.

In our flora northeastern bladderwort (*U. resupinata*) has a similar growth habit, but it is smaller overall and has purple flowers. Northeastern bladderwort is known from a single historical collection at Presque Isle in 1879 and is believed to be extirpated.

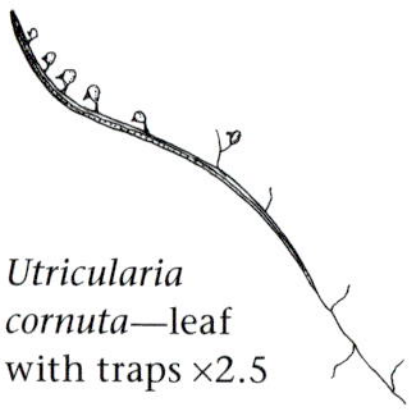

Utricularia cornuta—leaf with traps ×2.5

Utricularia cornuta—leaves with traps ×1/2

Utricularia cornuta—flowering stem ×3/4

Flowers of *Utricularia cornuta*

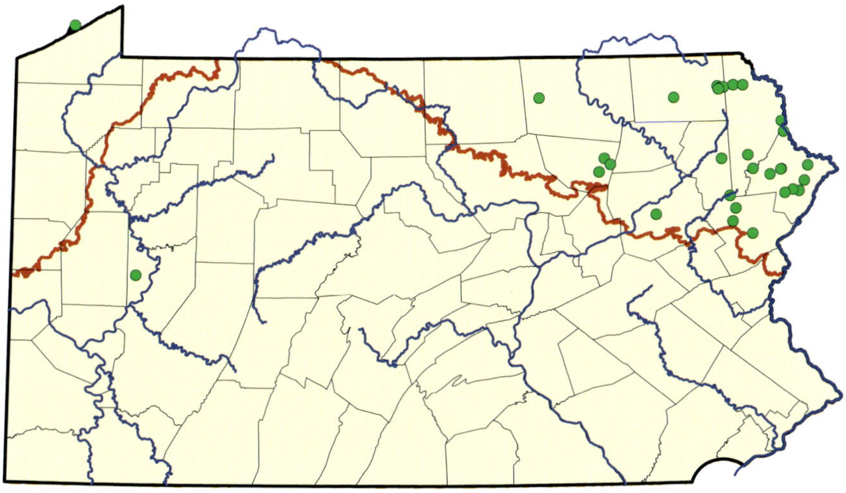

Utricularia cornuta

Bladderwort *Utricularia geminiscapa* Benj.

Native
Free-floating perennial

Utricularia geminiscapa resembles common bladderwort (*Utricularia macrorhiza*), but is smaller overall. The erect flowering stalk bears 2–8 yellow flowers crowded near the tip. Another distinguishing feature is the frequent presence of closed (cleistogamous) underwater flowers, which look like stalked buds. Turions are also produced.

This species ranges from Newfoundland to Quebec, northern Michigan, and northern Wisconsin, south to Pennsylvania and North Carolina. In Pennsylvania *Utricularia geminiscapa* is scattered in the eastern and central regions of the state. It often grows in small pools in the surfaces of sphagnum bogs or in shallow water of ditches or marshes.

In addition to its smaller size, *Utricularia geminiscapa* can be distinguished from common bladderwort by the presence of cleistogamous flowers.

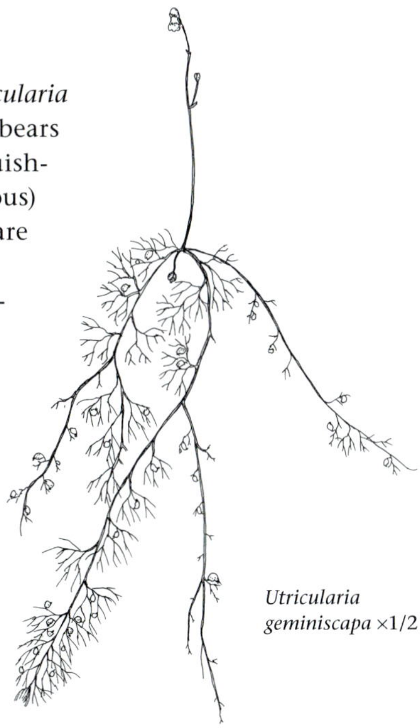

Utricularia geminiscapa ×1/2

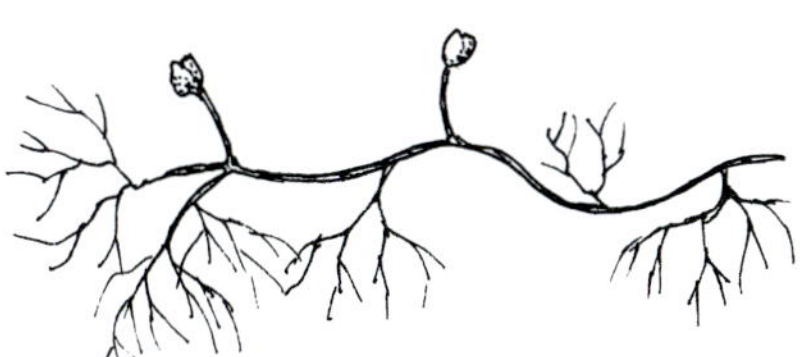

Utricularia geminiscapa—stem with cleistogamous flowers ×1

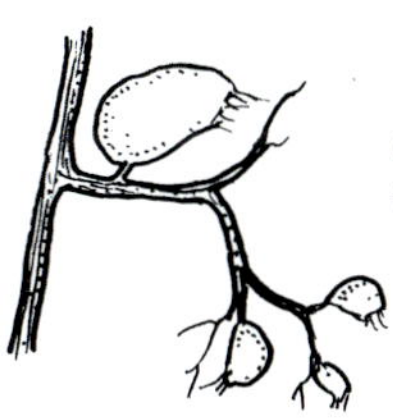

Utricularia geminiscapa—traps ×5

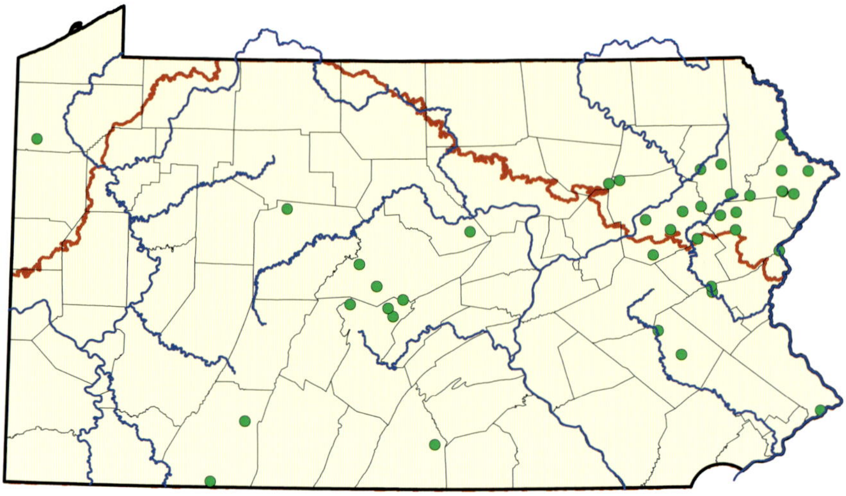

Utricularia geminiscapa

Humped bladderwort ***Utricularia gibba* L.**

Native

Free-floating annual or perennial

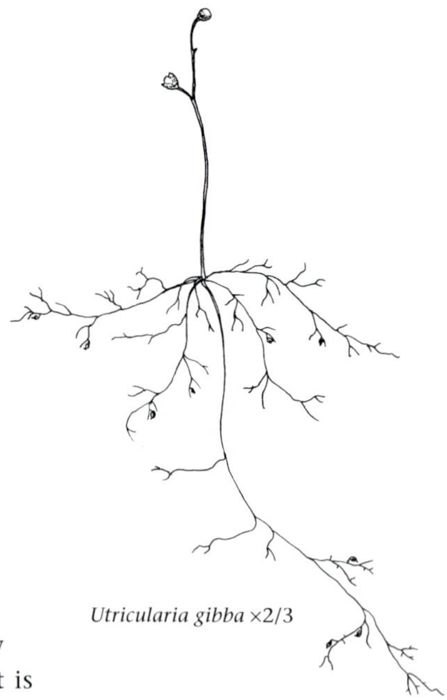

Utricularia gibba ×2/3

The slender strands of humped bladderwort are often draped over other submergent plants or may form tangled mats that float just below the water surface in lakes and ponds. The lightly branched stems bear small leaves that are only once or twice divided. Bladders are borne at the tips of short lateral leaf segments.

Flowering is rare in deep water. Flowering stalks are more often produced by plants growing on peat or in water only a few centimeters deep. The inflorescence consists of 2–6 yellow flowers on a slender, emergent peduncles 1–20 cm tall. Below each flower is a semicircular bract that is about 1 mm long, and has a rounded or truncate apex. Very reduced, submersed inflorescences with cleistogamous (closed) flowers are sometimes present. Humped bladderwort does not produce turions, and may grow as an annual in the northern part of its range.

A wide-ranging and very variable species, humped bladderwort grows throughout the Old and New World tropics and north as far as eastern and western Canada and the United States. In Pennsylvania it occurs primarily in the glaciated northeast, with scattered sites elsewhere across the state.

The lightly branched form of humped bladderwort stems makes this species easy to identify, even in the absence of flowers.

Stems and flowers of *Utricularia gibba* growing in Bucks County, Pennsylvania

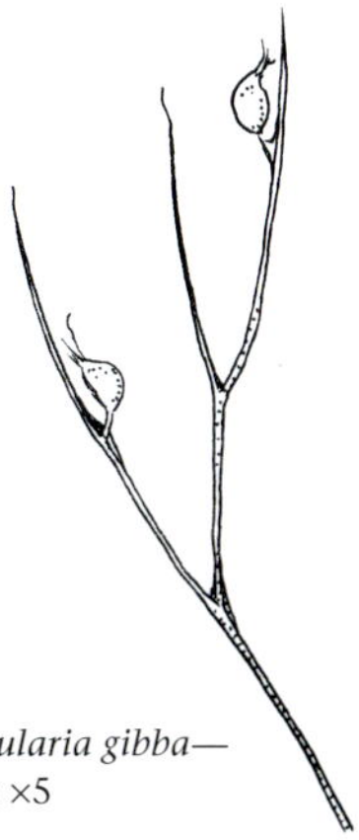

Utricularia gibba—traps ×5

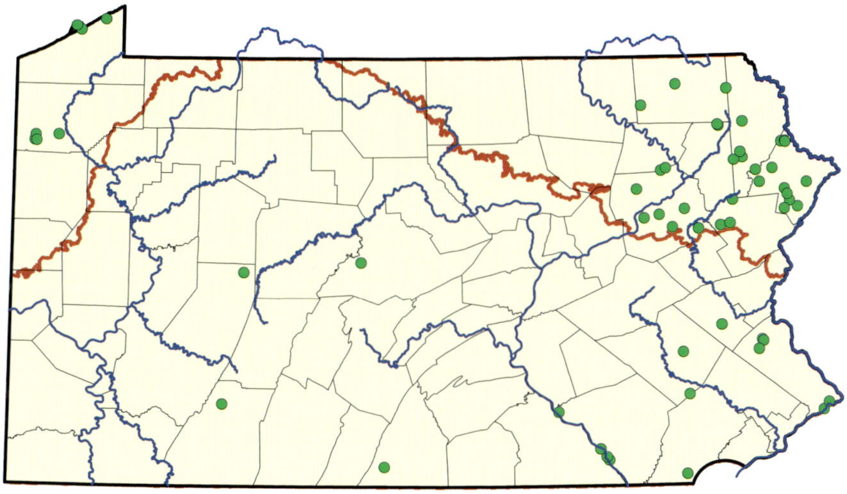

Utricularia gibba

Inflated bladderwort *Utricularia inflata* Walter

Native
Free-floating perennial

Inflated bladderwort is a very unusual plant, even for a bladderwort. Not only does it trap tiny aquatic animals and digest their bodies to obtain nutrients, it also grows its own personal flotation device (pfd) to keep its flowers above water. The flowering stem, bearing 5–8 yellow flowers, extends above the water surface, held aloft by a radiating ring of inflated branches.

Inflated bladderwort is a recent addition to the flora of Pennsylvania. It was first noted in two lakes in Sullivan County, Pennsylvania in the early

Utricularia inflata ×1/4

Utricularia inflata growing in a lake in Monroe County, Pennsylvania

1990s. Only in 2000, during a botanical survey we conducted of natural areas at Tobyhanna State Park, was it found in the Pocono region. We have subsequently documented its presence at 21 sites in four counties in northeastern Pennsylvania. Botanists in New York and Massachusetts also report its recent appearance in those states.

A plant of lakes and ponds, inflated bladderwort was formerly limited to the Atlantic Coastal Plain from New Jersey south to Florida and Texas. How it made the jump to the northeast is unknown, perhaps as seeds adhering to the feathers of migrating waterfowl, or possibly stuck to the surface of a boat. We will probably never know for sure.

Floating bladderwort (*Utricularia radiata*) is very similar to inflated bladderwort (*U. inflata*), but smaller. In addition, the inflated leaves that support the flowering stem are cylindrical, rather than tapering to the base as do those of inflated bladderwort.

Non-flowering plants of inflated bladderwort can be distinguished from common bladderwort (*Utricularia macrorhiza*) by the alternate leaves and reddish color and crowded form of the stem tip, or by microscopic examination of the trigger hairs on the bladders.

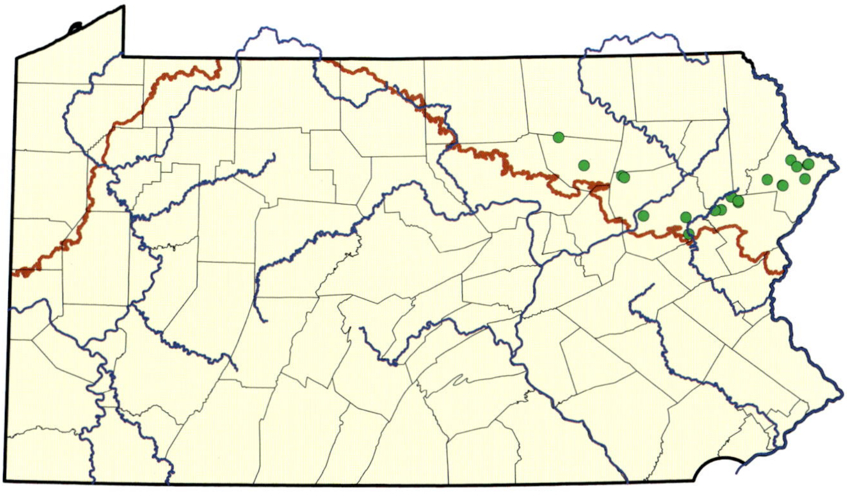

Utricularia inflata

Flat-leaved bladderwort *Utricularia intermedia* Hayne

PNHP
Native
Anchored, submergent perennial

Flat-leaved bladderwort has two distinguishing characteristics that make it easy to separate from other species. Its leaf segments are flattened and have a distinct midrib, in contrast to the thread-like leaf segments of most bladderworts. The leaves are fan-shaped in outline; in addition, the bladders are borne on separate branches from the leaves. The bladder-bearing branches bear very reduced leaves and are often white rather than green, due to being buried in saturated peat below the water surface. At flowering, two to four yellow blooms are borne on emergent peduncles. Dense turions are formed.

Flat-leaved bladderwort has a worldwide distribution in northern latitudes (circumboreal); in North America it extends south to New Jersey, Pennsylvania, Indiana, and California. In Pennsylvania it is found primarily in natural ponds and lakes in the glaciated northeastern and northwestern regions. Early records from the coastal plain have not been relocated in over 100 years.

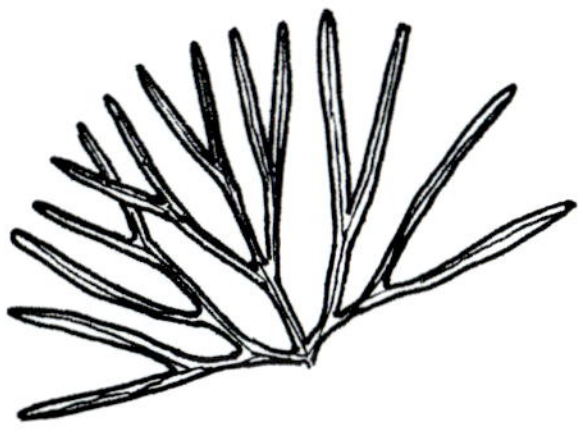

Utricularia intermedia—leaf ×3

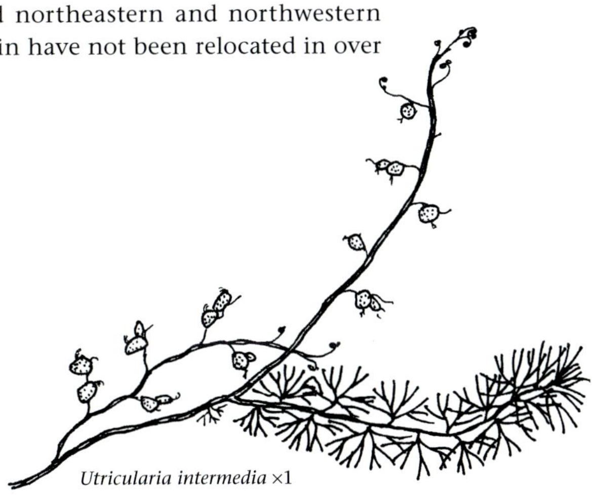

Utricularia intermedia ×1

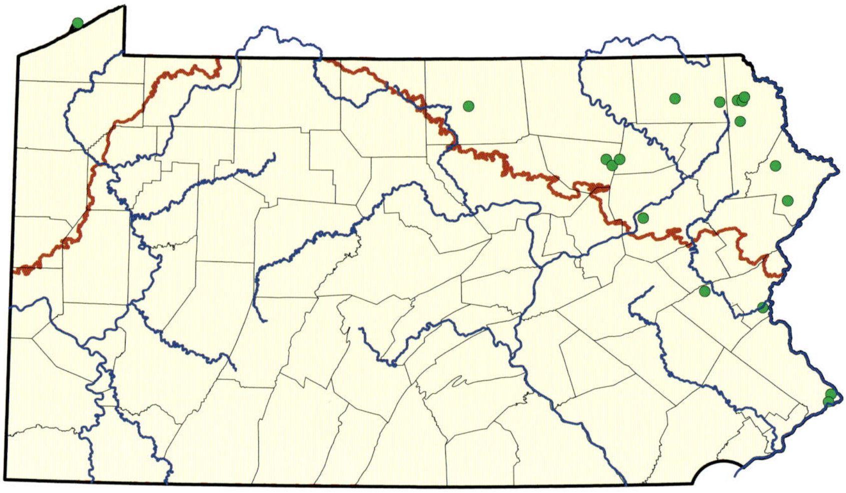

Utricularia intermedia

Common bladderwort *Utricularia macrorhiza* LeConte

Native
Free-floating perennial

Common bladderwort is the largest of our native bladderworts. Leafy stems are free-floating below the water surface. Yellow flowers arise on leafless stalks elevated above the water surface, creating an inflorescence 25–35 cm in length. The flowers are strongly bilabiate and spurred; the upper lip is 2-lobed and smaller than the lower, which is 3-lobed with reddish-brown streaks. The inflorescence of this plant has 3–14 flowers, with each flower subtended by a bract, and 1–5 empty bracts below the lowest flower. Turions, dense leafy stem tips up to 3 cm or more long, are frequently visible late in the growing season.

Common bladderwort is frequent in lakes, ponds, swamps, marshes, and ditches. It occurs in North America throughout the United States and Canada, in western Mexico, and in eastern Asia. In Pennsylvania it is found primarily in the glaciated regions.

Common bladderwort is the most frequently encountered bladderwort species in Pennsylvania; however, it can be confused with *U. geminiscapa*. If the plants are not flowering, distinguishing the species can be difficult. Common bladderwort is generally larger and more robust than *U. geminiscapa*, but depauperate plants may be similar in size. If the plants are flowering, more characters are available to distinguish them. The upper corolla lip of *U. geminiscapa* is larger than the lower; in addition, the bracts that subtend the flowers taper to

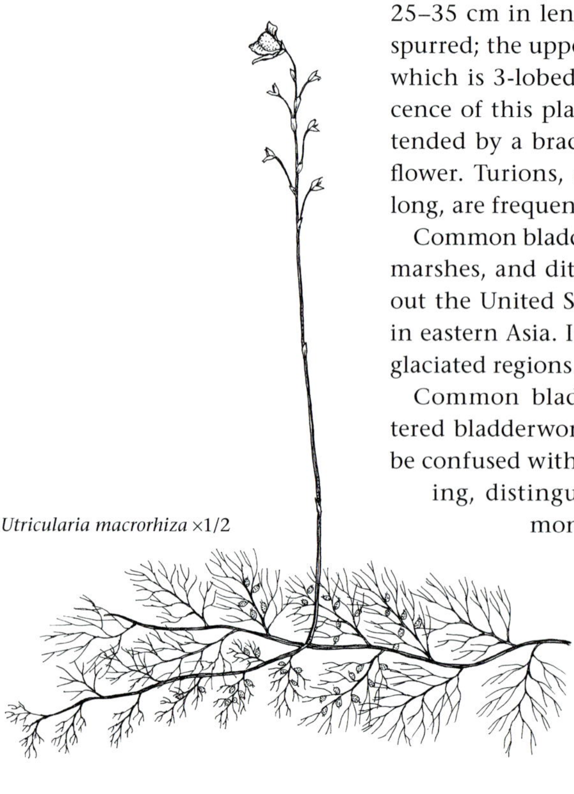

Utricularia macrorhiza ×1/2

a clasping base, while the bracts of common bladderwort are not clasping at the base. Another difference is that *U. geminiscapa* often bears small cleistogamous (closed) flowers that look like stalked buds on the underwater stems. Cleistogamous flowers are never present on common bladderwort.

Non-flowering plants of common bladderwort are also difficult to differentiate from those of inflated bladderwort; see additional comments under that species.

Emergent flowering stem of *Utricularia macrorhiza* with leaves of *Nymphaea odorata* in a lake in Pike County, Pennsylvania

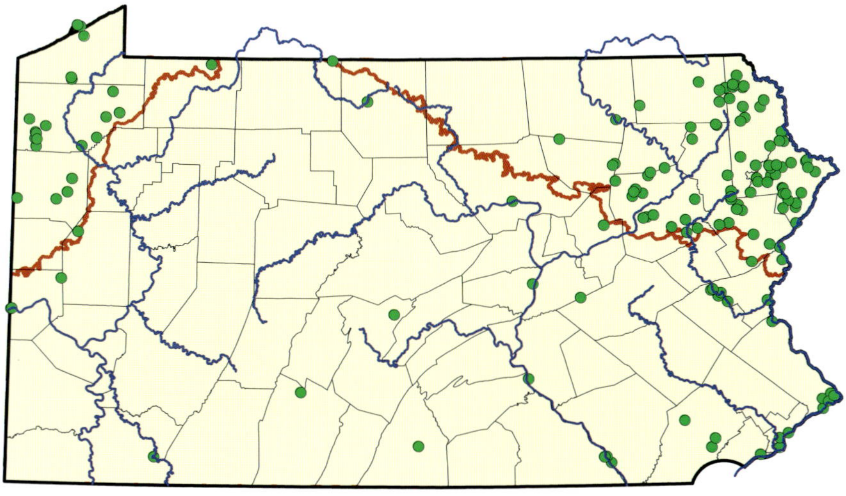

Utricularia macrorhiza

Lesser bladderwort *Utricularia minor* L.

Native

Anchored, submergent perennial

Utricularia minor—stem with bladders ×5

Lesser bladderwort is the smallest of our species. It grows with its stolons and many of the leaves and traps imbedded in a substrate of saturated peat or floating leaf compost. Its leaves are of two forms; those produced above the substrate or on floating stems are roughly circular in outline and palmately divided into many narrow, somewhat flattened segments. Leaves imbedded within the substrate are reduced to a few filiform segments, which bear most of the traps. Turions are formed. Flowering stalks are rarely as much as 8 cm tall and bear 3–9 yellow flowers. Lesser bladderwort is very easy to overlook when it is not in flower.

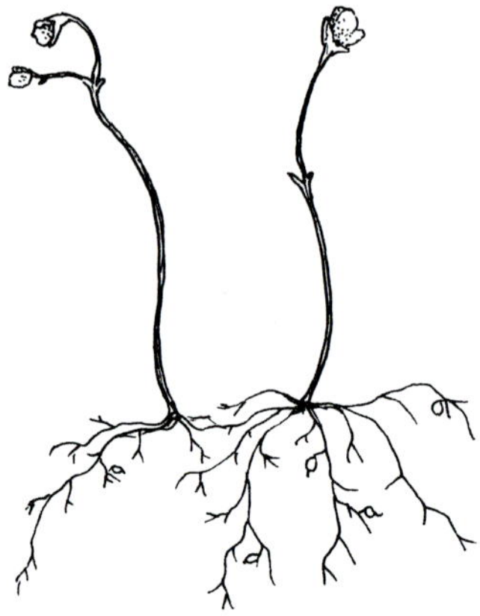

Utricularia minor ×1

Lesser bladderwort is a plant of exposed soft peat and other saturated organic substrates on lake or pond margins or peaty islands. It is a circumboreal plant, meaning that it occurs throughout the northern latitudes; its southernmost occurrences in North America are in New Jersey, Pennsylvania, Indiana, and California. In Pennsylvania it is found only in the glaciated northeastern and northwestern corners of the state.

Floating sterile fragments of lesser bladderwort can be confused with flat-leaved bladderwort.

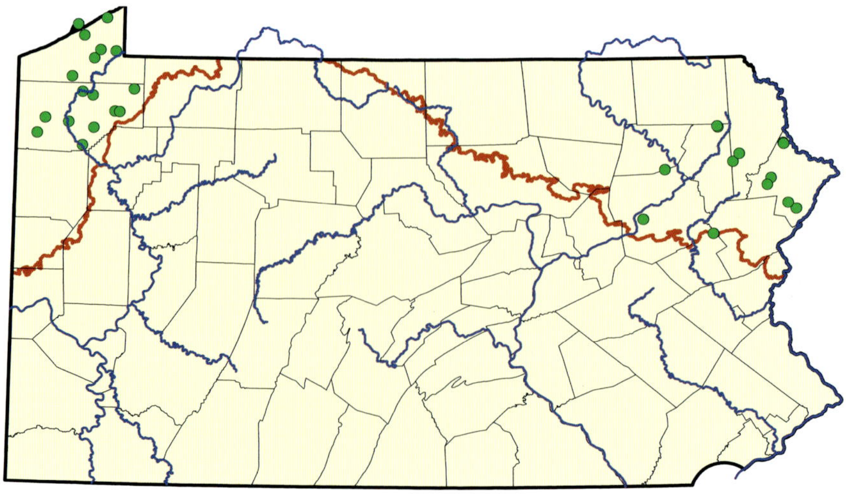

Utricularia minor

Purple bladderwort — *Utricularia purpurea* Walter

Native
Free-floating perennial

A plant of lakes and ponds, this species is often so abundant that at flowering time a haze of purple is visible over the water surface as thousands of stems bloom synchronously. Individual flowers are pinkish-purple with a yellow blotch on the palate; each peduncle holds 3–5 blooms. Throughout the growing season the free-floating stems with whorled leaves grow at one end and gradually decompose at the other.

Purple bladderwort grows from Nova Scotia to Quebec, Minnesota, Michigan, Indiana, and Illinois, and south along the Atlantic Coastal Plain to Florida and Louisiana. Its range also includes Mexico, Central America, and the West Indies. In Pennsylvania it is found in the glaciated northeast. It can become very abundant under certain circumstances.

Purple bladderwort is the only species of our native bladderworts with whorled leaves, and traps at the tips of the leaf segments, making it easy to distinguish at all times.

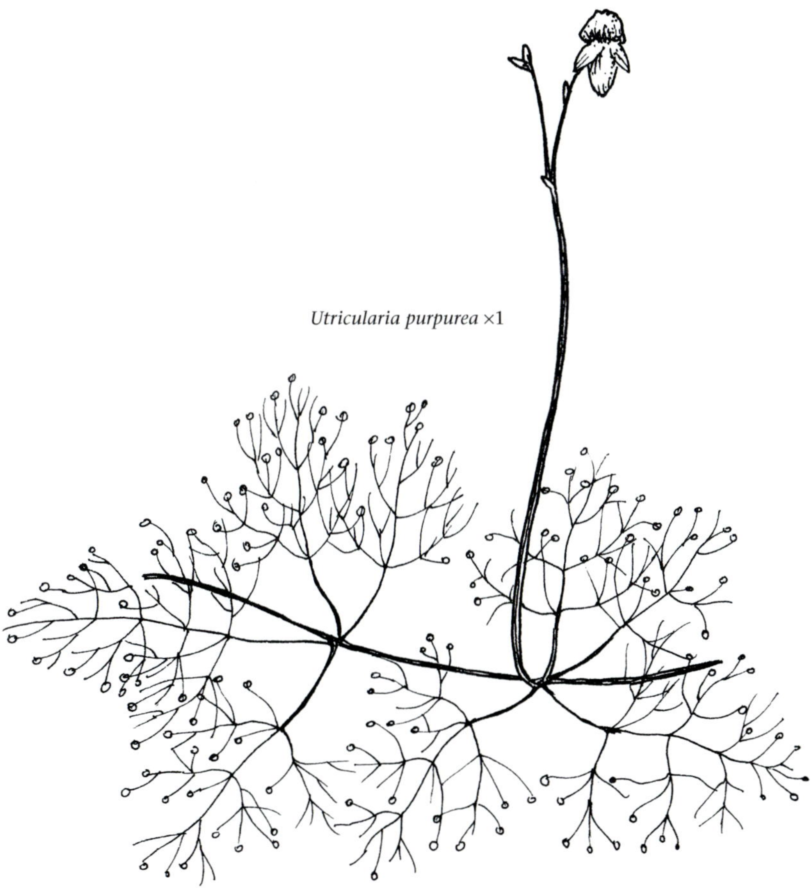

Utricularia purpurea ×1

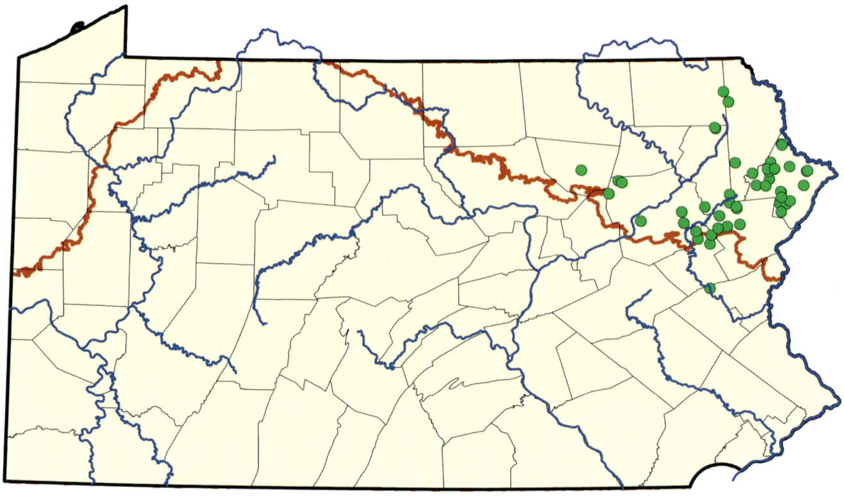

Utricularia purpurea

Emergent flowering stems of *Utricularia purpurea* among *Nymphaea odorata* leaves at a lake in Pike County, Pennsylvania

COONTAIL, HORNWORT

CERATOPHYLLUM L.

Hornwort Family—Ceratophyllaceae

The coontails or hornworts are native submergent perennial plants. They do not have roots or rhizomes, but the lower portion of the stem is sometimes imbedded in the substrate. The branching stems, which can reach lengths of 1–1.3 m, bear whorls of leaves. Each leaf is palmately divided into narrow segments. Flowers are unisexual, solitary, and sessile in the leaf axils; they are water-pollinated. The fruits are flattened disks with a spiny margin. However, fruits are often hard to find, and vegetative reproduction is the more common method of increase. Stem segments frequently break off and float. These detached segments can be transported to a new location where they continue to grow.

Both morphology and recent DNA studies indicate that *Ceratophyllum* is a very ancient angiosperm, among the oldest extant lineages of flowering plants (Judd et al. 2002).

Fruits and leaves of coontail are eaten by ducks; however, they are not a highly preferred food of waterfowl.

Key to the Species of Coontail

A. leaf segments somewhat flattened, divided once or at most twice; marginal teeth on the segments conspicuous; fruit with 3 long spines .... *Ceratophyllum demersum*

A. leaf segments threadlike, divided 3 or 4 times; marginal teeth on the segments inconspicuous or absent; fruit with 8–10 long and short spines .... *Ceratophyllum echinatum*

Coontail *Ceratophyllum demersum* L.

Native

Submergent, suspended perennial

Coontail is a stiff plant, which, unlike many aquatics, holds its shape out of the water. The name is derived from the dense, bushy stem tips caused by overlapping whorls of finely divided leaves. The leaves of coontail are palmately divided into narrow segments with small teeth on the sides. Typically 1 or 2 divisions occur in which similar-size narrow flattened segments are produced. The fruits are flattened disks about 6 mm across with 3 protruding spines.

Ceratophyllum demersum ×1

Ceratophyllum demersum occurs worldwide; occurrences in Pennsylvania are mainly in

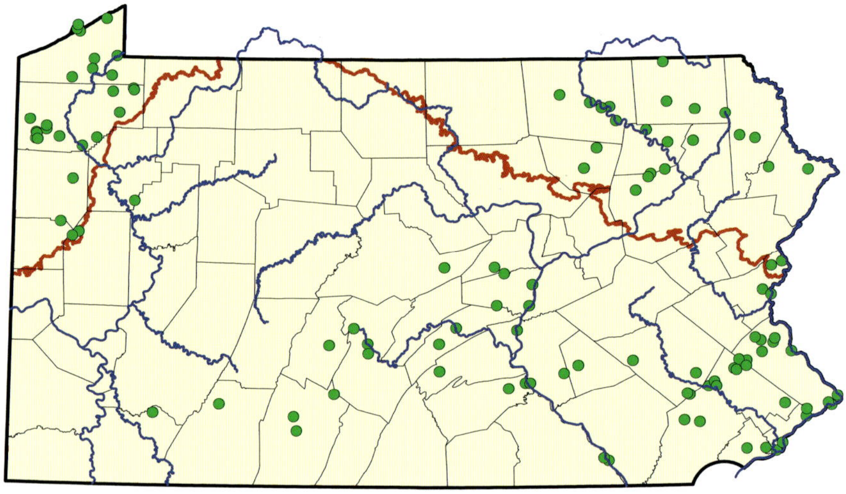

Ceratophyllum demersum

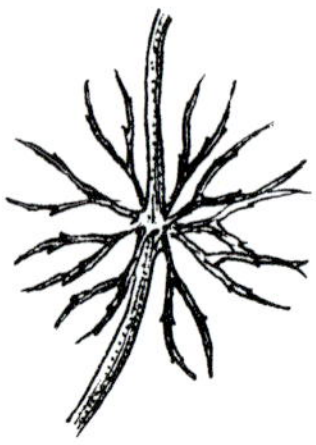

Ceratophyllum demersum—whorl of leaves ×1

the eastern, central, and northwestern regions. Coontail can be very abundant in quiet water of mildly to strongly eutrophic lakes and ponds. It usually grows in water bodies with soft bottom sediments.

Ceratophyllum demersum is distinguished from *C. echinatum* by its stiffer overall habit, leaf segments that fork only 2 times as compared to 3–4 times in *C. echinatum*, more conspicuous leaf teeth, and fruits that lack shorter spines between the 3 longer ones.

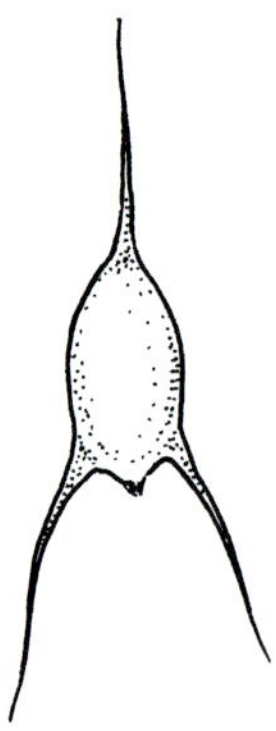

Ceratophyllum demersum—fruit ×2

Hornwort, coontail — *Ceratophyllum echinatum* A. Gray [synonym: *C. muricatum* Cham.]

Native
Submergent, suspended perennial

Hornwort is very similar to coontail (*Ceratophyllum demersum*). Both have whorls of finely palmately divided leaves at the nodes of the elongate stems. *Ceratophyllum echinatum* is not as stiff as *C. demersum* when removed from the water; in addition, its leaf segments are more slender and more finely divided. *Ceratophyllum echinatum* fruits are disk-shaped and about 6 mm in diameter with several short spines between the 3 longer ones around the margin. Fruiting can be infrequent in some populations.

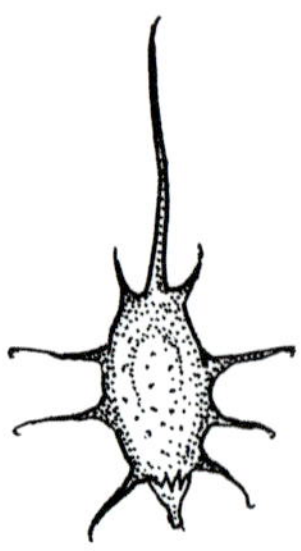

Ceratophyllum echinatum—fruit ×2

Ceratophyllum echinatum is a North American plant; it grows from Newfoundland to British Columbia and south to Florida, Texas, and Mexico. In Pennsylvania it is found mostly north of the glacial boundary. Hornwort occurs in lower-nutrient, more acidic lakes than *C. demersum*, and is less likely to become a problem due to overabundance.

See additional comments under *Ceratophyllum demersum* for hints on distinguishing the two species.

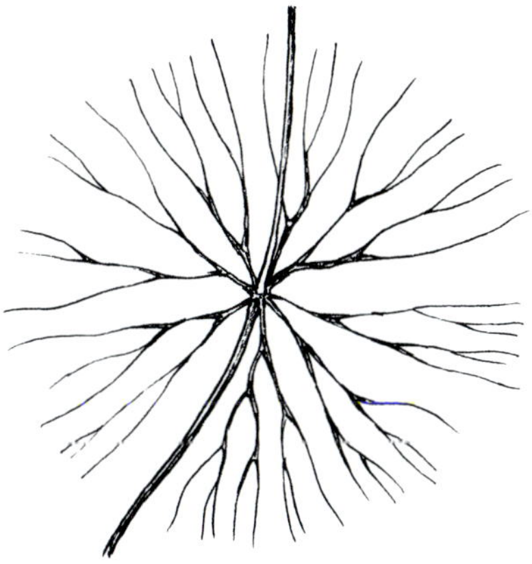

Ceratophyllum echinatum—whorl of leaves ×1

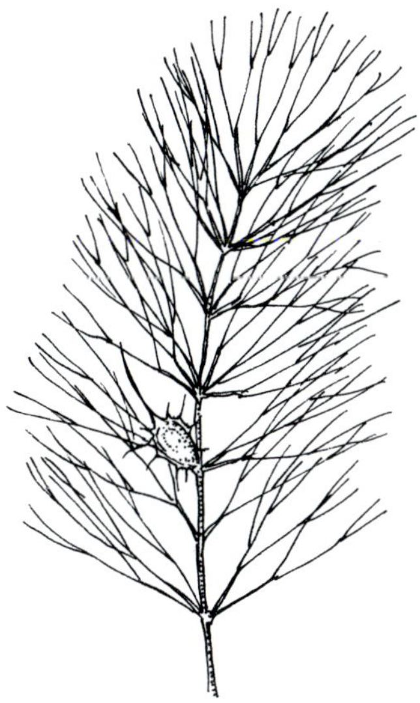

Ceratophyllum echinatum ×1

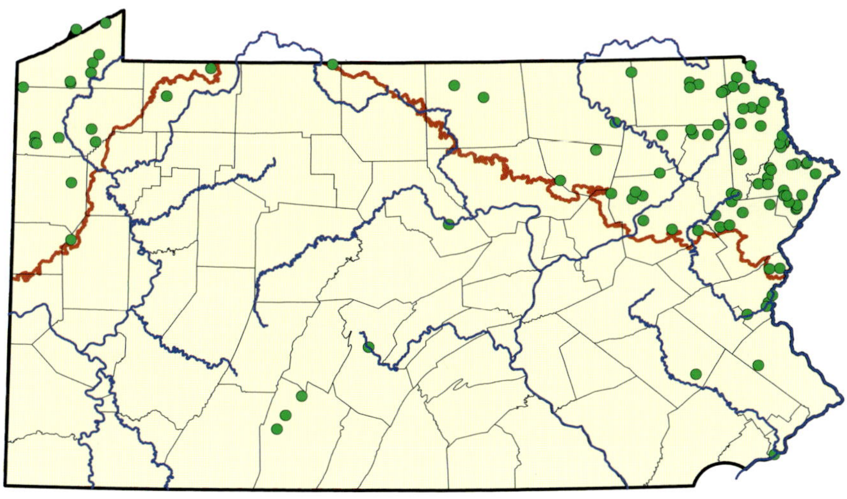

Ceratophyllum echinatum

FANWORT

CABOMBA AUBL.

Watershield Family—Cabombaceae

Fanwort ***Cabomba caroliniana*** **A.Gray**

Introduced

Rooted, submergent and floating-leaf perennial

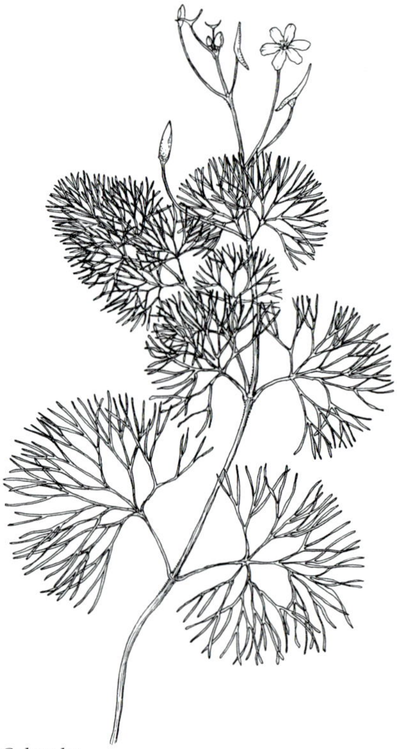

Cabomba caroliniana ×1/2

Fanwort is a rooted, mostly submergent aquatic plant with finely, palmately divided underwater leaves that occur in pairs (opposite) on the stem. Unlike the very similar leaves of coontail and water buttercup, fanwort leaves have a distinct petiole. Small, elliptical floating leaves are produced at the tips of the stems in association with flowering. The flowers extend above the water surface; they are 1–1.8 cm in diameter, have 3 white petals and 3 similar sepals, and are insect-pollinated.

Fanwort is native to the southeastern United States from Virginia to Florida and Texas; its popularity in the aquarium trade is believed to have led to introduction and subsequent naturalization north to Massachusetts as well as in Oregon. In Pennsylvania it is recorded from sites in the eastern and western regions of the state.

Fanwort has displayed weedy, aggressive tendencies in some areas.

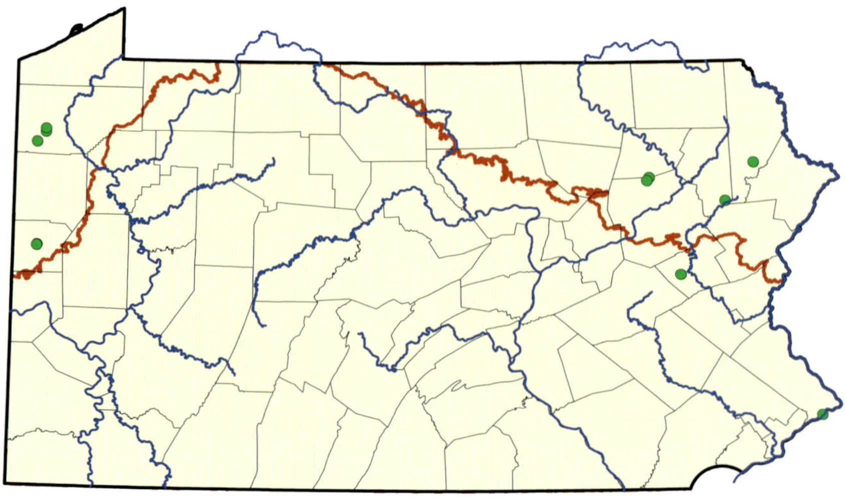

Cabomba caroliniana

MERMAID-WEED

PROSERPINACA L.

Water-milfoil Family—Haloragaceae

Mermaid-weeds are amphibious perennial plants that grow from a rhizomatous base. Stems are prostrate and often produce roots at the lower nodes; the stem tips are emergent. Leaves are alternate throughout. Leaf form varies with water depth, from deeply pinnately divided to merely dentate. Flowers are produced in the upper leaf axils; they are 3-merous. The fruit is a 3-angled nutlet.

Key to the Species of Mermaid-weed

A. emersed leaves merely serrate, submersed leaves pinnatifid
  B. fruits 4–6 mm wide with the angles winged .... *Proserpinaca palustris* var. *palustris*
  B. fruits 2–4 mm wide with the angles lacking wings .... *Proserpinaca palustris* var. *crebra*
A. leaves all pinnatifid, including those subtending flowers and fruits .... *Proserpinaca pectinata*

Common mermaid-weed — *Proserpinaca palustris* L.

Native
Rooted, emergent or submergent perennial

The submersed leaves of common mermaid-weed are 2–4 cm long and deeply pinnately divided into linear segments; they may be completely absent in very shallow water or on mud flats. Leaves of the erect flowering branches are larger and more shallowly lobed or merely toothed. Fruits are 2–6 mm and 3-angled.

The range of common mermaid-weed extends from Nova Scotia to Wisconsin and south to Cuba and Guatemala. In Pennsylvania it is an occasional plant of swamps, bogs, ponds, and marshes; it occurs primarily in the southeastern and northwestern corners of the state.

Two varieties are found in Pennsylvania: *Proserpinaca palustris* var. *palustris* has fruits 4–6 mm wide with the angles winged. *Proserpinaca palustris* var. *crebra* has fruits 2–4 mm wide with the angles lacking wings.

Proserpinaca palustris ×1/2

Proserpinaca palustris var. *crebra*—fruit ×4, note winged angles

Proserpinaca palustris var. *palustris*—fruit ×3

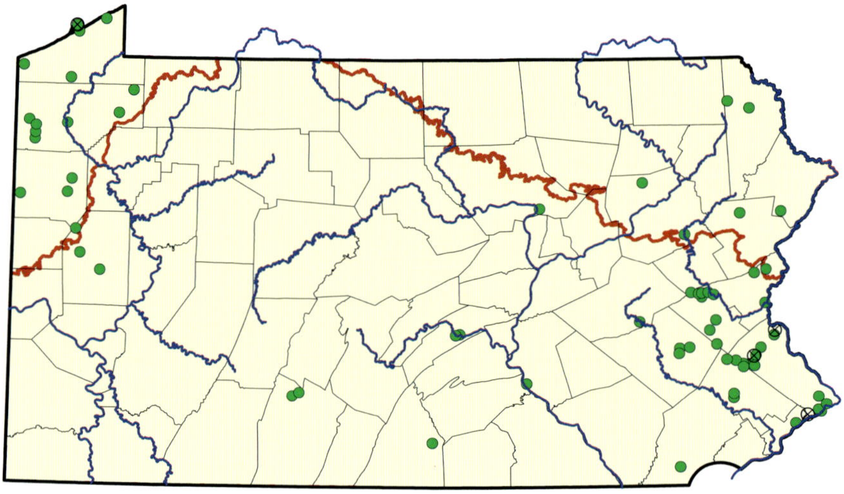

Proserpinaca palustris with varieties

Comb-leaved mermaid-weed *Proserpinaca pectinata* Lam.

PNHP

Native

Rooted, emergent or submergent perennial

Proserpinaca pectinata—flower ×3

The emersed leaves of comb-leaved mermaid-weed are 1.5–3 cm long and deeply pinnately lobed. The nutlets are 2.4–5 mm wide and obtusely angled.

Comb-leaved mermaid-weed occurs near the coast from Nova Scotia to Florida and Louisiana. In Pennsylvania it is a rare plant of the Atlantic Coastal Plain in the southeastern corner of the state.

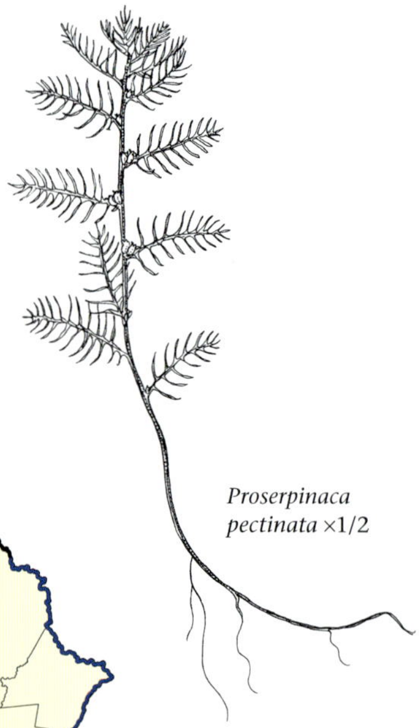

Proserpinaca pectinata ×1/2

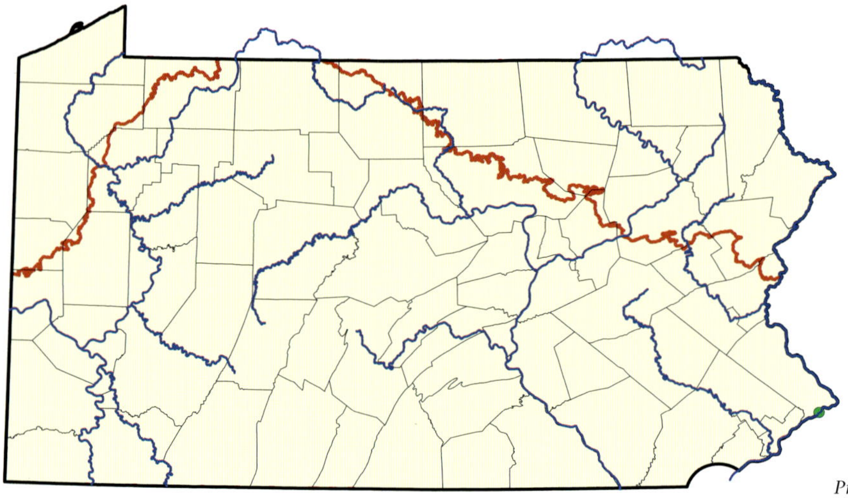

Proserpinaca pectinata

WATER-CROWFOOT
Ranunculus L.
Buttercup Family—Ranunculaceae

Members of the Buttercup Family are herbaceous annuals or perennials with alternate, often lobed or compound leaves. The flowers have 3–6 sepals, 5–6 petals, numerous stamens, and numerous, spirally arranged separate carpels. Each carpel matures to form an achene. Petals are often modified as nectar-producing structures.

Most crowfoots (or buttercups) are terrestrial or wetland plants; of the 20 species known to occur in Pennsylvania, only 2 are truly aquatic. Both have finely divided submersed leaves and often some merely lobed leaves near the stem tips.

Key to the Species of Water-crowfoot

A. flowers white .... *Ranunculus aquatilis*
A. flowers yellow .... *Ranunculus flabellaris*

White water-crowfoot — *Ranunculus aquatilis* L. var. *diffusus*

PNHP
Native
Rooted, submergent annual or perennial

White water-crowfoot has long, flexible stems with alternate, finely palmately divided leaves and white flowers that extend above the water surface. A few of the uppermost leaves may be merely lobed, but this character is highly variable. White water-crowfoot sometimes grows in fairly fast-moving streams where the clumps of stems become aligned with the current.

The flowers arise singly from upper leaf axils and have 5 whitish petals, each with a yellowish base; they are insect-pollinated. Flowering occurs May through July.

A plant of lakes, ponds, and rivers, white water-crowfoot is rare and scattered in eastern, central, and northwestern Pennsylvania, but it can be locally abundant. Its total range includes all of the United States and Canada with the exception of South Carolina, Georgia, Florida, and the Gulf Coast states.

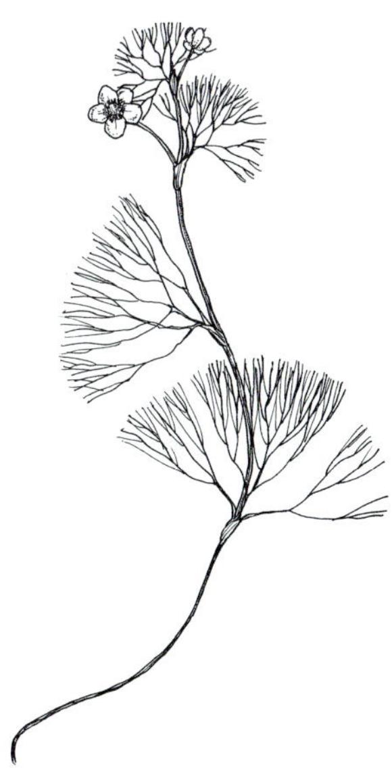

Ranunculus aquatilis ×1/2

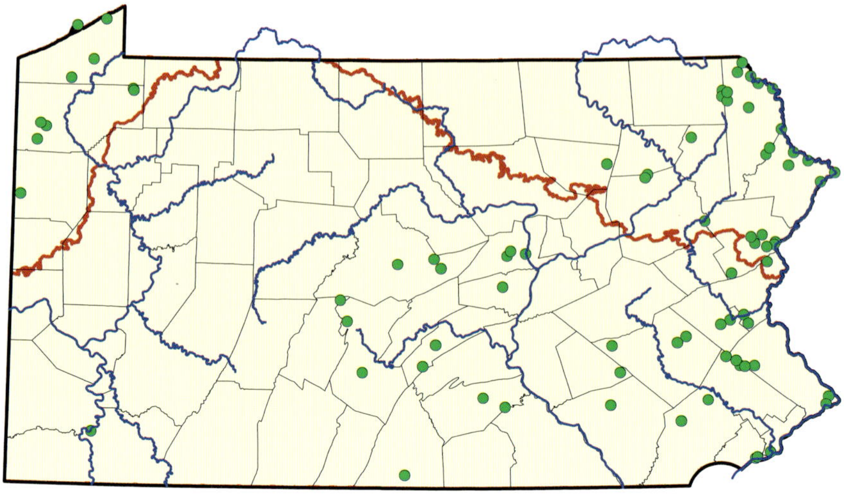

Ranunculus aquatilis

Ranunculus aquatilis growing in a stream in Pike County, Pennsylvania

Yellow water-crowfoot *Ranunculus flabellaris* Raf.

PNHP
Native
Rooted, emergent perennial

The stems of yellow water-crowfoot are floating or prostrate, and often form roots at the nodes. Submersed leaves are finely palmately divided, but leaves that are merely lobed may be present near the stem tips. The flowers extend above the water surface; they have 5–6 yellow petals and 5 sepals and are insect-pollinated. Flowering occurs May through August.

Yellow water-crowfoot is a rare plant of shallow water and muddy shores at scattered locations across Pennsylvania. Its total range includes the northeastern and north central United States with additional sites in the Pacific Northwest.

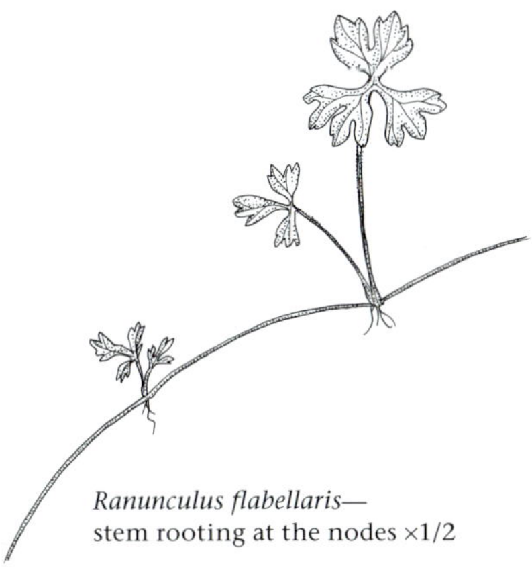

Ranunculus flabellaris—
stem rooting at the nodes ×1/2

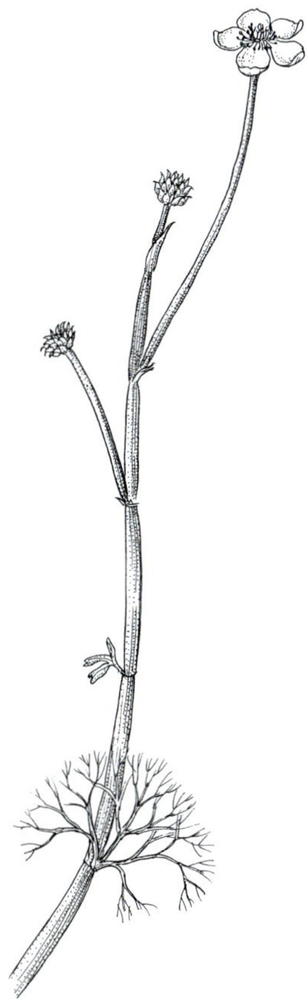

Ranunculus flabellaris—
showing finely divided submergent leaf ×1/2

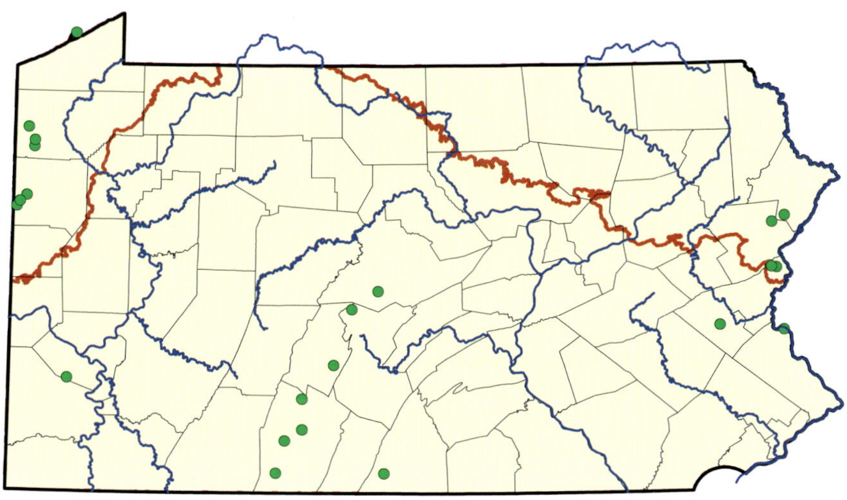

Ranunculus flabellaris

WATER-MARIGOLD
Bidens L.
[synonym: *Megalodonta beckii* (Torr. and Spreng.) Greene]
Aster Family—Asteraceae

Beck's water-marigold, water beggar-ticks

***Bidens beckii* Torre [synonym: *Megalodonta beckii* (Torr. and Spreng.) Greene]**

PNHP
Native
Rooted, submergent perennial

Beck's water-marigold has conspicuous flowering heads that extend above the water surface; they are about 2.5 cm in diameter with yellow rays. The long branching stems have opposite, sessile leaves. Underwater leaves are finely divided into slender, palmately arranged segments; emergent stem tips bear broader, simple leaves with toothed margins. Flowering occurs in mid- to late summer.

Beck's water-marigold grows in sites with a soft substrate and quiet, calcium-rich water; it can tolerate a range of pH values from 6 to nearly 9. It grows in water to 2 m deep.

Although locally abundant at a few sites, water-marigold is very rare in Pennsylvania; it is known from only a handful of sites in the glaciated northwestern and northeastern corners of the state. Its total range extends from Quebec to New Jersey and west to Manitoba and Missouri. Water-

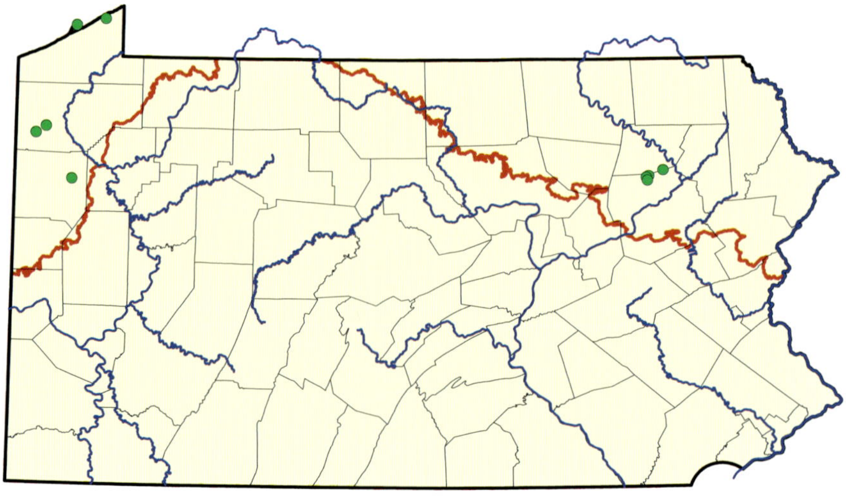

Bidens beckii

marigold also occurs in the Pacific Northwest from Oregon to British Columbia.

Other aquatic species with elongate stems and finely divided underwater leaves include coontail (*Ceratophyllum echinatum* and *C. demersum*), water-crowfoot (*Ranunculus aquatilis*), and fanwort (*Cabomba caroliniana*). Water-crowfoot can be distinguished by its alternate leaf arrangement. The leaves of fanwort, while opposite, are distinctly stalked, unlike the sessile leaves of Beck's water-marigold (see Figure 2.8, p. 28). Coontail has opposite, sessile leaves but it is best distinguished by the lack of emergent stem tips. The inconspicuous flowers of coontail occur in the axils of submersed leaves.

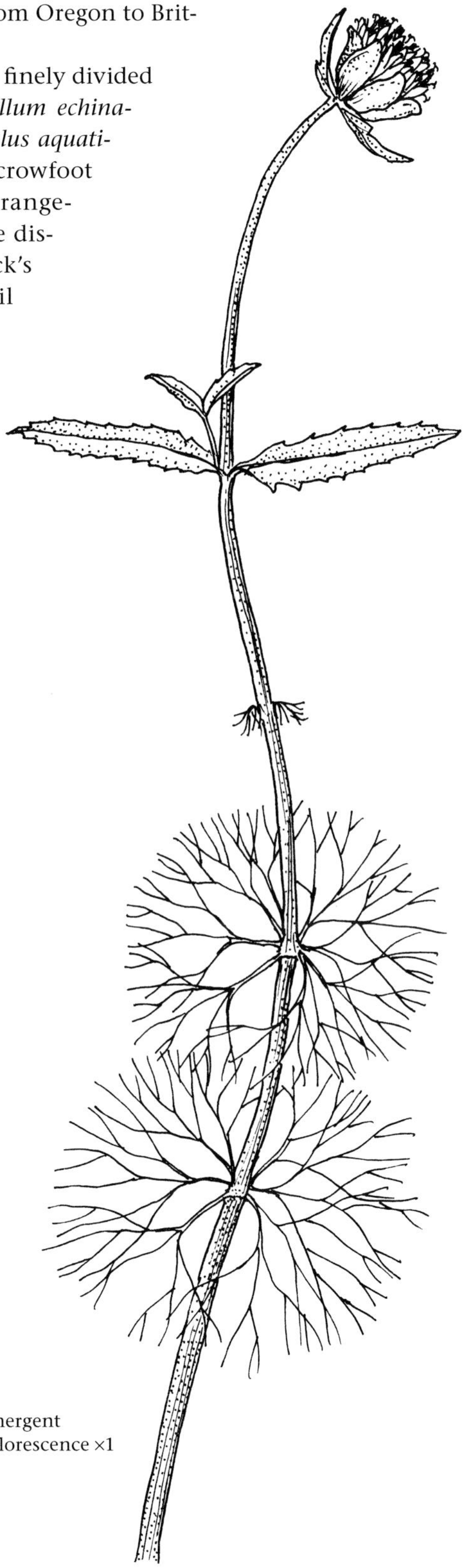

Bidens beckii—showing submergent and emergent leaves and inflorescence ×1

WATER-MILFOIL

MYRIOPHYLLUM L.

Water-milfoil Family—Haloragaceae

The water-milfoils are rooted submergent, or occasionally emergent, perennials. Leaves are alternate, opposite, or whorled; they are finely pinnately divided in most species. Flowers are in the axils of submersed leaves or floral bracts (reduced leaves) that develop on emersed stem tips. The flowers are small, perfect, and have a 4-parted, inferior ovary, 4 stamens, 4 tiny sepals, and 4 petals. The fruits, which are necessary for an accurate identification in some species, consist of a cluster of 4 cylindrical to ovate nutlets.

Eight species of water-milfoil are present in Pennsylvania; six are native and two are introduced. Four of our native species are classified as endangered, threatened, or rare by the Pennsylvania Natural Heritage Program (PNHP). Several species are highly invasive.

The fruits and foliage of water-milfoils are eaten by ducks and other marsh and shore birds, but are not a high quality food.

Key to the Species of Water-milfoil

A. leaves all alternate, reduced to minute bracts .... *Myriophyllum tenellum*
A. at least some of the leaves whorled, well-developed
  B. submersed leaves partly whorled and partly alternate or scattered; flowers and fruits borne in the axils of fully developed submersed leaves
    C. fruits smooth; emergent stem tips sometimes present .... *Myriophyllum humile*
    C. fruits tuberculate; emergent stem tips not produced .... *Myriophyllum farwellii*
  B. submersed and emersed leaves all whorled
    D. leafy vegetative branches wholly to partly emersed, with rigid, bluish-green leaves, submersed leaves similar in size, but with narrower segments and lax .... *Myriophyllum aquaticum*
    D. leafy vegetative branches submersed with only the flowering tips, with their reduced, bract-like leaves, extending above the water surface
      E. bracts (reduced leaves) subtending the flowers deeply pinnately lobed .... *Myriophyllum verticillatum*
      E. bracts (reduced leaves) subtending the flowers entire or serrate
        F. floral bracts serrate, much longer than the flowers .... *Myriophyllum heterophyllum*
        F. floral bracts entire, slightly longer than, equal to, or shorter than the flowers
          G. leaf segments 4–14 per side; stem diameter uniform .... *Myriophyllum sibiricum*
          G. leaf segments 5–24 per side; stem thicker below the inflorescence .... *Myriophyllum spicatum*

Parrot's-feather *Myriophyllum aquaticum* (Vell.) Veerdc.

Introduced
Rooted, emergent perennial

Parrot's-feather has whorled leaves on stems that are partly to completely emergent or sometimes rooted in exposed mud. On emergent stems the finely divided pinnate leaves are bluish-green with a hydrophobic surface that sheds water; the leaves are rigid, holding their shape out of water. Submergent foliage has thinner, more lax leaf segments. Flowering is infrequent, but when present, flowers are in the axils of the scarcely reduced upper leaves. Only female plants are known in North America.

Parrot's-feather is native to South America; it was introduced for use in aquaria and water gardens. In Pennsylvania it has become naturalized in ponds and canals at scattered locations; it has become a serious aquatic weed in the South.

Myriophyllum aquaticum ×1/2

Myriophyllum aquaticum

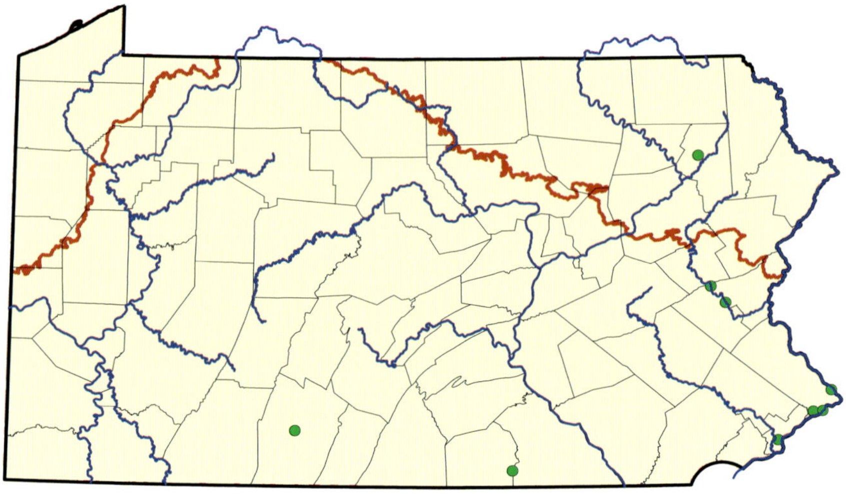

Myriophyllum aquaticum

Farwell's water-milfoil *Myriophyllum farwellii* Morong

PNHP
Native
Rooted, submergent perennial

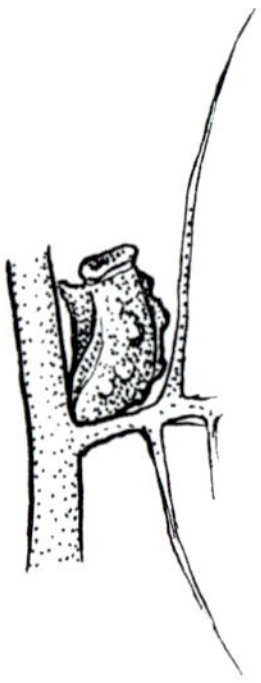

Myriophyllum farwellii—fruit in axil of submersed leaf ×5, note warty surface

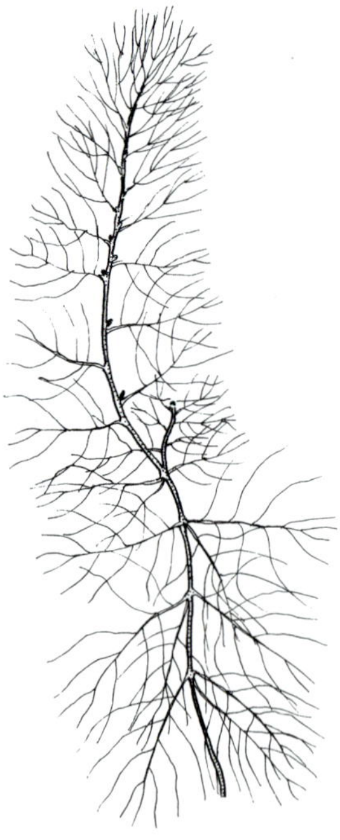

Myriophyllum farwellii ×1/2

Farwell's water-milfoil is a rooted submergent plant; it has alternate or subopposite, very fine, pinnately divided leaves that are crowded on the stems and limp when removed from the water. Stems arise in clusters, but are unbranched or only lightly branched. Terminal winter buds, known as turions, are produced. Flowers and fruits are formed in the axils of submersed leaves. The fruits have a warty surface (visible at 10X magnification).

Farwell's water-milfoil occurs from Ontario and Nova Scotia to New England, New York, and northern Michigan and Minnesota. It is at the southern limit of its range in northeastern Pennsylvania, where it is a plant of lakes and ponds, generally growing in water less than 2 m deep. However, detached stems often float and may accumulate at the down-wind side of a body of water.

All the sites where Farwell's water-milfoil has been collected in Pennsylvania are impoundments (lakes formed by dams). Like several other water-milfoil species, this plant can be sufficiently abundant in shallow water bodies to interfere with swimming, fishing, and boating.

Myriophyllum farwellii is very similar to *M. humile*. Mature fruits are needed to distinguish the two species; the fruits of *M. farwellii* have a warty surface in contrast to the smooth fruits of *M. humile*. The two species often grow together.

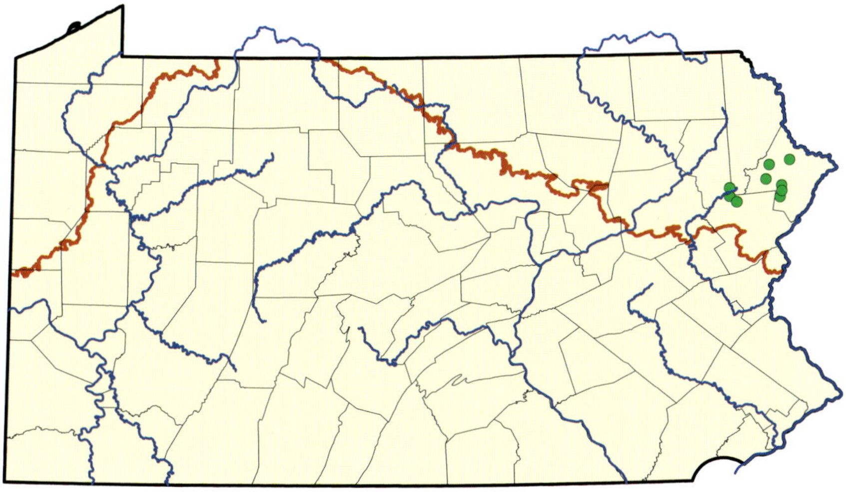

Myriophyllum farwellii

Broad-leaved water-milfoil *Myriophyllum heterophyllum* Michx.

PNHP
Native
Rooted, submergent perennial

In Pennsylvania broad-leaved water-milfoil was known from several mid to late-1800s collections from Bucks County, but has not been collected in the southeastern part of the state for many years. It is also known from several sites in western Pennsylvania that are within the natural range of the species (see below). Recently, though, broad-leaved water-milfoil has invaded a number of lakes and ponds in northeastern Pennsylvania very aggressively, rapidly developing dense monocultures, interfering with the growth of other plants, and restricting recreational use.

Broad-leaved water-milfoil is native from western New York and Ontario to North Dakota and south to Florida and Texas. It has recently invaded New England, where it is considered an exotic invasive in Maine, New Hampshire, Vermont, Massachusetts, and Connecticut (USDA Plants 2010; IPANE 2010). Studies in Connecticut have detected the presence of hybrids in invasive populations there (Moody and Les 2002). However, a preliminary study of Pennsylvania material failed to find evidence of hybridization (Treher et al. 2006).

The leaves of broad-leaved water-milfoil are whorled throughout; submersed leaves are finely pinnately divided and limp when removed from the water. Flowers and fruits, subtended by floral bracts, are produced on swollen stem tips that extend above the water surface. The fruits are subglobose with a rough surface.

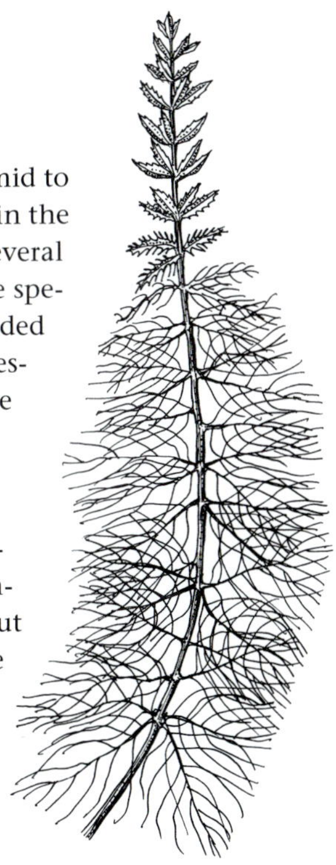

Myriophyllum heterophyllum—stem with emergent tip ×1/2

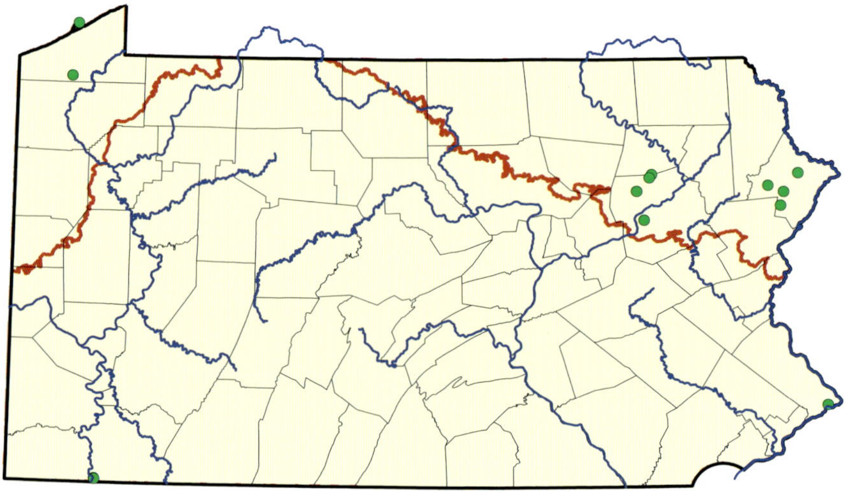

Myriophyllum heterophyllum

Myriophyllum heterophyllum with emergent, flowering stem tips

Water-milfoil ***Myriophyllum humile*** **(Raf.) Morong**

Native

Rooted, submergent perennial

Myriophyllum humile is normally a submergent species. Its finely divided leaves are whorled below and alternate near the stem tips; they are limp when removed from the water. Flowers and fruits are found in the axils of submersed leaves. However, during low water, *Myriophyllum humile* develops a semi-terrestrial or emergent form with rigid, simple leaves. The same growth form often develops at the tips of detached floating stems. Turions are not produced. The fruits are 0.7–1.2 mm long and smooth and rounded on the back of each nutlet.

The range of *Myriophyllum humile* extends from Nova Scotia and Vermont to Maryland and locally west to Minnesota. In Pennsylvania it is a common plant of lakes and ponds in the glaciated northeastern region of the state; it also occurs at widely scattered locations elsewhere.

See comments under *Myriophyllum farwellii* above for a discussion of how to distinguish these two very similar species.

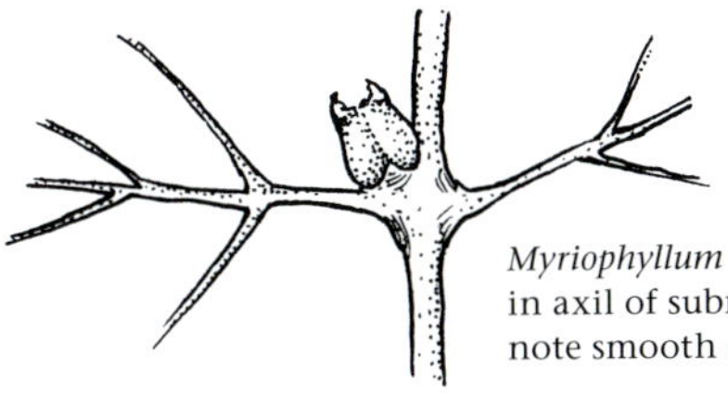

Myriophyllum humile—fruit in axil of submersed leaf ×5, note smooth surface

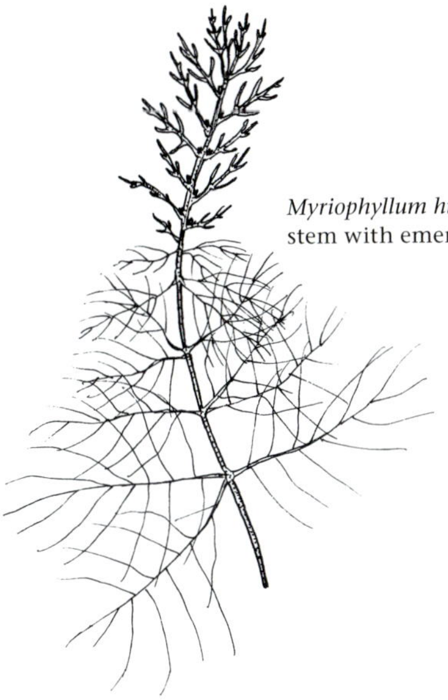

Myriophyllum humile—stem with emersed tip ×3/4

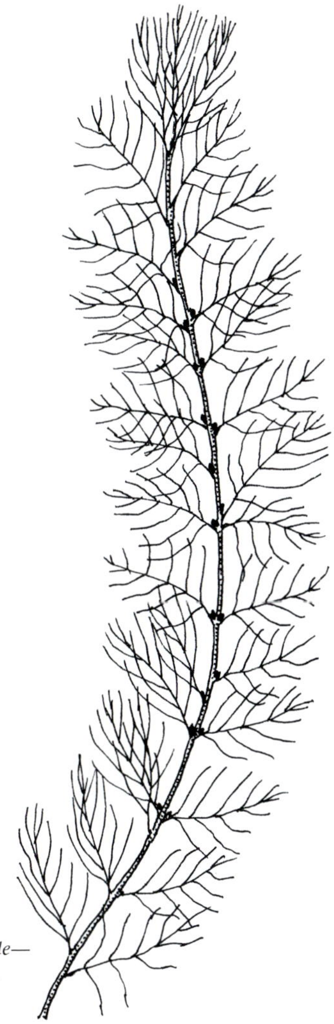

Myriophyllum humile—submersed stem ×1

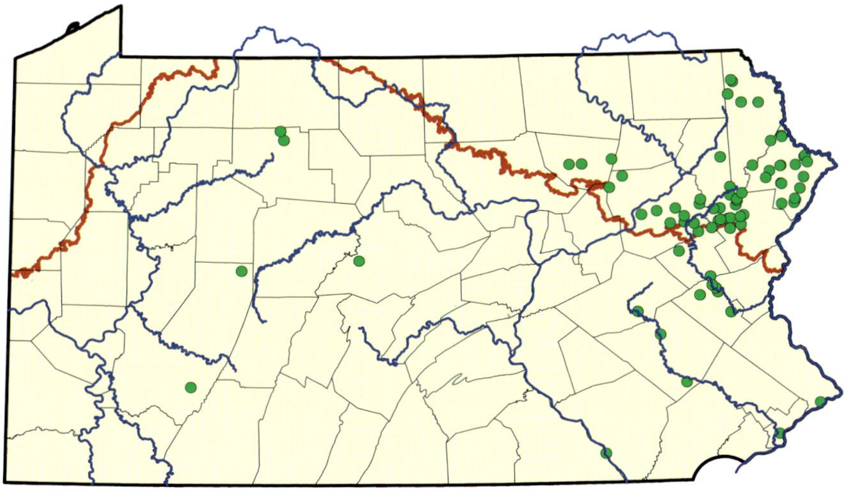

Myriophyllum humile

Northern water-milfoil — *Myriophyllum sibiricum* Komarov

PNHP
Native
Rooted, submergent perennial

Northern water-milfoil has leaves in whorls of 3–4; submersed leaves have 4–14 segments per side. Flowers and fruits are produced in the axils of small floral bracts on the emergent stem tips. Turions, consisting of condensed terminal stem segments, are produced in the fall.

The range of northern water-milfoil extends from Labrador to Alaska and south to Virginia, Missouri, and California. It also grows in Eurasia. In Pennsylvania it is a rare plant of rivers, lakes, ponds, and marshes.

Non-flowering plants of *Myriophyllum sibiricum* are very similar to *M. spicatum* and *M. verticillatum*. See Table 8.1 for a comparison.

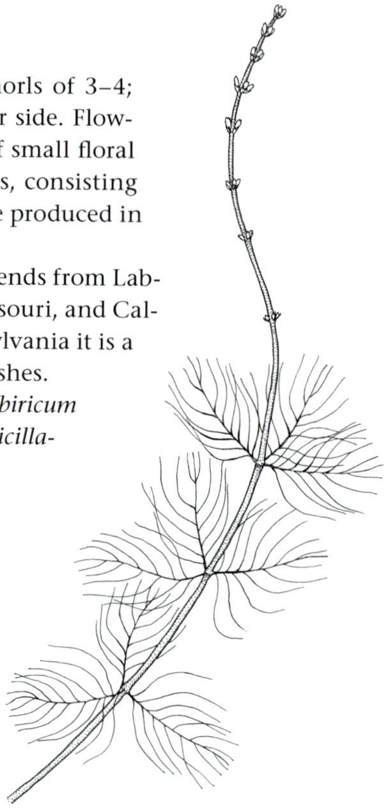

Myriophyllum sibiricum ×1/2

Myriophyllum sibiricum

Table 8.1. Comparison of Vegetative Characteristics of *Myriophyllum sibiricum*, *M. spicatum*, and *M. verticillatum*

	M. SIBIRICUM	*M. SPICATUM*	*M. VERTICILLATUM*
leaves per node	4, less often 5	4, rarely 5	4 or 5
leaf length	1.5–3 cm	2–4 cm	2–4.5 cm
pairs of leaf segments	4–14	5–24	9–17
stem width	constant throughout	twice as wide below the inflorescence	constant throughout
turions	cylindrical, tapering, leaves smaller than normal summer leaves	not produced	club-shaped, leaves similar to normal summer leaves

Adapted from Aiken et al. (1979).

Eurasian water-milfoil *Myriophyllum spicatum* L.

Introduced
Rooted, submergent perennial

One of the most abundant of the non-native, invasive aquatic plants, Eurasian water-milfoil occurs in lakes and rivers throughout Pennsylvania. It is most problematic in high-nutrient (eutrophic) waters. Although Eurasian water-milfoil may become established in low-nutrient, low pH lakes, it rarely becomes abundant in such situations. It is native to Eurasia.

Rooted stems of Eurasian water-milfoil are known in water as deep as 3 m; the stems typically branch profusely just below the water surface. Leaves are borne in whorls of 3–4; submersed leaves are pinnately divided, with 5–24 slender segments on each side of the midrib. Inflorescences are produced on emergent stem tips; the flowers and fruits are subtended by short bracts that are entire to slightly toothed. Turions are not produced; however, some shoots from the previous year remain green all winter and resume active growth in the spring.

Vegetative fragmentation and transport of stem sections by water currents or human activities are probably the most common means of dispersal of Eurasian water-milfoil from one water body to another.

The branching stems of Eurasian water-milfoil can form a dense canopy just below the water surface which inhibits the growth of native plants, interferes with recreation, and impedes navigation. Efforts to control its growth have included herbicides and mechanical harvesting; both methods have unintended consequences, including effects on non-target organisms.

A more specific control strategy utilizes a native submersed aquatic weevil (*Euhrychiopsis lecontei*) which is a water-milfoil specialist. Fortunately, the weevil has been shown to prefer Eurasian water-milfoil over its native host, the closely related *Myriophyllum sibiricum* (Solarz and Newman 2001; Gettys et al. 2009).

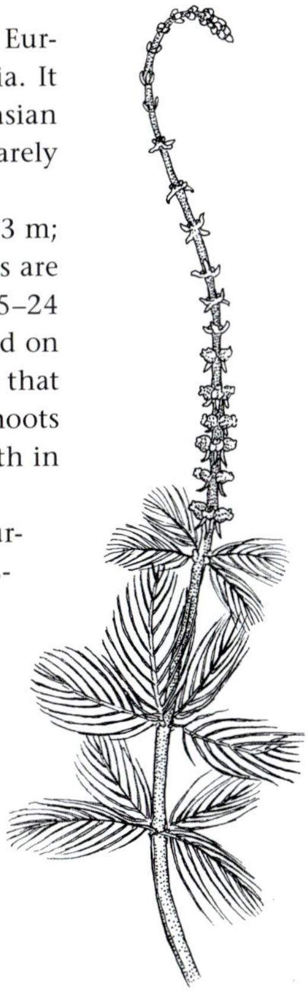

Myriophyllum spicatum ×3/4

Eurasian water-milfoil is very similar to the rare native species *Myriophyllum sibiricum*. See Table 8.1 for a comparison of vegetative characteristics of *Myriophyllum sibiricum*, *M. spicatum*, and *M. verticillatum*.

Myriophyllum spicatum—fruit ×4

Myriophyllum spicatum—flower ×4

Myriophyllum spicatum—emergent stem tip with flowers

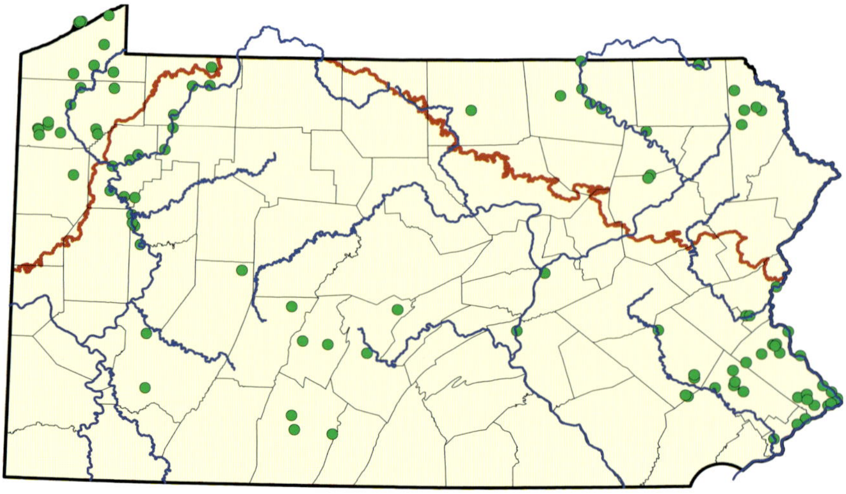

Myriophyllum spicatum

Slender water-milfoil *Myriophyllum tenellum* Bigelow

PNHP
Native
Rooted, submergent perennial

Slender water-milfoil is unlike any of the other *Myriophyllum* species that occur in Pennsylvania. It is a rooted, submergent plant with stiffly upright stems 2–20 cm tall, which arise singly from rhizomes. Tiny scale-like leaves are alternate on the stems. Vegetative plants often form a turf in deep water. In shallow water flowers are produced in the upper axils of emergent stems.

Slender water-milfoil is a northern plant ranging from Newfoundland and Quebec to New Jersey and west to Minnesota. In Pennsylvania it is a rare, but locally abundant, plant of natural glacial lakes in the northeastern corner of the state.

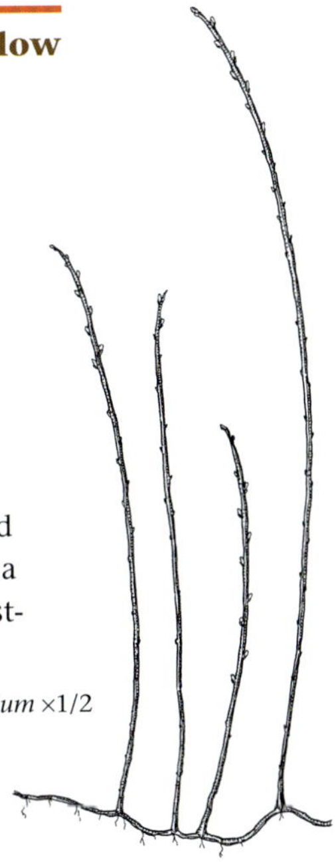

Myriophyllum tenellum ×1/2

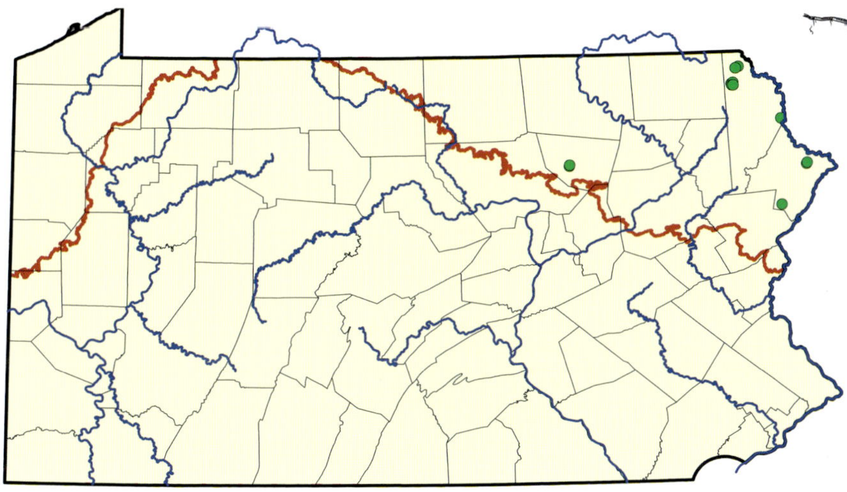

Myriophyllum tenellum

Whorled water-milfoil *Myriophyllum verticillatum* L.

PNHP
Native
Rooted, submergent perennial

Myriophyllum verticillatum occurs throughout the boreal regions of the northern hemisphere. In eastern North America it reaches its southern limit in New York and western Pennsylvania. In Pennsylvania it is a very rare plant known only from ponds and marshes in the northwestern corner of the state.

The submersed leaves of whorled water-milfoil are in whorls of 4–5 per node; each leaf has 9–17 segments per side. The inflorescences, on emersed stem tips, have floral bracts that are also

Myriophyllum verticillatum ×1/2

Myriophyllum verticillatum—stem tip with turion ×3/4

pinnately lobed or at least strongly toothed and as long as, or longer than, the flowers. Turions are formed in the fall and remain dormant until spring.

In vegetative condition, whorled water-milfoil can easily be confused with *M. spicatum* and *M. sibiricum*. See Table 8.1 for differences.

Myriophyllum verticillatum—emergent stem tip with flowers

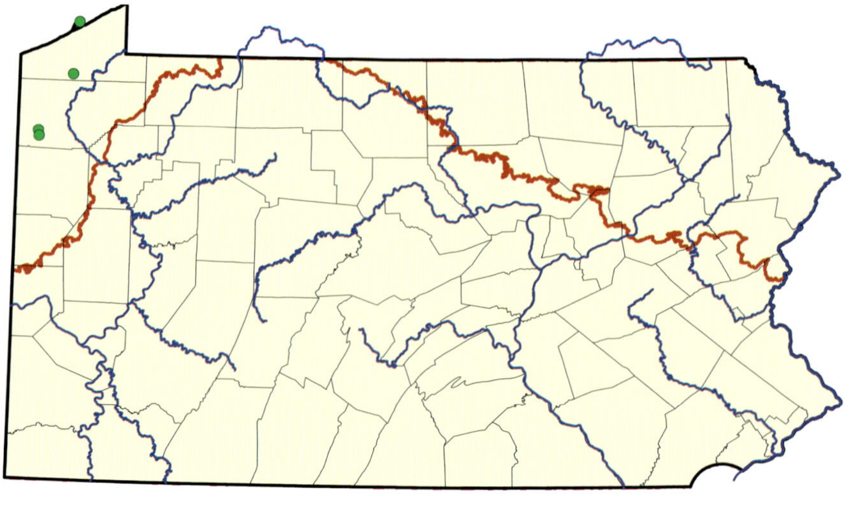

Myriophyllum verticillatum

Glossary

achene—dry, indehiscent, one-seeded fruit with the seed coat not fused to the fruit wall
abaxial—the side away from the stem axis
acuminate—forming an angle of less than 45 degrees
adaxial—the side toward the stem axis
adventitious roots—roots that develop at nodes on a stem
aerenchyma—spongy tissue in the stems or roots of aquatic plants that allows for buoyancy and gas exchange
aereolae—the areas between the veinlets of a leaf
alternate-leaved—having only 1 leaf at each node on the stem
annual—a plant that completes its entire life cycle in a single growing season
anther—sac-like part of the stamen where pollen is produced
appressed—flat against a plant organ
auriculate—having an ear-shaped appendage
awn—a long, bristle-like appendage
axil—the junction of leaf and stem
axillary—occurring in the leaf axil

bract—a small, leaf-like structure, often subtending a flower

calyx—collective term for the sepals
capillary leaves—very slender, fine, or finely-divided leaves
carpel—the organ that bears the ovules and seeds of an angiosperm, the basic unit of the pistil
cauline—of the stem
ciliate—bearing long, stiff hairs along an edge or keel
cleistogamous—flowers that do not open, but undergo self-fertilization and produce seeds
compound leaf—a leaf divided into leaflets
corm—a vertical, enlarged, solid, underground stem
cuticle—natural waxy coating on the surface of leaves and/or stems

decumbent—reclining on the surface, but with the tip ascending
depauperate—growing in reduced form, often as a result of marginal habitat conditions
dichotomous branching—stems repeatedly dividing into 2 equal branches
dioecious—species in which male and female flowers are borne on different plants
dystrophic—low-nutrient, acidic lakes characterized by peat accumulation and water stained brown by organic acids and tannins

emergent—growing with the roots below the water surface, but most of the shoot above the water
entire margin—leaf edge that lacks teeth or lobes

epiphyte—a plant that uses another plant for physical support, but does not derive any nutrition from its host
eutrophic—high-nutrient aquatic systems

filiform—thread-like
frond—a leaf of a fern, also sometimes applied to the segments of duckweed plants

glaucous—having a waxy coating, often resulting in a bluish appearance
glumes—the pair of bracts at the base of each grass spikelet

herbaceous—remaining soft and green during the growing season, not becoming woody

imbricate—made up of scales whose edges overlap
inferior ovary—condition in which the stamens and perianth are attached above the ovary
inflorescence—a grouping or cluster of flowers
introduced—a non-native plant that is deliberately brought in to an area
invasive species—a species with aggressive growth capable of disrupting normal ecosystem structure or function; many, but not all, invasive species are non-native (exotic)
involucre—a set of usually small, overlapping bracts subtending an inflorescence or flower

keel—a longitudinal ridge

lanceolate—much longer than wide, with the widest part below the middle; lance-shaped
lemma—the lower of the 2 bracts that enclose each floret in a grass spikelet
ligule—a tongue- or strap-shaped projection present at the junction of the blade and sheath of a grass or sedge leaf

mesotrophic—lakes with moderate nutrient levels
monoecious—a species with unisexual flowers but with both male and female flowers borne on the same plant
mucronate—abruptly tapered to a short, sharp point

nerve—a prominent vein in a leaf or other organ
node—specific region of a stem at which leaves or buds are produced

ocrea—a stipular sheath formed at the nodes of plants in the Smartweed Family
ocreolae—a small stipular sheath that subtends flowers or groups of flowers in the Smartweed Family
oligotrophic—low-nutrient aquatic systems
opposite-leaved—bearing 2 leaves at each node

ovary—the portion of the pistil that contains the ovules and seeds; the ovary of a flower matures into a fruit

palea—the upper of the 2 bracts that enclose each grass flower in a grass spikelet
palmate leaf—a compound leaf with the leaflets radiating from a common point of attachment
panicle—a branched inflorescence with individually pedicellate flowers
papillose—a surface covered with small rounded projections
pedicel—the stalk of an individual flower
peduncle—a stalk bearing an inflorescence
perennial—a plant that lives for more than 2 seasons
perianth—a collective term for the sepals and petals of a flower
petiole—the stalk of a leaf
pinnate leaf—a compound leaf with leaflets attached along an elongate central axis
pinnatifid—a leaf that is pinnately cleft or lobed half or more of the distance to the midrib
pistil—the female organ of the flower, includes the ovary, style and stigma; made up of one or more carpels

raceme—an unbranched inflorescence with individually pedicellate flowers arrayed along an elongate axis
radial symmetry—radiating equally in all directions
receptacular bracts—bracts located between the flowers in the flowering head of a member of the Aster Family.
rhizoid—a small, root-like structure without vascular tissue
rhizome—a horizontal, underground stem that produces roots and shoots
rhizomatous—producing rhizomes
rosette—a radiating cluster of leaves at ground level

sagittate—arrowhead-shaped with the basal lobes pointed downward
sepal—the outer whorl of flower parts, usually green and smaller than the petals; sepals are collectively known as the calyx
sessile—attached directly to the stem, without a stalk
sheath—the basal portion of many monocot leaves, that surrounds the stem
spadix—a fleshy, spike-like inflorescence, usually surrounded by a sheathing bract
spathe—the sheathing bract that surrounds the spadix in inflorescences of the Arum Family
spike—an inflorescence in which individual flowers are attached directly to an elongate central axis
spikelet—a small spike-like inflorescence or subunit of an inflorescence, especially in the grass and sedge families
stamen—the male organ of the flower; consists of a filament and an anther and produces pollen
stigma—the portion of the pistil that receives pollen
stolon—horizontal stem at ground level, produces roots and/or shoots at the nodes

stoloniferous—having stolons
strobilus—a cone-like cluster of spore sacs on an axis
style—the portion of the pistil that connects the stigma to the ovary
submergent—growing completely below the water surface; refers to plant growth habit
submersed—occurring below the water surface; refers to specific plant organs
subulate—tipped by a fine, sharp point
superior ovary—an ovary that is attached above the petals and sepals

tepals—a term applied to petals and sepals when the two are similar in color and texture
terete—rounded when viewed in cross section
thalloid—see thallus
thallus—a flattened plant body that lacks distinct stems or leaves
tuber—an underground storage organ of stem origin
tuberculate—having small swellings or projections
turion—overwintering short, vegetative shoot produced by some aquatic plants

umbel—an inflorescence in which the pedicels arise from a common point

References and Sources

Aiken, S. G., P. R. Newroth, and I. Wile. 1979. The biology of Canadian weeds. 34. *Myriophyllum spicatum* L. *Canadian Journal of Plant Science* 59: 201–205.

Bissell, James K. and Charles W. Bier. 1987. Botanical Survey and Natural Community Classification for Presque Isle State Park, Erie County, Pennsylvania. Report submitted to the Pennsylvania Department of Natural Resources, Bureau of Forestry, Harrisburg.

Bowes, George, and Michael E. Salvucci. 1989. Plasticity in the photosynthetic carbon metabolism of submersed aquatic macrophytes. *Aquatic Botany* 34: 233–266.

Browne, Janet. 1995. *Charles Darwin, Voyaging*. Princeton University Press, Princeton.

Brunton, Daniel F. and Donald M. Britton. 1996. Taxonomy and distribution of *Isoetes valida*. *American Fern Journal* 86(1): 16–25.

Brunton, Daniel F. and Donald M. Britton. 1997. Appalachian quillwort (*Isoetes appalachiana, sp. nov.;* Isoetaceae), a new Pteridophyte from the eastern United States. *Rhodora* 99(898): 118–133.

Casati, Paula, Maria V. Lara, and Carlos S. Andreo. 2000. Induction of a C_4-like mechanism of CO_2 fixation in *Egeria densa,* a submersed aquatic species. *Plant Physiology* 123: 1611–1621.

Charalambidou, and Luis Santamaria. 2002. Waterbirds as endozoochorous dispersers of aquatic organisms: a review of experimental evidence. *Acta Oecologica* 23: 165–176.

Commonwealth of Pennsylvania, Department of Conservation and Natural Resources. 1993. Pennsylvania Code, Title 17, Chapter 45. Conservation of Pennsylvania Native Wild Plants. Harrisburg.

Cook, Christopher D.K. 1988. Wind pollination in aquatic angiosperms. *Annals of the Missouri Botanical Garden* 75: 768–777.

Cook, Christopher D.K. 1990. Seed dispersal of *Nymphoides peltata* (J.G. Gmel.) Kuntze (Menyanthaceae). *Aquatic Botany* 37: 325–340.

Darwin, Charles. 1898. *Insectivorous Plants*. D. Appleton and Company, New York.

Dillard, Gary E.1999. *Common Freshwater Algae of the United States*. J. Cramer, Berlin.

Elakovich, S.D. and J.W. Wooten. 1987. An examination of the phytotoxicity of the watershield, *Brasenia schreberi*. *Journal of Chemical Ecology* 13: 1935–1940.

Ellmore, George. 1981. Root dimorphism in *Ludwigia peploides* (Onagraceae): structure and gas content of mature roots. *American Journal of Botany* 68(4): 557–568.

Flora of North America Editorial Committee. 1993. *Flora of North America Vol. 2*. Oxford University Press, New York.

Flora of North America Editorial Committee. 1997. *Flora of North America Vol. 3*. Oxford University Press, New York.

Flora of North America Editorial Committee. 2000. *Flora of North America Vol. 22*. Oxford University Press, New York.

Flora of North America Editorial Committee. 2002. *Flora of North America Vol. 26*. Oxford University Press, New York.

Flora of North America Editorial Committee. 2006. *Flora of North America Vol. 21*. Oxford University Press, New York.

Gettys, Lyn A., William T. Haller, and Marc Bellaud. 2009. *Biology and Control of Aquatic Plants, a Best Management Practices Handbook*. Aquatic Ecosystem Restoration Foundation, Marietta, GA.

Givens, Florence M. and Wayne Ferren. 1972. *Marsilea quadrifolia* L. in Delaware County, Pennsylvania. *Bartonia* 41: 60.

Gleason, Henry A. and Arthur Cronquist. 1991. *Manual of Vascular Plants of Northeastern United States and Adjacent Canada*. New York Botanical Garden, Bronx.

Grund, Steve. Undated. Documenting the Native and Introduced Flora Associated with Glacial Lakes in Northwest Pennsylvania. Western Pennsylvania Conservancy, Pittsburgh.

Haynes, Robert R. 1988. Reproductive biology of selected aquatic plants. *Annals of the Missouri Botanical Garden* 75: 805–810.

Invasive Plant Atlas of New England. http://nbii-nin.ciesin.columbia.edu/ipane/, accessed 1/21/2010.

Judd, Walter S., Christopher S. Campbell, Elizabeth A. Kellogg, Peter F. Stevens, and Michael J. Donoghue. 2002. *Plant Systematics, A Phylogenetic Approach, Second Edition*. Sinauer Associates, Inc. Sunderland, MA.

Kaplan, Zdenek, Judith Fehrer, and C. Barre Hellquist. 2009. New hybrid combinations revealed by molecular analysis: the unknown side of North American pondweed diversity (*Potamogeton*). *Systematic Botany* 34(4): 625–642.

Kibriya, Sadia, and J. Iwan Jones. 2007. Nutrient availability and the carnivorous habit in *Utricularia vulgaris*. *Freshwater Biology* 52: 500–509.

Les, Donald H. 1988. Breeding systems, population structure, and evolution in hydrophilous angiosperms. *Annals of the Missouri Botanical Garden* 75: 819–835.

Les, Donald H., Robert S. Capers, and Nicholas P. Tippery. 2006. Introduction of *Glossostigma* (Phrymaceae) to North America: a taxonomic and ecological overview. *American Journal of Botany* 93(6): 927–939.

Manville, G. Christine. 2001. An annotated checklist of liverworts and hornworts (hepatics and Anthocerotes) of Pennsylvania. Wild Resources Conservation Fund, Pennsylvania Department of Conservation and Natural Resources, Harrisburg.

Martin, Alexander C., Herbert S. Zim, and Arnold L. Nelson. 1961. *American Wildlife and Plants*. Dover Publications, New York.

McCourt, Richard M., Charles F. Delwiche, and Kenneth G. Karol. 2004. Charophyte algae and land plant origins. *Trends in Ecology and Evolution* 19(12): 661–666.

Moerman, Daniel E. 2000. *Native American Ethnobotany*. Timber Press, Portland, OR.

Moody, Michael L. and Donald H. Les. 2002. Evidence of hybridity in invasive water-milfoil (*Myriophyllum*) populations. *Proceedings National Academy of Science* 99(23): 14867–14871.

NatureServe Explorer. http://www.natureserve.org/explorer/, accessed 1/21/2010.

Neinhuis, C. and W. Barthlott. 1997. Characterization and distribution of water-repellent, self-cleaning plant surfaces. *Annals of Botany* 79: 667–677.

Osborn, Jeffrey M. and Edward L. Schneider. 1988. Morphological studies of the Nymphaeaceae *sensu lato*. XIV. The floral biology of *Brasenia schreberi*. *Annals of the Missouri Botanical Garden* 75: 778–794.

Padget, Donald J. 1997. A biosystematic monograph of the genus *Nuphar* Sm. (Nymphaeaceae). PhD Dissertation, University of New Hampshire.

Pennsylvania Flora Database. Morris Arboretum of the University of Pennsylvania, Philadelphia. http://www.paflora.org, accessed 1/11/2009.

Pennsylvania Natural Heritage Program, rare plant listings. http://www.naturalheritage.state.pa.us/PlantsPage.aspx, accessed 6/16/2010.

Petroutka, Marianne, Wolfram Adlassnig, Michael Volgger, Thomas Lendl, Walter G. Url, and Irene K. Lichtscheidl. 2008. *Utricularia*: a vegetarian carnivorous plant? *Plant Ecology* 199: 153–162.

Preston, C. D. 1995. *Pondweeds of Great Britain and Ireland*, B.S.B.I. Handbook No. 8. Botanical Society of the British Isles, London.

Rhoads, Ann Fowler, and Timothy A. Block. 2000. *The Plants of Pennsylvania: An Illustrated Manual, First Edition*. University of Pennsylvania Press, Philadelphia.

Rhoads, Ann Fowler, and Timothy A. Block. 2007. *The Plants of Pennsylvania: An Illustrated Manual, Second Edition*. University of Pennsylvania Press, Philadelphia.

Sirová, Dagmara, Jakub Borovec, Hana Šantrůčková, Jiří Šantrůček, Jaroslav Vrba, and Lubomír Adamec. 2010. *Utricularia* carnivory revisited: plants supply photosynthetic carbon to traps. *Journal of Experimental Botany* 61(1): 99–103.

Solarz, Susan L. and Raymond M, Newman. 2001. Variation in host plant preference and performance by the milfoil weevil, *Euhrychiopsis lecontei* Deitz, exposed to native and exotic milfoils. *Oecologia* 126: 66–75.

Stuckey, Ronald L. 1974. The introduction and distribution of *Nymphoides peltata* (Menyanthaceae) in North America. *Bartonia* 42: 14–23.

Stuckey, Ronald L. 1979. Distributional history of *Potamogeton crispus* (curly pondweed) in North America. *Bartonia* 46: 22–42.

Stuckey, Ronald L. 1985. Distributional history of *Najas marina* (spiny naiad) in North America. *Bartonia* 51: 2–16.

Taylor, Peter. 1989. *The Genus Utricularia: A Taxonomic Monograph*. Kew Bulletin Additional Series XIV. Royal Botanic Gardens, Kew, London.

Titus J.G., D.E. Hudgens, D.L. Trescot, M. Craghan, W.H. Nuckols, C.H. Hershner, J.M. Kassakian, C. J. Linn, P. G. Merritt, T. M. McCue, J. F. O'Connell, J. Tanski, and J. Wang. 2009. State and local governments plan for development of most land vulnerable to rising sea level along the US Atlantic coast. *Environmental Research Letters* 4 (2009) 044008 (7pp) doi:10.1088/1748-9326/4/4/044008.

Treher, Amanda, Timothy Block, Ann Rhoads, Lucinda McDade, James Macklin, and Carrie Kiel. 2006. A study of *Myriophyllum heterophyllum* in Pennsylvania and Delaware. *Journal of the Pennsylvania Academy of Science* 79:128, abstract.

USDA Plants Database, Natural Resources Conservation Service. Washington. http://plants.usda.gov/.

Wetzel, Robert G. 2001. *Limnology, Third Edition*. Academic Press, New York.

Wiersema, John. 1988. Reproductive biology of *Nymphaea* (Nymphaeaceae). *Annals of the Missouri Botanical Garden* 75: 795–804.

Appendix
Taxonomic List of Plants Included

NON-FLOWERING PLANTS

Charophyte Algae

Chara spp.	musk-grass	Native
Nitella spp.	stonewort	Native

Bryophytes—Liverworts

Riccia fluitans	slender riccia	Native
Ricciocarpus natans	purple-fringed riccia	Native

Bryophytes—Mosses

Amblystegium tenax	moss	Native
Fontinalis sullivantii	water moss	Native
Sphagnum sp.	sphagnum moss	Native

Equisetaceae (Horsetail Family)

Equisetum x litorale	shore horsetail	Native
Equisetum fluviatile	water horsetail	Native

Isoetaceae (Quillwort Family)

Isoetes appalachiana	Appalachian quillwort	Native
Isoetes echinospora	spiny-spored quillwort	Native
Isoetes engelmannii	Engelmann's quillwort	Native
Isoetes riparia	shore quillwort	Native
Isoetes valida	quillwort	Native

Ferns

Azolla caroliniana	eastern mosquito fern	Introduced?
Marsilea quadrifolia	European water-clover	Introduced

FLOWERING PLANTS

Acanthaceae (Acanthus Family)

Justicia americana	water-willow	Native

Alismataceae (Water-plantain Family)

Alisma subcordatum	southern water-plantain	Native
Alisma triviale	northern water-plantain	Native
Sagittaria australis	Appalachian arrowhead	Native
Sagittaria calycina	long-lobed arrowhead	Native
Sagittaria cuneata	northern arrowhead	Native
Sagittaria filiformis	arrowhead	Native
Sagittaria graminea	grass-leaved arrowhead	Native

Sagittaria latifolia	duck-potato, wapato	Native
Sagittaria rigida	sessile-fruited arrowhead	Native
Sagittaria subulata	subulate arrowhead	Native

Amaranthaceae (Amaranthus Family)

Amaranthus cannabinus	water-hemp ragweed	Native

Apiaceae (Parsley Family)

Sium suave	water-parsnip	Native

Araceae (Arum Family)

Landoltia punctata	eastern water-flaxseed	Introduced
Lemna minor	duckweed	Native
Lemna obscura	little water duckweed	Native
Lemna perpusilla	minute duckweed	Native
Lemna trisulca	star duckweed	Native
Lemna turionifera	winter duckweed	Native
Lemna valdiviana	pale duckweed	Native
Orontium aquaticum	goldenclub	Native
Peltandra virginica	arrow-arum	Native
Spirodela polyrhiza	water-flaxseed, greater duckweed	Native
Wolffia borealis	dotted watermeal	Native
Wolffia brasiliensis	pointed watermeal	Native
Wolffia columbiana	watermeal	Native
Wolffiella gladiata	bogmat	Native

Asteraceae (Aster Family)

Bidens beckii	water-marigold	Native
Bidens bidentoides	swamp beggar-ticks	Native
Bidens discoidea	small beggar-ticks	Native
Bidens laevis	showy bur-marigold	Native

Brassicaceae (Mustard Family)

Nasturtium officinale	watercress	Introduced

Butomaceae (Flowering-rush Family)

Butomus umbellatus	flowering-rush	Introduced

Cabombaceae (Watershield Family)

Brasenia schreberi	watershield	Native
Cabomba caroliniana	fanwort	Introduced?

Campanulaceae (Bellflower Family)

Lobelia chinensis	Chinese lobelia	Introduced
Lobelia dortmanna	water lobelia	Native

Ceratophyllaceae (Hornwort Family)

Ceratophyllum demersum	coontail, hornwort	Native
Ceratophyllum echinatum	hornwort	Native

Cyperaceae (Sedge Family)

Cladium mariscoides	twig-rush	Native
Dulichium arundinaceum	three-way sedge	Native
Eleocharis acicularis	needle spike-rush	Native
Eleocharis obtusa var. *obtusa*	blunt spike-rush	Native
Eleocharis obtusa var. *peasei*	Wright's spike-rush	Native
Eleocharis palustris	creeping spike-rush	Native
Eleocharis parvula	dwarf spike-rush	Native
Eleocharis quadrangulata	four-angled spike-rush	Native
Eleocharis robbinsii	Robbins' spike-rush	Native
Schoenoplectus acutus	great bulrush, hard-stemmed bulrush	Native
Schoenoplectus fluviatilis	river bulrush	Native
Schoenoplectus pungens	three-square, Chairmaker's-rush	Native
Schoenoplectus purshianus	Pursh's bulrush	Native
Schoenoplectus smithii	Smith's bulrush	Native
Schoenoplectus subterminalis	water bulrush	Native
Schoenoplectus tabernaemontani	great bulrush, soft-stemmed bulrush	Native
Schoenoplectus torreyi	Torrey's bulrush	Native
Scirpus ancistrochaetus	northeastern bulrush	Native

Elatinaceae (Waterwort Family)

Elatine americana	American waterwort	Native
Elatine minima	small waterwort	Native
Elatine triandra	long-stem waterwort	Introduced

Eriocaulaceae (Pipewort Family)

Eriocaulon aquaticum	seven-angle pipewort	Native
Eriocaulon decangulare	ten-angle pipewort	Native
Eriocaulon parkeri	Parker's pipewort	Native

Haloragaceae (Water-milfoil Family)

Myriophyllum aquaticum	parrot-feather	Introduced
Myriophyllum farwellii	Farwell's water-milfoil	Native
Myriophyllum heterophyllum	broad-leaved water-milfoil	Native
Myriophyllum humile	water-milfoil	Native
Myriophyllum sibiricum	northern water-milfoil	Native
Myriophyllum spicatum	Eurasian water-milfoil	Introduced
Myriophyllum tenellum	slender water-milfoil	Native
Myriophyllum verticillatum	whorled water-milfoil	Native
Proserpinaca palustris	common mermaid-weed	Native
Proserpinaca pectinata	comb-leaved mermaid-weed	Native

Hydrocharitaceae (Frog's-bit Family)

Egeria densa	Brazilian waterweed	Introduced
Elodea canadensis	common waterweed	Native
Elodea nuttallii	waterweed	Native
Elodea schweinitzii	Schweinitz' waterweed	Native

Hydrilla verticillata	hydrilla	Introduced
Najas flexilis	northern waternymph	Native
Najas gracillima	slender waternymph	Native
Najas guadalupensis	southern waternymph	Native
Najas marina	holly-leaved naiad	Native
Najas minor	waternymph	Introduced
Vallisneria americana	water-celery, eel-grass	Native

Iridaceae (Iris Family)

Iris pseudacorus	yellow iris	Introduced
Iris versicolor	northern blue flag	Native
Iris virginica	southern blue flag	Native

Juncaceae (Rush Family)

Juncus militaris	bayonet rush	Native

Lamiaceae (Mint Family)

Lycopus rubellus	water-horehound	Native

Lentibulariaceae (Bladderwort Family)

Utricularia cornuta	horned bladderwort	Native
Utricularia geminiscapa	bladderwort	Native
Utricularia gibba	humped bladderwort	Native
Utricularia inflata	inflated bladderwort	Native
Utricularia intermedia	flat-leaved bladderwort	Native
Utricularia macrorhiza	common bladderwort	Native
Utricularia minor	lesser bladderwort	Native
Utricularia purpurea	purple bladderwort	Native
Utricularia radiata	floating bladderwort	Native
Utricularia resupinata	northeastern bladderwort	Native
Utricularia subulata	slender bladderwort	Native

Lythraceae (Loosestrife Family)

Decodon verticillatus	water-willow	Native
Trapa natans	European water-chestnut	Introduced

Menyanthaceae (Buckbean Family)

Nymphoides cordata	floating-heart	Native
Nymphoides peltata	yellow floating-heart	Introduced

Nelumbonaceae (Lotus-lily Family)

Nelumbo lutea	American lotus	Native
Nelumbo nucifera	sacred lotus	Introduced

Nymphaeaceae (Water-lily Family)

Nuphar advena	yellow water-lily, spatterdock, cow-lily	Native
Nuphar microphylla	small yellow water-lily	Native
Nuphar variegata	yellow water-lily, spatterdock, cow-lily	Native

Nymphaea (pink)	pink water-lily	Introduced
Nymphaea odorata	fragrant water-lily	Native

Onagraceae (Evening-primrose Family)

Ludwigia hexapetala	water-primrose	Introduced?
Ludwigia palustris	water-purslane	Native
Ludwigia peploides ssp. *glabrescens*	primrose-willow	Introduced?

Phrymaceae (Lopseed Family)

Glossostigma cleistanthum	mudmat	Introduced

Plantaginaceae (Plantain Family)

Callitriche heterophylla	water-starwort	Native
Callitriche palustris	water-starwort	Native
Callitriche stagnalis	water-starwort	Introduced
Callitriche terrestris	water-starwort	Native
Limosella australis	awl-shaped mudwort	Native
Micranthemum micranthemoides	Nuttall's mud-flower	Native

Poaceae (Grass Family)

Echinochloa walteri	Walter's barnyard-grass	Native
Glyceria acutiflora	creeping mannagrass	Native
Glyceria borealis	northern mannagrass	Native
Leersia oryzoides	rice cutgrass	Native
Torreyochloa pallida	pale meadowgrass	Native
Zizania aquatica	wild-rice	Native

Podostemaceae (Riverweed Family)

Podostemum ceratophyllum	riverweed	Native

Polygonaceae (Smartweed Family)

Persicaria amphibia	water smartweed	Native
Persicaria hydropiper	smartweed, water-pepper	Introduced
Persicaria hydropiperoides	mild water-pepper, water smartweed	Native
Persicaria punctata	dotted smartweed	Native
Persicaria robustior	large dotted smartweed	Native

Pontederiaceae (Pickerel-weed Family)

Heteranthera dubia	water star-grass	Native
Heteranthera multiflora	mud-plantain	Native
Heteranthera reniformis	mud-plantain	Native
Pontederia cordata	pickerel-weed	Native

Potamogetonaceae (Pondweed Family)

Potamogeton amplifolius	bigleaf pondweed	Native
Potamogeton bicupulatus	pondweed	Native
Potamogeton confervoides	Tuckerman's pondweed	Native
Potamogeton crispus	curly pondweed	Introduced
Potamogeton diversifolius	snailseed pondweed	Native

Potamogeton epihydrus	ribbonleaf pondweed	Native
Potamogeton filiformis	threadleaf pondweed	Native
Potamogeton foliosus	leafy pondweed	Native
Potamogeton friesii	Fries' pondweed	Native
Potamogeton gramineus	grassy or variable pondweed	Native
Potamogeton hillii	Hill's pondweed	Native
Potamogeton illinoensis	Illinois pondweed	Native
Potamogeton natans	floating-leaf pondweed	Native
Potamogeton nodosus	longleaf pondweed	Native
Potamogeton oakesianus	Oakes' pondweed	Native
Potamogeton obtusifolius	blunt-leaved pondweed	Native
Potamogeton pectinatus	sago pondweed	Native
Potamogeton perfoliatus	perfoliate pondweed, redhead-grass	Native
Potamogeton praelongus	white-stem pondweed	Native
Potamogeton pulcher	spotted or heartleaf pondweed	Native
Potamogeton pusillus	pondweed	Native
Potamogeton richardsonii	redhead pondweed	Native
Potamogeton robbinsii	flat-leaved or fern pondweed	Native
Potamogeton spirillus	snailseed pondweed	Native
Potamogeton strictifolius	narrow-leaved or straight-leaved pondweed	Native
Potamogeton tennesseensis	Tennessee pondweed	Native
Potamogeton vaseyi	Vasey's pondweed	Native
Potamogeton zosteriformis	flat-stem pondweed	Native
Zannichellia palustris	horned pondweed	Native

Ranunculaceae (Buttercup Family)

Ranunculus aquatilis var. *diffusus*	white water-crowfoot	Native
Ranunculus flabellaris	yellow water-crowfoot	Native

Saururaceae (Lizard-tail Family)

Saururus cernuus	lizard's-tail	Native

Sparganiaceae (Bur-reed Family)

Sparganium americanum	bur-reed	Native
Sparganium androcladum	branching bur-reed	Native
Sparganium angustifolium	narrow-leaf bur-reed	Native
Sparganium emersum	bur-reed	Native
Sparganium eurycarpum	bur-reed	Native
Sparganium fluctuans	bur-reed	Native
Sparganium natans	small bur-reed	Native

Typhaceae (Cat-tail Family)

Typha angustifolia	narrow-leaved cat-tail	Introduced?
Typha latifolia	common cat-tail	Native
Typha x *glauca*	hybrid cat-tail	Introduced?

Index

Acanthaceae, 137
Acanthus Family, 137
Alisma, 129
  subcordatum, 129
  triviale, 131
Alismataceae, 31, 129
Amaranthaceae, 125
Amaranthus cannabinus, 125
Amaranthus Family, 125
Amblystegium tenax, 183
Apiaceae, 128
Araceae, 29, 77, 166
arrow-arum, 29
arrowhead, 31
  Appalachian, 31
  grass-leaved, 35
  long-lobed, 33
  northern, 34
  sessile-fruited, 38
  subulate, 40
Arum Family, 29, 77, 166
asexual reproduction, 6
Aster Family, 42, 278
Asteraceae, 42, 278
Azolla caroliniana, 178
Azollaceae, 178

barnyard-grass, Walter's, 79
beggar-ticks, 42
  small, 44
  swamp, 42
  water, 278
Bellflower Family, 74, 251
Bidens, 42, 278
  beckii, 278
  bidentoides, 42
  discoidea, 44
  laevis, 45
bladderwort, 255, 259
  common, 264
  flat-leaved, 263
  floating, 262, 266
  horned, 258
  humped, 260
  inflated, 261
  lesser, 266
  northeastern, 257, 258
  purple, 267
  slender, 257
Bladderwort Family, 255
blue flag, 92
  northern, 92
  southern, 93
bogmat, 166, 177
bogs, 11
Brasenia schreberi, 161
Brassicaceae, 141
Buckbean Family, 143
bulrush, 47
  great, 48, 56
  hard-stemmed, 48
  northeastern, 59
  Pursh's, 51
  river, 49
  Smith's, 52
  soft-stemmed, 56
  Torrey's, 58
  water, 53
bur-marigold, showy, 45
bur-reed, 61, 62, 68, 69, 70
  branching, 64
  narrow-leaf, 66
  small, 17
Bur-reed Family, 61
Butomaceae, 76
Butomus umbellatus, 76
Buttercup Family, 275

Cabomba caroliniana, 272
Cabombaceae, 161, 272
Callitriche, 229
  heterophylla, 229
  palustris, 230
  stagnalis, 231
  terrestris, 232
caltrop, 149
Campanulaceae, 74, 251
cat-tail, 71
  common, 71
  hybrid, 72
  narrow-leaved, 72
Cat-tail Family, 71
Ceratophyllaceae, 269
Ceratophyllum, 269
  demersum, 269
  echinatum, 270
  muricatum, 270
Chairmaker's-rush, 50
Chara spp., 164
Charophyte Algae, 164
Cladium mariscoides, 123
coontail, 269, 270
cow-lily, 153
cutgrass, rice, 83
Cyperaceae, 47, 111, 121, 123

Darwin, Charles, 7, 255
Decodon verticillatus, 137
decomposers, 9
Delaware Estuary, 13–15
dispersal, 6–7
ditch-moss, 234
duck-potato, 36
duckweed, 166, 168
  common, 168
  greater, 172
  little water, 168
  minute, 169
  pale, 172
  star, 170
  winter, 171
Dulichium arundinaceum, 121
dystrophic, 11

Echinochloa walteri, 79
ecology, 1–19
eel-grass, 253
Egeria densa, 233
Elatinaceae, 237
Elatine, 237
  americana, 237
  minima, 238
  triandra, 239
Eleocharis, 111
  acicularis, 111, 258
  dulcis, 149
  obtusa var. *obtusa*, 113
  obtusa var. *peasei*, 114
  palustris, 115
  parvula, 116
  quadrangulata, 117
  robbinsii, 118
Elodea, 233
  canadensis, 234
  nuttallii, 235
  schweinitzii, 233
endangered species, 15–17

Endangered Species Act, 15
Equisetaceae, 88
Equisetum, 88
  fluviatile, 88
  x *litorale*, 88
Eriocaulaceae, 243
Eriocaulon, 243
  aquaticum, 243
  decangulare, 243
  parkeri, 14, 243
eutrophic, 13
Evening-primrose Family, 132, 136
evolution, 1–4
exotic species, 17–19

fanwort, 272
floating-heart, 143
  yellow, 145
flowering-rush, 76
Flowering-rush Family, 76
Fontinalis sullivantii, 183
freshwater tidal marsh, 14–15
Frog's-bit Family, 185, 223, 233, 253

glacial lakes, 11–13
Glossostigma
  cleistanthum, 241
  diandrum, 241
Glyceria, 79, 81
  acutiflora, 81
  borealis, 82
goldenclub, 77
Grass Family, 79

habitat structure, 9–10
Haloragaceae, 273, 280
Heteranthera, 96, 221
  dubia, 221
  multiflora, 96
  reniformis, 98
hornwort, 269, 270
Hornwort Family, 269
Horsetail Family, 88
horsetail, 88
  shore, 88
  water, 88
hydrilla, 185
Hydrilla verticillata, 185
Hydrocharitaceae, 185, 223, 233, 253

invasive species, 14, 17–19
Iridaceae, 90
Iris Family, 90
Iris, 90
  pseudacorus, 90
  versicolor, 92
  virginica, 93
iris, yellow, 90
Isoetaceae, 245
Isoetes, 245
  appalachiana, 245
  echinospora, 246
  engelmannii, 247
  engelmannii var. *caroliniana*, 250
  engelmannii var. *georgiana*, 245
  engelmannii var. *valida*, 250
  riparia, 248
  valida, 250

Juncaceae, 102
Juncus militaris, 102
Justicia americana, 139

lakes, 11
Lamiaceae, 127
Landoltia punctata, 167
Leersia oryzoides, 83
Lemna, 166, 168
  minor, 168
  obscura, 168
  perpusilla, 169
  trisulca, 170
  turionifera, 171
  valdiviana, 172
Lentibulariaceae, 255
Limosella australis, 14
littoral zone, 9
liverworts, aquatic, 163
lizard's-tail, 94
Lizard's-tail Family, 94
Lobelia, 74, 251
  chinensis, 74
  dortmanna, 251
lobelia, 74, 251
  Chinese, 74
  water, 251
Loosestrife Family, 137, 149
Lopseed Family, 241
lotus, 147
  American, 147
  sacred, 147
Lotus-lily Family, 147
Ludwigia, 132, 136
  hexapetala, 132
  palustris, 136
  peploides ssp. *glabrescens*, 134
Lycopus rubellus, 127
Lythraceae, 137, 149

mannagrass, 81
  creeping, 81
  northern, 82
Marsilea quadrifolia, 151
Marsileaceae, 151
meadowgrass, pale, 84
Megalodonta beckii, 278
Menyanthaceae, 143
mermaid-weed, 273
  comb-leaved, 274
  common, 273
mesotrophic, 11
Micranthemum micranthemoides, 14
Mint Family, 127
mosquito fern, eastern, 178
Mosquito Fern Family, 178
moss, 183
  sphagnum, 183
  water, 183
mudmat, 241
mud-plantain, 96, 98
mudwort, 14
  awl-shaped, 14
  Nuttall's, 14
musk-grass, 164
Mustard Family, 141
Myriophyllum, 280
  aquaticum, 281
  farwellii, 282
  heterophyllum, 283
  humile, 284
  sibiricum, 286
  spicatum, 287
  tenellum, 289
  verticillatum, 289

naiad, 227
  bushy, 224
  holly-leaved, 226
  southern, 225
Najas, 223
  flexilis, 223
  gracillima, 224
  guadalupensis, 225
  marina, 226
  minor, 227
Nasturtium officinale, 141
Nelumbo, 147
  lutea, 147
  nucifera, 147
Nelumbonaceae, 147

Nitella spp., 164
non-native species, 17–19
Nuphar, 153
  advena, 154
  microphylla, 155
  variegata, 156
Nymphaea, 153
  pink, 158
  odorata, 158
Nymphaeaceae, 153
Nymphoides, 143
  cordata, 143
  peltata, 145

oligotrophic, 11, 15
Onagraceae, 132, 136
Orontium aquaticum, 77
over-abundant native
    species, 19

parrot's-feather, 281
Parsley Family, 128
Peltandra virginica, 29
Pennsylvania Natural Heri-
    tage Program, 15
Persicaria, 104
  amphibia, 104
  hydropiper, 105
  hydropiperoides, 107
  punctata, 108
  robustior, 110
photosynthesis, 7–9
Phrymaceae, 241
pickerel-weed, 100
Pickerel-weed Family, 96, 100,
    221
pipewort, 243
  Parker's, 243
  seven-angle, 243
  ten-angle, 243
Pipewort Family, 243
Plantaginaceae, 229
Plantain Family, 229
Poaceae, 79
Podostemaceae, 13, 180
Podostemum ceratophyllum, 180
Polygonaceae, 104
Polygonum
  amphibium, 104
  hydropiper, 105
  hydropiperoides, 107
  punctatum, 108
  robustius, 110
pond-lily, 153
pondweed, 186, 191, 211
  bigleaf, 189
  blunt-leaved, 206
  curly, 193
  fern, 213
  flat-leaved, 213
  flat-stem, 219
  floating-leaf, 203
  Fries', 198
  grassy, 200
  heartleaf, 210
  Hill's, 201
  horned, 220
  Illinois, 202
  leafy, 197
  longleaf, 204
  narrow-leaved, 215
  Oakes', 205
  perfoliate, 208
  redhead, 212
  ribbonleaf, 195
  Richardson's, 212
  sago, 207
  snailseed, 194, 214
  spotted, 210
  straight-leaved, 215
  Tennessee, 216
  threadleaf, 197
  Tuckerman's, 192
  variable, 200
  Vasey's, 217
  white-stem, 209
Pondweed Family, 186
Pontederia cordata, 100
Pontederiaceae, 96, 100, 121
Potamogeton, 186
  amplifolius, 189
  bicupulatus, 191
  confervoides, 192
  crispus, 193
  diversifolius, 194
  diversifolius var. *tricho-*
    phyllus, 191
  epihydrus, 195
  filiformis, 197
  foliosus, 197
  friesii, 198
  gramineus, 200
  hillii, 201
  illinoensis, 202
  natans, 203
  nodosus, 204
  oakesianus, 205
  obtusifolius, 206
  pectinatus, 207
  perfoliatus, 208
  praelongus, 209
  pulcher, 210
  pusillus, 211
  richardsonii, 212
  robbinsii, 213
  spirillus, 214
  strictifolius, 215
  tennesseensis, 216
  vaseyi, 217
  zosteriformis, 219
Potamogetonaceae, 186
primrose-willow, 134
problem vegetation, 17–19
Proserpinaca, 273
  palustris, 273
  pectinata, 274

quillwort, 245, 250
  Appalachian, 245
  Engelmann's, 247
  shore, 248
  spiny-spored, 246
Quillwort Family, 245

ragweed, water-hemp, 125
Ranunculaceae, 275
Ranunculus, 275
  aquatilis var. *diffusus*, 275
  flabellaris, 277
rare species, 15–17
redhead-grass, 208
reproduction, 4–6
Riccia fluitans, 163
riccia, 163
  purple-fringed, 163
  slender, 163
Ricciocarpus natans, 163
rivers, 13
riverweed, 13, 180
Riverweed Family, 180
Rush Family, 102
rush, bayonet, 102

Sagittaria, 31
  australis, 31
  calycina, 33
  cuneata, 34
  filiformis, 17
  graminea, 35
  latifolia, 36
  montevidensis ssp.
    calycina, 33
  rigida, 38
  subulata, 40
Saururaceae, 94

Saururus cernuus, 94
Schoenoplectus, 47
  acutus, 48
  fluviatilis, 49
  pungens, 50
  purshianus, 51
  smithii, 52
  subterminalis, 53
  tabernaemontani, 56
  torreyi, 58
Scirpus, 47
  ancistrochaetus, 59
  torreyi, 58
  validus, 56
sea level rise, 15
Sedge Family, 47, 111, 121, 123
sedge, three-way, 121
sexual reproduction, 4–5
Sium suave, 128
smartweed, 104, 105
  dotted, 108
  large dotted, 110
  water, 104, 107
Smartweed Family, 104
Sparganiaceae, 61
Sparganium, 61
  americanum, 62
  androcladum, 64
  angustifolium, 66
  chlorocarpum, 68
  emersum, 68
  eurycarpum, 69
  fluctuans, 70
  natans, 17
spatterdock, 153, 154, 156
Sphagnum sp., 183
spike-rush, 111
  blunt, 113
  creeping, 115
  dwarf, 116
  four-angled, 117
  needle, 111
  Robbin's, 118
  Wright's, 114
Spirodela, 167, 172
  polyrhiza, 172
  punctata, 167
star-grass, water, 221
stonewort, 164
streams, 13
Stuckenia
  filiformis, 197
  pectinata, 207

threatened species, 15–17
three-square, 50
Torreyochloa pallida, 84
Trapa natans, 149
tuckahoe, 29
twig-rush, 123
Typha, 71
  angustifolia, 72
  latifolia, 71
  x *glauca*, 72
Typhaceae, 71

Utricularia, 255
  cornuta, 258
  geminiscapa, 259, 264
  gibba, 260
  inflata, 261
  intermedia, 263
  macrorhiza, 264
  minor, 266
  purpurea, 267
  radiata, 262
  resupinata, 17, 258
  subulata, 17, 257

Vallisneria americana, 4–5, 253

wapato, 36
water-celery, 4–5, 253
water-chestnut, 149
water-chickweed, 231
water-clover, European, 151
Water-clover Family, 151
watercress, 141
water-crowfoot, 275
  white, 275
  yellow, 277
water-flaxseed, 166, 172
  eastern, 167
water-hemp, saltmarsh, 125
water-horehound, 127
water-lily, 153
  fragrant, 158
  pink, 158
  small yellow, 155
  yellow, 153, 154, 156
Water-lily Family, 153
water-marigold, Beck's, 278
watermeal, 166, 173, 176
  dotted, 174
  pointed, 175
water-milfoil, 280, 284
  broad-leaved, 283
  Eurasian, 287
  Farwell's, 282
  northern, 286
  slender, 289
  whorled, 289
Water-milfoil Family, 273, 280
waternymph, 223, 227
  northern, 223
  slender, 224
  southern, 225
water-parsnip, 128
water-pepper, 105
  mild, 107
water-plantain, 129
  northern, 131
  southern, 129
Water-plantain Family, 129
water-primrose, 132
water-purslane, 136
watershield, 161
Watershield Family, 161, 272
water-starwort, 229, 230, 231, 232
waterweed, 233, 235
  Brazilian, 233
  common, 234
  Schweinitz', 233
water-willow, 137, 139
waterwort, 237
  American, 237
  long-stem, 239
  small, 238
Waterwort Family, 237
wild-rice, 86
Wolffia, 166, 173
  borealis, 174
  brasiliensis, 175
  columbiana, 176
Wolffiella gladiata, 177

Zannichellia palustris, 220
Zizania aquatica, 86
Zosterella dubia, 221